C0-AUO-962

Cytogenetics of Aneuploids

CYTOGENETICS OF ANEUPLOIDS

GURDEV S. KHUSH

Plant Breeder and Head
Varietal Improvement Department
International Rice Research Institute
Los Baños, Philippines

ACADEMIC PRESS New York and London 1973
A Subsidiary of Harcourt Brace Jovanovich, Publishers

ACADEMIC PRESS, INC.
111 Fifth Avenue, New York, New York 10003

United Kingdom Edition published by
ACADEMIC PRESS, INC. (LONDON) LTD.
24/28 Oval Road, London NW1

Library of Congress Cataloging in Publication Data

Khush, Gurdev S
Cytogenetics of aneuploids.

Bibliography: p.
1. Aneuploidy. 2. Cytogenetics.
I. Title. [DNLM: 1. Aneuploidy. 2. Cytogenetics.
QH605 K45c 1973]
QH461.K48 574.2 72-12199
ISBN 0-12-406250-4

PRINTED IN THE UNITED STATES OF AMERICA

To

Charles and Martha Rick

and

Ledyard and Barbara Stebbins

Contents

Preface

Ever since the first discoveries of trisomic *Datura* in 1915 and monosomic *Drosophila* in 1920, a great body of literature dealing with the aneuploids has been published. These studies have helped elucidate some of the most important principles of heredity. No wonder then that introductory and advanced courses in genetics and cytogenetics regularly devote several lectures to this topic. Because of the limitations placed by the number of other topics to be covered, cytogenetics textbooks have not treated aneuploids adequately. This book has been written to supplement those works that can be used for advanced courses in genetics and cytogenetics.

Aneuploids have been particularly useful in studying the basic genetics of many plant species, particularly the crop species. In genetically well-known species such as *Datura,* maize, tomato, barley, wheat, and tobacco, aneuploids have been thoroughly investigated. This book reviews the literature on aneuploidy in an integrated manner for the first time. It should serve as a standard reference for research workers and teachers.

During the last decade a great deal of interest has developed in the aneuploids of man and other animals, resulting in numerous publications. Scientists working with animal species, however, are generally unfamiliar with studies on aneuploids of plant species, and many working with plant species are similarly unfamiliar with research on animal aneuploids. This book should help bridge this gap.

Over the years, different terms have been used to describe the same aneuploid; conversely, the same terms have been used for entirely different ones. To remedy this situation and to avoid further confusion, I have suggested a standard terminology for all aneuploids. I hope my fellow research workers concerned about the confusion in terminology will find the proposed system useful.

The first chapter, in addition to dealing primarily with terminology, gives a brief history of the development of this field of enquiry. The next five chapters review the entire literature on trisomics. Chapter 2 discusses sources of various kinds of trisomics. Subsequent chapters deal with the cytology, transmission rates, genetics, and morphology of trisomics. Chapters 7, 8, and 9 are devoted to monosomics and nullisomics. The two chapters that follow deal with intervarietal substitutions and alien additions and substitutions, respectively. This allows sequential reading of these related subjects. The last chapter was added to introduce plant science workers to the problems and progress made in studying the aneuploids of man and animals, but is not intended as an authoritative review.

I have tried to present in each chapter the basic principles and theoretical considerations of each topic discussed, followed by experimental results and reviews of available studies.

I first conceived of writing this book when I was engaged in investigating the aneuploids of the tomato in Professor C. M. Rick's laboratory at the University of California, Davis. These studies were supported by grant GM 06209 from the National Institutes of Health, U.S. Public Health Service. Five chapters on trisomics were completed before I moved to the International Rice Research Institute. Dr. R. F. Chandler, then director of IRRI, kindly consented to my returning to the University of California for a one month's leave of absence enabling me to complete the book.

It is my pleasure to acknowledge my gratitude for the invaluable suggestions and comments made on the various chapters by Professors C. M. Rick, G. L. Stebbins, S. R. Snow, E. R. Sears, L. M. Sears, G. Kimber, and D. R. Ramirez. However, I am solely responsible for any errors and omissions which may be present in this book. I am indebted to many colleagues for providing photographs for the figures. Thanks are due Dr. R. A. McIntosh for checking Table 9.6; Miss Dora Hunt for editing the entire manuscript; Mrs. Leni P. Nazarea for typing and retyping the manuscript; Mrs. Marilyn Stein for drawing Figures 4.1 and 5.1; Mr. Noli del Rosario for Figures 2.1, 2.2, 6.3, 6.4, 6.5, 6.8, and 7.1; Mrs. Lina M. Vergara and her staff for help in bibliographical matters; Mr. S. A. Breth for counsel on editorial matters; Mrs. Nancy Perez for preparing the Subject Index; and numerous publishers and individuals for permission to cite published data and reproduce figures from their copyrighted material. Special thanks are due to the immediate members of my family who willingly allowed me to spend innumerable evenings and weekends in libraries.

GURDEV S. KHUSH

Cytogenetics of Aneuploids

1

Introduction and Terminology

Phenotypic changes in plant species, especially the crop species, and in man which were later associated with aneuploidy have been recorded for over a century. For example, morphological variants called fatuoids in oats described by Buchman (1857) were found to be deficient for a chromosome pair by Huskins (1927). Langdon-Down (1866) described an anomaly in man which he called mongolian idiocy. Several authors such as Waardenburg (1932) and Bleyer (1934) suggested that this anomaly may be due to an extra chromosome. However, almost a century was to pass before Lejeune *et al.* (1959a) showed conclusively that mongolian idiots had an extra chromosome.

The first observation of plants with aneuploid chromosome number was made by Lutz (1909) when she reported that two plants from the cross of *Oenothera lata* × *Oenothera gigas* had fifteen chromosomes. She also found that several plants of *O. lata* had fifteen chromosomes and a few sibs had fourteen chromosomes. Gates (1909) similarly observed aneuploid chromosome numbers in the *Oenotheras*. However, in my opinion the single event which gave birth to the cytogenetics of aneuploids took place in 1915. In that year a mutant of *Datura stramonium* was discovered by Mr. B. T. Avery, Jr., an assistant of Dr. A. F. Blakeslee, at the Connecticut Agricultural College, Storrs, Connecticut. This mutant was called Globe from the globose shape of its capsules. *Datura stramonium* was used as demonstration material for students in genetics. Two single gene mutants, white flower and smooth capsules, were used for explaining the genetic ratios. The inheritance of Globe was investigated in subsequent years at the Station for Experimental Evolution at Cold Spring Harbor where Dr. Blakeslee had moved in 1915. It was found that Globe character did not segregate in the conventional manner expected of Mendelian genes. When the Globe parent was selfed, only about one-fourth of the progeny was Globe. About same proportion of the progeny was Globe when the Globe parent was crossed with pollen from the normal plants. The trait was transmitted to less than 2% of the progeny when the Globes were used as pollen parents in crosses with normals. Thus true breeding Globes could not be obtained by any breeding scheme. Between 1915 and 1919 Blakeslee and associates identified twelve mutants whose breeding behavior was similar to that of Globe (Blakeslee, 1921c).

In 1920, Blakeslee invited an eminent cytologist, Mr. John Belling, who was a visiting scientist at the Station for Experimental Evolution at Cold Spring Harbor, to carry out the cytological examination of these mutants. His critical observation led to the discovery that each of these mutants was distinguished by having an extra chromosome (Blakeslee *et al.*, 1920). This finding clarified the mystery of peculiar breeding behavior of these mutants and led to the rapid development of a field of inquiry dealing with the genetics and cytology of these mutants.

Blakeslee (1921b) coined the term "trisome" to signify the addition of one member of the basic chromosome complement. He called these mutants simple "trisomic" mutants. Since *Datura* has twelve pairs of chromosomes, only twelve such mutants were expected. However, another group of mutants were isolated which had considerable resemblance to the simple trisomic mutants and seemed to be related to them. Belling and Blakeslee (1924) showed that these mutants had an extra isochromosome instead of whole chromosome. Blakeslee (1924) named these mutants "secondary" mutants in contrast with the original or "primary" mutants. When mutants of another kind in which the extra chromosome consists of two arms of nonhomologous chromosomes were discovered, they were designated "tertiary" mutants by Belling and Blakeslee (1926). One mutant, which originated in 1921 among the offspring of a plant that had been exposed to radium emanation did not fit into any of the three groups. Blakeslee (1927a) showed that in this plant one chromosome was missing but it was compensated for by two tertiary chromosomes. One tertiary chromosome had one arm of the missing chromosome while the second tertiary had the other arm. Blakeslee (1927b) called it a "compensating" mutant.

Although terms primary, secondary, tertiary, and compensating trisomics are now widely used, Blakeslee and his associates never used these terms. They always referred to these aneuploids as primary, secondary, tertiary, or compensating mutants or primary $2n + 1$ types, etc., or simply as secondary types or compensating types, etc. Rhoades (1933b) made a distinction between primary, secondary, and tertiary trisomes. However, Goodspeed and Avery (1939) were the first to use the names primary, secondary, or tertiary trisomics. Khush and Rick (1967a) used the name compensating trisomic for the first time.

The fifth type of trisomic, e.g., the individual having only one chromosome arm (a telocentric) as extra was first recorded by Belling (1927) in *Datura*. Rhoades (1936) and Blakeslee and Avery (1938) called these fragment types. Burnham (1962) referred to them as "telosomic trisomics." Khush and Rick (1968b) suggested calling them "telotrisomics." Kimber and Sears (1968) called them "monotelotrisomics." However, the prefix "mono" in this term is redundant and the preferred name for these trisomics is telotrisomics. It may be noted that according to the nomenclature of Kimber and Sears (1968) secondary trisomic and monoisotrisomic are synonymous but the term secondary trisomic has been used so widely in the literature that it should be retained.

The missing chromosome in a compensating trisomic may be compensated for by (1) two tertiary chromosomes, (2) two isochromosomes, (3) one tertiary and one telocentric chromosome, (4) one iso and one telocentric chromosome, and (5) one iso and one tertiary chromosome. I have called these trisomics as "ditertiary," "diiso," "telotertiary," "isotelo," and "isotertiary compensating" trisomics, respectively, in this book.

The term "monosome" was coined by Blakeslee (1921b) to signify the absence of one chromosome of the basic chromosome complement. The first monosomic individual to be discovered was a fruit fly *(Drosophila melanogaster)* which lacked one chromosome 4. Bridges (1921a) called it "haplo-4." Clausen and Goodspeed (1926a) were the first to use the term "monosomic" to describe a plant of *Nicotiana tabacum* with a missing chromosome. *Drosophila*, tomato, and tobacco workers (notably Clausen and Cameron, 1944) have used the term "haplo" in combinations such as haplo-4 and haplo-p, while wheat, oat, and cotton workers have used the term "mono," for example, mono-5A, mono-14, and mono-M, to denote the missing chromosome. However, term monosomic has priority and I have followed the wheat workers in this respect.

The term "nullo" was coined by Bridges (1921a) to refer to a gamete with a missing chromosome. Plants with a pair of missing chromosomes were called "nullosomic" by Sears (1939), although he later changed the designation to "nullisomic" (Sears, 1941). Kihara (1924) and Winge (1924) were the first to report the nullisomic plants of wheat.

A chromosome arm with a terminal centromere was called a telocentric by Darlington (1939) and chromosome with two identical arms was named isochromosome by Darlington (1940). Sears (1954) and Kimber and Sears (1968) have coined descriptive terms for the individuals having telocentric(s) and/or isochromosome(s). The nomenclature used above to describe the aneuploids as well as the nomenclature suggested by Kimber and Sears (1968) is given below.

Disomic: An individual with euploid chromosome complement.

Trisomic: An individual with an extra chromosome (aneuploid chromosome number).

Tetrasomic: An individual having an extra pair of homologous chromosomes.

Primary trisomic: A trisomic in which the extra chromosome is one of the normal chromosomes of the complement.

Secondary trisomic: A trisomic in which the extra chromosome is the isochromosome for one of the chromosome arms of the complement. Also called "monoisotrisomic" by Kimber and Sears (1968).

Tertiary trisomic: A trisomic in which the extra chromosome consists of two arms of two nonhomologous chromosomes.

Telotrisomic: A trisomic in which the extra chromosome is the telocentric chromosome. Also called "monotelotrisomic" by Kimber and Sears (1968).

Compensating trisomic: An individual in which one chromosome is missing but is compensated for by two other modified chromosomes.

Diiso compensating trisomic: A compensating trisomic in which the missing chromosome is compensated for by two isochromosomes, one for each arm of the missing chromosome. Also called "doubleisotrisomic" by Kimber and Sears (1968).

Ditertiary compensating trisomic: A compensating trisomic in which the missing chromosome is compensated for by two tertiary chromosomes, one having one arm of the missing chromosome and the other second arm.

Isotertiary compensating trisomic: A compensating trisomic in which the missing chromosome is compensated for by one isochromosome and one tertiary chromosome.

Isotelo compensating trisomic: A compensating trisomic in which the missing chromosome is compensated for by one isochromosome and one telocentric chromosome. Also called "monotelo-monoisotrisomic" by Kimber and Sears (1968).

Telotertiary compensating trisomic: A compensating trisomic in which the missing chromosome is compensated for by one telocentric chromosome and one tertiary chromosome.

Ditelotrisomic: An individual deficient for one chromosome but which has a pair of homologous telocentric chromosomes for one arm of the missing chromosome.

Diisotrisomic: An individual deficient in one chromosome but which has a pair of homologous isochromosomes for one arm of the missing chromosome.

Doubletelotrisomic: An individual deficient in one chromosome but which has two telocentric chromosomes, one for each arm of the missing chromosome (this is really a "pseudo-trisomic").

Teloisotrisomic: An individual deficient in one chromosome but which has a telocentric chromosome and an isochromosome for the same arm of the missing chromosome.

Monosomic: An individual with one missing chromosome.

Nullisomic: An individual with one missing chromosome pair.

Tertiary monosomic: An individual in which two arms of two nonhomologous chromosomes are missing but the remaining two arms are joined to constitute a tertiary chromosome.

Monotelodisomic: An individual deficient in one chromosome arm.

Monoisodisomic: An individual deficient in one chromosome but which has an isochromosome for one of the arms of the missing chromosome. Also called "haplo-triplo-disomic" by Khush and Rick (1967d). In this volume, I have used the term "monoisodisomic" which is more descriptive.

Monotelosomic: An individual deficient in one entire chromosome and one arm of the other homolog.

Ditelosomic: An individual deficient in two homologous arms.

Tritelosomic: An individual which is missing one chromosome pair but has three homologous telocentric chromosomes.

Monoisosomic: An individual which is missing one chromosome pair but has an isochromosome for one arm of the missing pair.

Diisosomic: An individual which is missing one chromosome pair but has two homologous isochromosomes for same arm of the missing pair.

Triisosomic: An individual which is missing one chromosome pair but has three homologous isochromosomes for same arm of the missing pair.

Teloisodisomic: An individual in which one chromosome pair is missing but a telocentric for one arm of the missing pair and an isochromosome for the same arm are present.

Monotelomonoisosomic: An individual in which one chromosome pair is missing but a telocentric for one arm of the missing pair and an isochromosome for the other arm are present.

Doublemonotelosomic: An individual in which one chromosome pair is missing but two telocentrics, one for each arm of the missing pair, are present.

Ditelomonotelosomic: An individual in which one chromosome pair is missing but which has a pair of telocentric chromosomes for one arm and an unpaired telocentric chromosome of the other arm of the missing chromosome pair.

Doublemonoisosomic: An individual which is missing one chromosome pair but which has two isochromosomes, one for each arm of the missing chromosome pair.

Chromosomal Formulas

Blakeslee (1921b) suggested that designation $2n$ be used to indicate disomic chromosome number and $2n + 1$ and $2n - 1$ to indicate trisomic and monosomic individuals, respectively. These formulas have been used rather widely in referring to euploid and aneuploid chromosome numbers. Specific monosomics have been referred to as mono-5A or mono-M, etc., meaning thereby that the plant is monosomic for chromosome 5A (as in wheat) or chromosome M (as in cotton). The trisomic individuals have been designated by laboratory names which generally signify some phenotypic modification of the trisomic. Thus, the well-known trisomic of *Datura* with globose capsules was called Globe by *Datura* workers. Other names such as Poinsettia, Buckling, etc., were used for trisomics of other chromosomes. By 1930, the individual chromosomes of the *Datura* chromosome complement were cytologically identified and numbered. The longest chromosome was designated 1·2 chromosome, ·1 referring to one arm and ·2 to the other. Thus the shortest chromosome was labeled 23·24. When it was found that in the Globe trisomic 21·22 chromosome was extra, this trisomic could simply be referred to as $2n + 21 \cdot 22$. This nomenclature was universally used by *Datura* workers after 1930. The main merit of this system is that it did away with the necessity of remembering the trisomic names and the corresponding extra chromosomes. The nature of the trisomic can be readily determined by looking at the formula. For example, $2n + 21 \cdot 22$ is clearly a primary trisomic for the 21·22 chromosome; $2n + 9 \cdot 9$, a secondary in which the extra is the isochromosome for ·9 arm of the 9·10 chromosome; $2n + 1 \cdot 15$, a tertiary trisomic whose extra chromosome consists of ·1 arm of 1·2 chromosome and ·15 arm of 15·16 chromosome. Similarly, $2n - 1 \cdot 2 + 1 \cdot 9 + 2 \cdot 5$ is a compensating trisomic in which chromosome 1·2 is missing but is compensated for by ·1 arm of 1·9 tertiary chromosome and ·2 arm of 2·5 tertiary

TABLE 1.I

Chromosomal Formulas for Various Types of Aneuploids after Tomato and the Wheat Systems and Taking Chromosome 5 as an Example

Aneuploid	Chromosome no.	Tomato system	Wheat system
Disomic	$2n$	$2n$	$21''$
Trisomic	$2n + 1$	$2n$ + 5S·5L	$20'' + 1'''$
Tetrasomic	$2n + 2$	$2n$ + 2(5S·5L)	$20'' + 1''''$
Primary trisomic	$2n + 1$	$2n$ + 5S·5L	$20'' + 1'''$
Secondary trisomic (monoisotrisomic)	$2n + 1$	$2n$ + 5S·5S	$20'' + i2'''$
Tertiary trisomic	$2n + 1$	$2n$ + 5S·7L	$19'' + 1'''''$
Telotrisomic	$2n + 1$	$2n$ + ·5S	$20'' + t2'''$
Compensating trisomic	$2n + 1$		
Diiso compensating trisomic (doubleisotrisomic)	$2n + 1$	$2n$ − 5S·5L + 5S·5S + 5L·5L	$20'' + (i + i)\ 1'''$
Ditertiary compensating trisomic	$2n + 1$	$2n$ −5S·5L + 1L·5S + 6S·5L	
Isotertiary compensating trisomic	$2n + 1$	2n − 5S·5L +5S·5S + 1L·5L	
Isotelo compensating trisomic (monotelo-monoisotrisomic)	$2n + 1$	$2n$ − 5S·5L + 5S·5S + ·5L	$20'' + (t + i)\ 1'''$

Telotertiary compensating trisomic	$2n + 1$	$2n$ − 5S·5L + 5S·7S + ·5L	
Ditelotrisomic	$2n + 1$	$2n$ − 5S·5L + 2(·5L)	20″ + (t″) 1″ ′
Diisotrisomic	$2n + 1$	$2n$ − 5S·5L + 2(5L·5L)	20″ + (i″) 1″ ′
Doubletelotrisomic	$2n + 1$	$2n$ − 5S·5L + ·5S + ·5L	20″ + (t + t) 1″ ′
Teloisotrisomic	$2n + 1$	$2n$ − 5S·5L + ·5L + 5L·5L	20″ + (ti) 1″ ′
Monosomic	$2n - 1$	$2n$ − 5S·5L	20″ + 1′
Nullisomic	$2n - 2$	$2n$ − 2(5S·5L)	20″
Tertiary monosomic	$2n - 1$	$2n$ − 5S·6S	
Monotelodisomic	$2n$	$2n$ − ·5L	20″ + t1″
Monoisodisomic	$2n$	2n − 5S·5L + 5L·5L	20″ + i1″
Monotelosomic	$2n - 1$	$2n$ − 2(5S·5L) + ·5L	20″ + t′
Ditelosomic	$2n$	$2n$ − 2(5S·5L) + 2(·5L)	20″ + t″
Tritelosomic	$2n + 1$	$2n$ − 2(5S·5L) + 3(·5L)	20″ + 1″ ′
Monoisosomic	$2n - 1$	$2n$ − 2(5S·5L) + 5L·5L	20″ + i′
Diisosomic	$2n$	$2n$ − 2(5S·5L) + 2(5L·5L)	20″ + i″
Triisosomic	$2n + 1$	$2n$ − 2(5S·5L) + 3(5L·5L)	20″ + i″ ′
Teloisodisomic	$2n$	$2n$ − 2(5S·5L) + 5L·5L + ·5L	20″ + ti″
Monotelomonoisosomic	$2n$	$2n$ − 2(5S·5L) + ·5L + 5S·5S	20″ + t′ + i′
Doublemonotelosomic	$2n$	$2n$ − 2(5S·5L) + ·5L + ·5S	20″ + t′ + t′
Ditelomonotelosomic	$2n + 1$	$2n$ − 2(5S·5L) + 2(·5L) + ·5S	20″ + t″ + t′
Doublemonoisosomic	$2n$	$2n$ − 2(5S·5L) + 5S·5S + 5L·5L	20″ + i′ + i′

chromosome. In this volume, I have used the laboratory names and chromosomal formulas for *Datura* trisomics at different places.

Workers in several other species, most notably tomato (Rick and Barton, 1954), have referred to specific trisomics by the prefix triplo, such as triplo-7 for trisomic in which the extra chromosome is No. 7 of the chromosome complement. Khush and Rick (1967a) initiated the use of chromosomal formulas for tomato trisomics. In tomato, chromosomes are numbered from 1 to 12 in the decreasing order of their length at the pachytene stage of meiosis and each chromosome has a short and a long arm. Thus chromosome 1 is referred to as 1S·1L and the primary trisomic for chromosome 1 as $2n$ + 1S·1L. The formula for a secondary trisomic for long arm of chromosome 8 is $2n$ + 8L·8L. The tertiary trisomic having an extra chromosome composed of short arm of chromosome 7 and long arm of chromosome 5 is written as $2n$ + 7S·5L. A primary monosomic for chromosome 11 reads as $2n$ − 11S·11L while a tertiary monosomic, in which short arms of chromosomes 1 and 4 are missing, reads as $2n$ − 1S·4S (see Table 1.I for formulas of complex aneuploids). This system has been followed in this volume for tomato aneuploids.

I see a lot of merit in using this nomenclature in other species. However, it can be used only in those species in which individual chromosomes of the complement as well as the two arms of each chromosome can be identified.

Kimber and Sears (1968) have suggested a system of symbols to indicate the chromosomal constitution and the pairing situation in the aneuploids of Triticinae. The chromosomes are designated by Arabic numerals and short and long arms of the chromosomes by capital letters S and L, respectively. Isochromosome is designated by small letter i, the telocentric by small letter t, and the complete monosome by numeral 1. The pairing status is indicated by superscripts following the chromosome symbols. For example, an euploid plant of wheat ($2n = 42$) is designated as 21″ and a monosomic as 20″ + 1′. When the chromosome involved in the deficiency or duplication is to be identified, the number and letter indicating the homoeologous group and the genome of the chromosome is added to the symbol indicating the configuration. For example, a monosomic deficiency of chromosome 5A is symbolized as 20″ + 1′5A and a plant with chromosome 6B represented by the telocentric bivalent of its long arm and a telocentric univalent of its short arm is represented as 20″ + t″6BL + t′6BS.

Table 1.I summarizes the chromosomal formulas for various kinds of aneuploids following the tomato system and assuming chromosome 5 to be involved in aneuploidy. The wheat system of symbolization is indicated in the last column of the table. The tomato system may appear complicated at first sight but is quite easy to read, and an exact idea of the amount of duplication and deficiency (the missing and duplicated arms) caused by the aneuploidy may be obtained by looking at the formulas. Some of the complex aneuploids are not expected to be viable in tomato and in other diploid species. However, the chromosomal formulas should help explain what is meant by various terms. This system may be used for other species, including wheat. Thus, by substituting 5A for 5, the monoisosomic for long arm of chromosome 5A of wheat would be written as $2n - 2(5AS \cdot 5AL) + 5AL \cdot 5AL$. These formulas may appear complicated as compared to the symbols given in the last column of Table 1.I, but when the number and letters indicating homoeologous group, the genome, and the arm of the chromosome involved are added to the symbol, the wheat system is not less cumbersome by any means.

The reader need not be unduly alarmed at the prospect of memorizing the terminology for complex aneuploids discussed in this chapter. By and large, the subject matter of the book deals with aneuploids with simple names and formulas. Complex names have been used only rarely and, at several places, in the text, I have given description along with the names of complex aneuploids.

Although substitution lines and alien additions and substitutions are not aneuploids in the strict sense of the word, yet aneuploids form the raw materials for the development of these types. Therefore, I have devoted one chapter to the discussion of the former and one to the latter.

In the substitution lines one chromosome pair is transferred from the donor variety into the recipient variety and the system of symbolization identifies both of these varieties and the chromosome pair substituted. Thus Chinese Spring (Hope-7B) indicates that homologous pair 7B from variety Hope has been substituted into variety Chinese Spring of wheat. In other words, a new line of wheat has been produced which has twenty pairs of chromosomes of Chinese Spring and one pair (7B) of Hope.

The alien additions have an extra chromosome (monosomic additions) or an extra chromosome pair (disomic additions) from a related species. These may be symbolized as $2n + 1$ I(R) (monosomic addition) or $2n + 1$ II(R) (disomic addition). Here the letter R refers to the alien rye chromosome. If the homoeologous relationships of the rye chromosome are known, these can be so indicated. Thus, if the alien chromosome is rye chromosome I, which is now known to be related to homoeologous group 5 of wheat, the disomic addition may be written as $2n + 1$ II(5R).

In the alien substitutions, an alien chromosome or a pair of alien chromosomes from a donor species replace one chromosome or a pair of chromosomes of the recipient species. If the alien chromosomes compensate for the missing chromosome they are considered homoeologous to the pair they replace. A 21-bivalent forming alien substitution of rye chromosome to wheat in which chromosome pair 6R of rye has replaced chromosome pair 6B of wheat may be written as $2n - 2(6B) + 2(6R)$. According to the nomenclature of Kimber and Sears (1968) it would be $20'' + 1''6R(6B)$.

2

Sources of Trisomics

The trisomics, like other chromosomal variants such as polyploids, deficiencies, translocations, and inversions, appear spontaneously in natural populations, but these variations are so rare that large populations have to be sampled to assemble a reasonable collection of chromosomal deviants. Cytogeneticists have employed various physical and chemical agents for the experimental production of trisomics and other chromosomal abnormalities. Individuals with abnormal chromosome complements have been further exploited to produce trisomics of various kinds. The sources of trisomics and the experimental techniques for inducing trisomy will be reviewed in this chapter.

Sources of Primary Trisomics

Of all the kinds of trisomics, the primaries are the easiest to obtain. The principal sources are (1) normal disomics: spontaneous occurrence or by treatment with physical and chemical agents; (2) asynaptic and desynaptic disomics; (3) polyploids, e.g., haploids, triploids, and tetraploids; (4) translocation heterozygotes; (5) primary trisomics, tetrasomics, multiple trisomics, and secondary, tertiary, and compensating trisomics; (6) monosomics; and (7) other chromosomal abnormalities.

Normal Disomics

Primary trisomics appear spontaneously among the progenies of normal disomics. The Globe trisomic of *Datura,* the first primary discovered and the forerunner of extensive studies on trisomics of this species, appeared spontaneously in 1915. By 1919 all the other primary trisomics of *Datura*, also of spontaneous origin, had been isolated by Blakeslee and his colleagues from *Datura* cultures. Between 1916 and 1920 the Globe trisomic appeared spontaneously nine times in normal disomic lines and fifteen times in various F_1 and backcross derivatives of *Datura* (Blakeslee, 1921a). Stubbe (1934) described four spontaneous trisomics of *Antirrhinum majus*. Among 2000 normal disomics of *Nicotiana sylvestris*, Goodspeed and Avery (1939) isolated six primary trisomic plants for four different chromosomes. Trisomics of spontaneous origin have been described in several other species (Table 2.I).

Spontaneous trisomics probably result from $n + 1$ gametes produced occasionally by the normal disomics as a result of nondisjunction in the germ line of the somatic tissues or during meiosis. The $n + 1$ gametes may also be produced by noncongression of a bivalent which fails to reach the metaphase plate and is included in one of the telophase I nuclei. Cytological evidence of nondisjunction was provided by Belling and Blakeslee (1924) who found eight cases of 11–13 disjunction in 1137 (PMC's) of normal disomic *Datura*, or about 0.4% $n + 1$ pollen grains.

Various physical and chemical agents can be employed in obtaining trisomics. Soriano (1957) and Dhillon and Garber (1960) induced trisomy in *Collinsia heterophylla* with colchicine. High and low temperatures and different kinds of radiation are effective in inducing aneuploidy. The high and low temperatures cause disturbances in the normal meiotic pairing leading to variable asynapsis (Matsuura, 1937; Sax, 1937b; Rana, 1965a; Wang *et al.*, 1965). The $n + 1$ gametes are produced from the sporocytes with univalents, and these gametes when fertilized give rise to trisomic progeny.

TABLE 2.I
Species in Which Trisomics of Spontaneous Origin Have Been Isolated

Species	Reference
Matthiola incana	Frost and Mann (1924)
Crepis capillaris	Babcock and Navashin (1930)
Secale cereale	Takagi (1935)
Corchorus olitorius	Nandi (1937)
Hordeum vulgare	Smith (1941); Kattermann (1939)
Lycopersicon esculentum	Rick (1945)
Avena sativa	McGinnis (1962a); Hacker and Riley (1963)
Gossypium hirsutum	Endrizzi *et al.* (1963); Brown (1966); Kohel (1966)
Humulus lupulus	Haunold (1968)

TABLE 2.II
Species in Which Primary Trisomics Have Been Induced by Radiation Treatments

Species	Type of radiation used	Reference
Datura stramonium	Radium rays	Gager and Blakeslee (1927)
Oryza sativa	X rays	Parthasarathy (1938)
Nicotiana sylvestris	X rays	Goodspeed and Avery (1939)
Crepis capillaris	X rays	Lewitsky (1940)
Hordeum vulgare	X rays	Tsuchiya (1960)
	Gamma rays	Das and Goswami (1967)
Oenothera berteriana	X rays	Arnold and Kressel (1965)
Arachis hypogea	X rays	Patil (1968)
Corchorus olitorius	Gamma rays	Iyer (1968)
Lycopersicon esculentum	Fast neutrons	G. S. Khush and C. M. Rick (unpublished)

Various types of radiations such as X rays, gamma rays, fast neutrons, and radium rays are known to increase the frequency of nondisjunction thus generating $2n + 1$ individuals in the irradiated progenies. Mavor (1924) showed statistically that X rays increased the rates of nondisjunction in *Drosophila*. The exact mechanism of induction of nondisjunction by radiation is not fully understood although Tikhomirova (1965) has presented data showing that chromosome breakage and reunion cause nondisjunction. Primary trisomics were induced in several species by different types of radiations (Table 2.II).

Asynaptic and Desynaptic Disomics

In asynaptic and desynaptic disomics a variable number of univalents are present at metaphase I of meiosis due to disturbances in normal pairing. Those univalents segregating at random to the two poles produce spores with extra chromosomes which produce gametes yielding simple primary trisomics and multiple trisomics upon fertilization by the haploid gametes. Koller (1938) found one trisomic plant in the progeny of asynaptic *Pisum sativum*, and Pal and Ramanujam (1940) isolated two trisomics from the progeny of asynaptic *Cap-*

sicum annuum. Of the progeny of asynaptic *Nicotiana sylvestris* studied by Goodspeed and Avery (1939), 4.4% were trisomic. Dyck (1964) and Dyck and Rajhathy (1965) isolated six of the seven possible primary trisomics from the progeny of desynaptic *Avena strigosa*, and Thomas and Rajhathy (1966) obtained several single and multiple trisomics in the progeny of an asynaptic genotype derived from the hybrid between *Avena barbata* and *A. abyssinica*. Shah (1964) obtained one trisomic in the progeny of an asynaptic plant of *Dactylis glomerata*.

Polyploids

Polyploids, especially the triploids, are the best and most dependable sources of primary trisomics. At meiosis, each trisomic group of chromosomes in an autotriploid may form a trivalent or a bivalent plus a univalent. During separation of chromosomes at anaphase I, each of the chromosomes in excess of the disomic number passes at random to either pole. Thus the triploids produce spores with all the chromosome numbers varying from n to $2n$ as reported by Belling (1921) for *Canna*, Belling and Blakeslee (1922) and Satina and Blakeslee (1937a) for *Datura*, Kihara (1924) for wheat, Lesley (1926) for tomato, Dermen (1931) for *Petunia*, Skovsted (1933) for diploid cotton, Sax (1937a) for *Tradescantia*, and Goodspeed and Avery (1939) for *Nicotiana sylvestris*. The selfed and backcrossed progenies of triploids therefore have variable chromosome numbers, generally in excess of $2n$. If all the gametophytes and zygotes with the different numbers of extra chromosomes were functional, individuals would be obtained in the triploid progeny with the range of extra chromosomes between $2n$ and $3n$. However, gametophytes and zygotes of most of the species with the higher numbers of extra chromosomes cannot survive due to the imbalance caused by these extra chromosomes. Therefore, most of the triploid progenies have a $2n$, $2n + 1$, $2n + 2$ or, rarely, $2n + 3$, chromosome number. Individuals with a higher number of extra chromosomes are obtained very rarely. This subject is considered further in Chapter 6.

The $2n + 1$ offspring of the triploids represent primary trisomics for the different chromosomes of the haploid complement of the species. If a large enough population is grown, all of the primary trisomics can be obtained from a triploid progeny. Usually more than one individual of a particular trisomic is obtained, and the $2n + 1$ individuals can be grouped into distinct morphological classes corresponding to the haploid chromosome number of species. The number of individuals of each trisomic in a triploid progeny generally varies (Table 2.III). For example, in the triploid progeny of tomato, the frequency of trisomics for longer chromosomes of the complement is lower than it is for the shorter chromosomes.

TABLE 2.III
Frequency of Primary Trisomics and Other Aneuploids in the Progenies of Triploids of Three Species: Tomato,[a] Barley,[b] and Rye[c]

Tomato			Barley			Rye		
No. chromosomes	No. plants	%	No. chromosomes	No. plants	%	No. chromosomes	No. plants	%
2n	303	37.9	2n	29	23.0	2n	12	20.7
2n + 1	342	42.7	2n + 1	59	46.7	2n + 1	23	39.7
2n + 2	131	16.4	2n + 2	22	17.5	2n + 2	12	20.7
2n + 3	7	0.9	2n + 3	6	4.8	2n + 3	6	10.3
3n	2	0.3	2n + 4	1	0.8	2n + 4	3	5.2
Not identified	14	1.7	Others[d]	9	7.2	Not identified	2	3.4
	799	99.9		126	100.0		58	100.0

Trisomic	No.	% of 2n + 1	Trisomic	No.	% of 2n + 1	Trisomic	No.	% of 2n + 1
1	4	1.1	Bush	8	13.6	Bush	9	39.1
2	26	7.6	Slender	11	18.6	Slender	5	21.7
3	13	3.8	Pale	16	27.2	Stout	4	17.4
4	79	23.1	Robust	4	6.8	Semistout	2	8.7
5	28	8.2	Pseudonormal	4	6.8	Dwarf	1	4.4
6	6	1.7	Purple	9	15.3	Feeble	1	4.4
7	50	14.6	Semierect	7	11.8	Pseudonormal	1	4.4
8	37	10.8						
9	22	6.4						
10	44	12.8						
11[e]	0	0.0						
12	32	9.3						

[a]From Rick and Barton, 1954.
[b]From Tsuchiya, 1960.
[c]From Kamanoi and Jenkins, 1962.
[d]There were two telotrisomics, three $2n + 1$ + telocentric, one triploid, one plant with twenty-five chromosomes and two were not identified.
[e]One plant identified by Rick and Barton (1954) as triplo-11 was later found to be a tertiary trisomic (Rick *et al.*, 1964; Reeves, 1968).

Because they are such rich sources of trisomics, any cytogeneticist searching for trisomics, inevitably, looks for triploids. In most species, triploids can be produced by crossing autotetraploids and diploids. However, in some, such as tomato, rice, and rye, reciprocal crosses between autotetraploids and diploids are difficult to make, so that triploids of spontaneous origin must be sought in natural populations or in cultivated fields. Spontaneous triploids are easily

TABLE 2.IV
Species in Which Trisomics Have Been Obtained from Progenies of Triploids

Species	Reference
Lycopersicon esculentum	Lesley (1926, 1928); Rick and Barton (1954)
Zea mays	McClintock (1929a)
Crepis capillaris	Babcock and Navashin (1930)
Oryza sativa	Nakamori (1932); Ramanujam (1937); Karibasappa(1961); Katayama (1963); Jackuck (1963); Sen (1965); Hu (1968); Watanabe *et al.* (1969); Iwata *et al.* (1970)
Frageria	Yarnell (1931)
Petunia	Levan (1937)
Nicotiana sylvestris	Goodspeed and Avery (1939)
Beta vulgaris	Levan (1942)
Oenothera blandina	Catcheside (1954)
Hordeum vulgare	Kerber (1954); Tsuchiya (1967)
Hordeum spontaneum	Tsuchiya (1954, 1958)
Clarkia unguiculata	Vasek (1956)
Antirrhinum majus	Rudorf-Lauritzen (1958); Sampson *et al.* (1961)
Spinacia oleracia	Tabushi (1958); Janick *et al.* (1959)
Arabidopsis thaliana	Steinitz-Sears (1963)
Secale cereale	Kamanoi and Jenkins (1962)
Collinsia heterophylla	Garber (1964)
Medicago sativa	McLennan and Kasha (1964); Kasha and McLennan (1967)
Sorghum vulgare	Price and Ross (1955, 1957); Schertz (1966); Lin and Ross (1969a)
Lotus pedunculatus	Chen and Grant (1968a)
Corchorus olitorius	Iyer (1968)
Verbena tenuisecta	Arora and Khoshoo (1969)
Solanum chacoense	Lam and Erickson (1971)
Humulus lupulus	Haunold (1970)

recognized by their sterility and semigigas characteristics. Rick (1945) reported that triploids appear spontaneously at the relatively high rate of 0.1% in all investigated varieties of tomato. Similarly, Nakamori (1932) and Morinaga and Fukushima (1935) found 70 and 150 triploid plants, respectively, in cultivated rice fields, and Watanabe *et al.* (1969) collected 27 spontaneous triploids from fields of the three varieties of rice.

All or some of the primary trisomics of the species listed in Table 2.IV have been obtained from the triploids.

The primary trisomics can sometimes be obtained from the autotetraploids. The $2n$ and $2n + 1$ or $2n + 2$ gametes produced by the autotetraploids occasionally develop parthenogenetically and give rise to disomics, trisomics, or double trisomics. Thus, out of 17,165 offsprings of $4n$ maize studied by Randolph and Fischer (1939), twenty-three were not tetraploids. Fifteen of the plants were disomics, one was trisomic, three were double trisomics, and three were not examined. Frandsen (1967) recovered trisomics arising by parthenogenesis in cultivated tetraploid potatoes. Among 1681 parthenotes obtained from tetraploid diploid crosses, 64 were trisomics ($2n = 25$), nine had 26 chromosomes, and one had 27. Hermsen (1969) obtained trisomics from the progeny of colchicine-induced autotetraploid *Solanum chacoense*. Hermsen *et al.* (1970) induced parthenogenesis by delayed pollination of tetraploid potatoes with pollen from a diploid species. The progeny consisted of disomics, trisomics, and multiple trisomics.

Haploids, when sufficiently fertile, are good sources of trisomics. Primary trisomics have been isolated from the progenies of haploids of several species. Sears (1939) obtained five trisomics in the progeny of haploid *Triticum aestivum* pollinated with the pollen from the normal plants. Among the 393 individuals in the immediate progeny of a haploid of *Datura stramonium*, 12 (3.05%) were trisomics (Blakeslee *et al.*, 1927). Endrizzi and Morgan (1955) obtained one trisomic plant from the progeny of haploid *Sorghum vulgare*. Schertz (1963) obtained five trisomics from the total progeny of 394 plants of haploid *Sorghum vulgare*. Twenty trisomic plants (6.0% of the progeny) were obtained from a haploid of *Nicotiana tabacum* by Rao and Stokes (1963). Endrizzi (1966) recovered two trisomics in the progeny of a haploid of *Gossypium barbadense*. Pochard (1968) isolated 49 trisomic plants, representing the twelve possible primary trisomics, from a progeny of 2500 plants of a haploid of *Capsicum annuum*.

The $n + 1$ spores from the haploids may originate in several ways depending upon the behavior of the univalents during meiosis (Fig. 2.1). According to one method (Fig. 2.1A), all of the univalents divide during the first division, and all but one disjoin and go to opposite poles. Both of the chromatids of the univalent which failed to disjoin are included with one of the anaphase I groups. The second division fails, and two spores, one with $n - 1$ and the other with $n + 1$ chromosomes, are produced. This type of nondisjunction of one univalent was observed by Gaines and Aase (1926) in the wheat haploid.

Another way in which $n + 1$ and $n - 1$ spores may be produced is by the failure of the first division and the inclusion of all the chromosomes in the restitution nucleus (Gaines and Aase, 1926). If one of the univalents has already divided when the restitution nucleus is formed, the two chromatids may be distributed independently at the next division and frequently pass to the same daughter nucleus and thus produce $n + 1$ and $n - 1$ spores (Fig. 2.1B). Conversely, these two chromatids may again divide equationally at second division and produce two $n + 1$ spores. Two successive divisions of univalents have

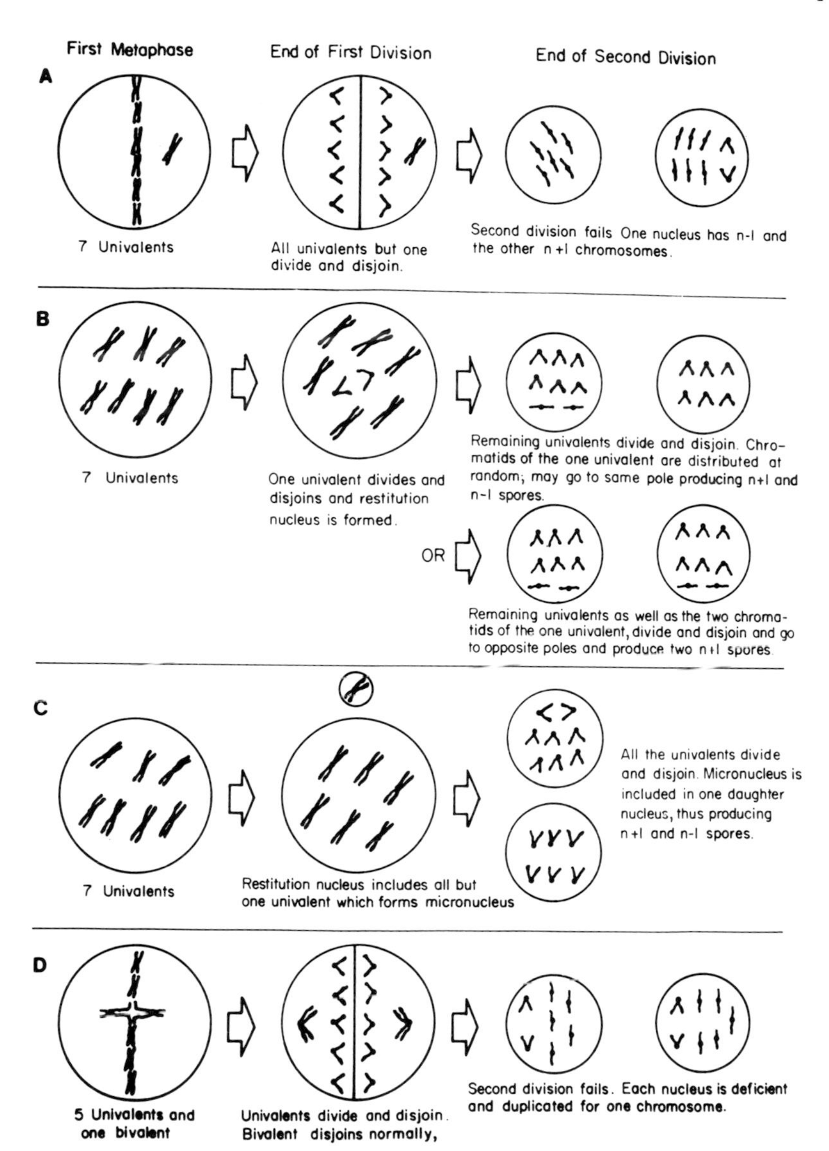

Fig. 2.1. Possible methods of origin of $n + 1$ and $n - 1$ gametes from a haploid of a species with seven pairs of chromosomes. (Modified from Sears, 1939.)

been reported by Bremer (1923) in *Saccharum* hybrids, Karpechenko (1927) in *Raphnobrassica* hybrids, Erlanson (1929) in *Rosa* hybrids, Meurman (1928) in *Ribes* hybrids, Darlington (1930) in *Prunus* hybrids, Federley (1931) in *Pygaera* hybrids, Ekstrand (1932) in asynaptic barley, and Avers (1954) in *Aster* hybrids.

A third method was suggested by Kihara and Nishiyama (1937). According to this method, a restitution nucleus, including all but one univalent, is formed at the end of first division. The excluded univalent forms a micronucleus. During second division, all of the univalents divide and disjoin. The univalent within the micronucleus also divides, but the two chromosomes thus formed are both included in one of the telophase groups, resulting in the formation of a $n + 1$ spore (Fig. 2.1C).

The fourth method, applicable to polyploid species only, was suggested by Von Berg (1935) from his observation on meiosis in a haploid plant of an intergeneric hybrid, *Triticum turgidum* × *Haynaldia villosa*. He observed that frequently all the univalents split and disjoined at the first division, the second division failed, and restitution nuclei were formed which included all the twenty-one chromatids. He pointed out that the presence and normal disjunction of a bivalent in a cell where all the univalents divide and go to opposite poles at first division would render the resulting spores deficient for one member of the bivalent and duplicated for the other (Fig. 2.1D). Such a spore cannot survive in the diploid species but may be functional in the polyploid species. The plants resulting from such gametes would have a disomic chromosome number but would be monosomic for one chromosome and trisomic for the other. The $2n + 1$ plant may be obtained from these individuals by selfing or by pollinating with pollen from normal disomics.

Translocation Heterozygotes

Due to multivalent formation, nondisjunction occurs regularly in the translocation heterozygotes. Any of the four chromosomes of a translocation tetravalent may give nondisjunction. Thus four kinds of $n + 1$ gametes are produced by a translocation heterozygote, two of which carry a primary extra chromosome. In this way, two primary trisomics may be obtained. This subject is treated in detail in the following section.

Primary Trisomics, Tetrasomics, Multiple Trisomics, and Secondary, Tertiary, and Compensating Trisomics

Various kinds of trisomics, tetrasomics, and multiple trisomics are good sources of primary trisomics. The primary trisomics produce nonparental primary trisomics among their progenies by nondisjunction. The rates of nondisjunction in the disomic chromosomes of different primary trisomics of *Datura* are higher than the rates of nondisjunction in normal disomics. For example, the disomics of *Datura* produced primary trisomics at the rate of 0.16% while 0.80% of the offspring of the primary trisomics were unrelated primaries (Blakeslee and

Avery, 1938). Why the presence of one extra chromosome increases the frequency of nondisjunction in other chromosomes is not clear. The increase may be caused by interference with meiotic divisions or, as suggested by Burnham (1962), a change in the physiology of trisomic plants may increase the rates of nondisjunction. The occurrence of nonparental primaries in the progenies of primaries have also been reported by Goodspeed and Avery (1939), Rick and Barton (1954), and Chen and Grant (1968b).

When available, tetrasomics yield primary trisomics at a very high frequency. A tetrasomic of the 15·16 chromosome ($2n + 15 \cdot 16 + 15 \cdot 16$) of *Datura* produced 70.5% trisomics in the $2n \times$ tetrasomic cross, 63.9% in the tetrasomic $\times\ 2n$ cross, and 49.3% in the tetrasomic × tetrasomic cross (Blakeslee and Avery, 1938).

The multiple trisomics ($2n + 2$, $2n + 3$, etc.), which are always present in the progenies of triploids, are good sources of primary trisomics. If all the primary trisomics do not appear in the immediate progeny of the triploid, the missing ones may be searched for in the progenies of the multiple trisomics. Goodspeed and Avery (1939) isolated eleven of the twelve primaries of *Nicotiana sylvestris* from various sources and found the twelfth in the progeny of a multiple trisomic (Goodspeed and Avery, 1941).

Most of the secondary, tertiary, and compensating trisomics regularly produce related primary trisomics in their progenies and are thus good sources of primaries. The cytological basis of this behavior and the experimental data are presented in the following two chapters.

Monosomics

A $2n - 1$ individual occasionally yields a trisomic for the monosomic chromosome. Müntzing (1930) and Uchikawa (1941) obtained trisomics for chromosome 5A in the progeny of mono-5A of *Triticum aestivum*, and Clausen and Cameron (1944) obtained trisomics in the progenies of monosomics of *Nicotiana tabacum*. The $n + 1$ spores probably result from the division and nondisjunction of the univalent chromosome at division I and passage to the same pole at division II.

Other Chromosomal Abnormalities

The production of $n + 1$ spores and $2n + 1$ individuals may result from other chromosomal abnormalities which cause disturbance of the normal meiotic process in the disomics. One variety of *Matthiola incana* has chromosomes which are several times longer than those of other varieties. A recessive gene probably interferes with the normal coiling of the chromosomes. Meiosis is disturbed and several univalents are present at metaphase I. Irregular distribution of univalents at anaphase I, produces $n + 1$ spores which give rise to trisomic individuals in the progeny (Lesley and Frost, 1927). A gene for long chromosomes has also been reported in barley (Burnham, 1946); the strain with long chromosomes yields trisomic individuals in its progeny (McLennan, 1947).

Clayberg (1959) has reported a gene in tomato causing precocious centromere division during meiosis. The precocity first appears at anaphase I in some bivalents which start to lag and divide after regular separation. The centromeres of those chromosomes which do not lag in the first division divide, in most instances, by prophase II. Irregular segregation of chromosomes during second anaphase produces aneuploid spores which result in trisomic progeny.

In many interspecific hybrids, reduced homology of the parental genomes results in incomplete meiotic pairing, and some univalents are present at metaphase I. Irregular distribution of these univalents produces aneuploid spores and progenies. Thus occasional trisomic individuals have been recorded in the progenies of numerous interspecific hybrids. Sometimes the frequency is quite high. Sachar *et al.* (1967) recorded 8.6% trisomics in the F_2 progeny of *Corchorus olitorius* × *C. capsularis*.

Sources of Tertiary Trisomics

The tertiary trisomics, like the primaries, may appear spontaneously in the progenies of normal disomics, but their frequency is extremely low. For example, among approximately two million untreated, normal *Datura* plants grown to maturity when a new tertiary could have been recognized, only six tertiary trisomics appeared spontaneously (Avery *et al.*, 1959). Asynaptic disomics or trisomics occasionally produce tertiary trisomics in their progenies. Sticky, a tertiary trisomic of *Nicotiana sylvestris*, appeared in the progeny of an asynaptic double trisomic (Goodspeed and Avery, 1939). Two or more univalents in an asynaptic may misdivide during the first meiotic division, and two telocentrics belonging to nonhomologous chromosomes join in the centromere region to produce a tertiary chromosome. This tertiary chromosome may be included in one of the telophase I nuclei and may separate at random to one of the poles during second division giving rise to an $n + 1$ spore with an extra tertiary chromosome. Similar events can take place in a triploid and a haploid where many univalents are also present. Thus one of the trisomics isolated from the triploids of tomato by Rick and Barton (1954) was a tertiary trisomic and was mistaken for primary triplo-11. However, some of the anomalous features of this trisomic which suggested its tertiary nature were pointed out by the authors, and later analysis by Rick *et al.* (1964) and Reeves (1968) showed it was indeed a tertiary trisomic. In the haploids, rearrangement may also result from pairing and crossing over

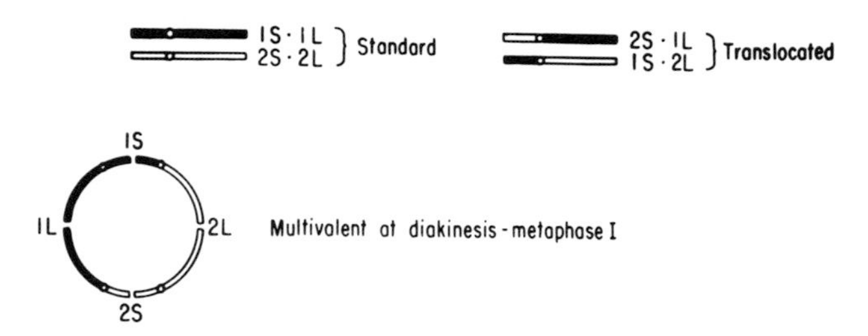

Functional gametes and trisomic progeny upon selfing

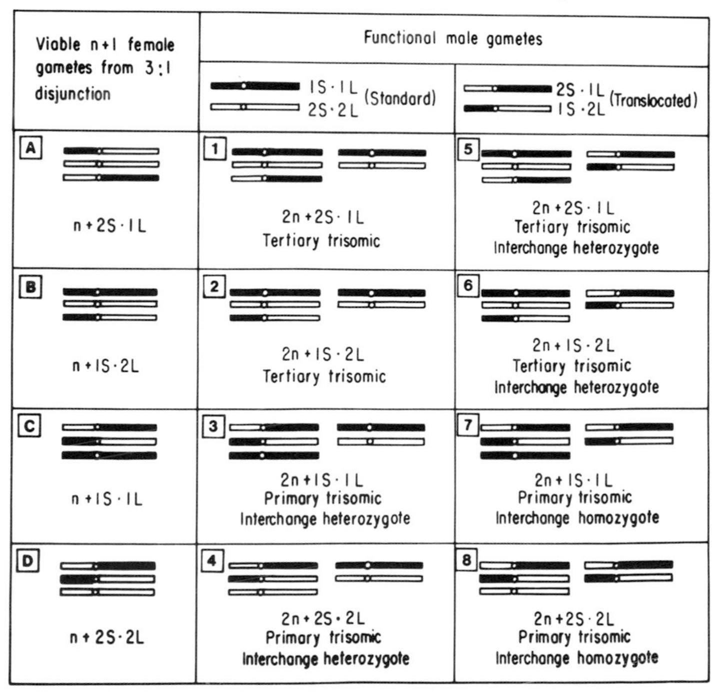

Fig. 2.2. Functional n and $n + 1$ gametes produced by a translocation heterozygote and the $2n + 1$ zygotes expected in its selfed progeny. (Modified from Khush and Rick, 1967b.)

between two nonhomologous chromosomes as reported by Endrizzi and Morgan (1955) in a haploid of *Sorghum vulgare*.

It may be noted that the appearance of tertiary trisomics in the above sources depends upon two events in the same meiotic cycle, namely, chromosome rearrangement (translocation or interchange) and production of $n + 1$ spores having an extra tertiary chromosome. The extremely low probability of these events occurring in the same meiotic cycle accounts for the rarity of tertiary trisomics from these sources. However, if a translocation is already present, a multivalent is formed at diakinesis-metaphase I and, as a result of nondisjunction from the multivalent, $n + 1$ spores are produced. Any of the four chromosomes from the multivalent which consists of two primary and two tertiary chromosomes may participate in the nondisjunction. Thus four different $n + 1$ spores and gametes may be produced by an individual heterozygous for one reciprocal translocation. Let us take a hypothetical translocation between chromosomes 1 and 2 in which 1S is translocated to 2L and 2S to 1L. Four different $n + 1$ gametes are expected from this translocation (first column, Fig. 2.2). Gametes A and B have standard chromosomes 1 and 2, but the extra chromosome in these gametes is translocated (tertiary). On the other hand, gametes C and D have translocated chromosome 2S·1L and 1S·2L, but the extra chromosome in these gametes is standard (primary).

In addition to the $n + 1$ gametes of four different types, two types of viable n gametes are produced by the translocation heterozygotes. One of these has standard 1S·1L and 2S·2L chromosomes and the other has translocated 2S·1L and 1S·2L chromosomes. All of the six types of gametes may be functional on the female side, but $n + 1$ gametes are not generally transmitted through the male side. Therefore, when an interchange heterozygote is selfed, three types of disomics are produced—standard homozygotes, interchange heterozygotes, and interchange homozygotes. In addition, eight different types of trisomics may also be expected in the progeny. These eight trisomics are shown in the second and third columns of Fig. 2.2. Trisomics 1 to 4 result from male gametes with a standard chromosome arrangement while trisomics 5 to 8 are sired by male gametes with translocated chromosomes. The extra chromosome in trisomics 1, 2, 5, and 6 is the tertiary chromosome while in trisomics 3, 4, 7, and 8 it is the primary chromosome which is extra. Trisomics 1 and 2 are homozygous for the standard arrangement, 3, 4, 5, and 6 are heterozygous for the interchange, while 7 and 8 are homozygous for the translocation.

To distinguish the different types of trisomics, it is necessary to grow and examine their selfed progenies and their progenies from crosses with normal disomics. Tertiary trisomics (1 and 2) and primary trisomic interchange homozygotes (7 and 8) when selfed produce disomics with bivalent chromosomes only. They may be further distinguished by crossing with normal disomics. The disomic F_1 progenies of tertiary trisomics would have only bivalents, but those of primary trisomic interchange homozygotes would have a multivalent of four chromosomes. The primary trisomic interchange heterozygotes (3 and 4) and tertiary trisomic interchange heterozygotes (5 and 6) produce selfed progenies that contain disomic individuals having only bivalents and disomic individuals having a multivalent of four chromosomes. They may be further distinguished by crossing them as females with normal disomics. The trisomic

F_1 progeny of primary trisomic interchange heterozygotes include primary trisomics exhibiting a multivalent of three chromosomes at diakinesis. All of the trisomic F_1 progeny of tertiary trisomic interchange heterozygotes exhibit multivalents of five chromosomes at diakinesis.

Barring position effects, the extra chromosome should have the same influence on the morphology of the individual irrespective of the chromosome arrangement of the disomic complement. Thus trisomic 1 in Fig. 2.2 should be identical to trisomic 5, and 2, 3, and 4 to 6, 7 and 8, respectively. If the morphology of the primary trisomics is known, they can be easily separated from the tertiaries.

Since the above procedure for identifying the trisomics is quite laborious, Khush and Rick (1967b) suggested backcrossing the translocation heterozygotes as females with the normal disomics instead of selfing. All the male gametes would be of the same type and have the standard chromosome arrangement. Therefore only four types of trisomics [1, 2, 3, and 4 (Fig. 2.2)] would be produced, two tertiary and two primary interchange heterozygotes, all having the maximum association of five chromosomes at diakinesis. In some cells a multivalent of four chromosomes would be present. In the tertiary trisomics, this multivalent would always occur as a chain and a closed ring of four chromosomes would never be observed. However, in primary trisomic interchange heterozygotes, rings of four would be quite frequent. Moreover, in the selfed progeny of tertiary trisomics, all the disomics would have only bivalents, while one-half the disomics in the selfed progeny of primary trisomic interchange heterozygotes would have a multivalent of four chromosomes. After identification, the primary trisomic interchange heterozygotes can be backcrossed as female to the normal disomics and made homozygous for the standard chromosome arrangement.

Tertiary and primary trisomics from selfed progenies of translocation heterozygotes have been obtained in several species. The first tertiary trisomic of any species, $2n + 1 \cdot 18$ (Wiry) of *Datura*, originated in the F_2 progenies of a cross of standard line 1A of *Datura* and another line "B white" from nature later found to be homozygous for one translocation. The F_1 plants of this cross were thus heterozygous for a translocation and yielded tertiary trisomic Wiry in their progenies. In fact, from the study of this trisomic, Belling and Blakeslee (1926) advanced for the first time the hypothesis of "segmental interchange." Several prime types of *Datura* differing in one or more translocations from the normal chromosome arrangement were later found in nature and several new translocations were induced by radiation treatment. According to Avery *et al.* (1959), thirty different tertiary trisomics of this species were isolated from translocation heterozygotes.

Burnham (1930) obtained two trisomics from a selfed progeny of a translocation heterozygote involving chromosomes 8 and 9 of maize. Both of these trisomics were later found to be primary trisomic interchange heterozygotes, one for chromosome 8 and the other for 9. In addition, two tertiary trisomics were isolated from this translocation (Burnham, 1934). Another translocation between chromosome 1 and 7 of maize yielded 0.8% $2n + 1$ individuals in its progeny (Burnham, 1948). Sansome (1933), Håkansson (1936), Lutkov (1937), and Sutton (1939) isolated primary and tertiary trisomics from interchange heterozygotes of *Pisum sativum*. Some of the trisomics obtained by Sutton (1939) were primary trisomics, others were tertiary. Lewitsky (1940) obtained a few tertiary trisomics from the translocation heterozygotes of *Crepis capillaris* Wallr. Primary and tertiary trisomics from the selfed progenies of translocation heterozygotes of barley have been obtained by several workers. For example, Burnham *et al.* (1954) isolated four and Hagberg (1954) seven trisomics from different translocations. Their exact nature, however, was not determined. Ramage (1960, 1963), Ramage and Day (1960), and Ramage and Humphrey (1964) obtained several trisomics from translocation heterozygotes of barley, but apparently only a few of them were identified with certainty. Tsuchiya (1960) obtained one tertiary and one primary trisomic in the X_2 progeny of X-rayed barley. The parental X_1 plant probably was heterozygous for a translocation. Das and Srivastava (1969) obtained five primary trisomic interchange heterozygotes from selfed progenies of translocation heterozygotes of barley. A tertiary trisomic has been reported for *Clarkia amoena* (Håkansson, 1940), *Oenothera blandina* (Catcheside, 1954), *Secale cereale* (Sybenga, 1966), and for mouse (Griffen, 1967). These appeared in the progenies of translocation heterozygotes.

Five tertiary trisomics of tomato were obtained from four translocations by Khush and Rick (1967b). Identification was simplified, as discussed earlier, by crossing translocation heterozygotes as females with pollen from the normal disomics. The resulting progenies were classified into disomics and trisomics, and the trisomics further grouped into primary and tertiary on the basis of morphological and cytological examinations. As the data of Table 2.V show, trisomics were present in all the families. However, none yielded trisomics belonging to all the four expected categories. Translocation 1-11 yielded only one of the expected tertiary types and neither of the two primaries. Translocation 5-7 produced one tertiary and both the expected primaries. Similarly, both primaries but only one tertiary appeared in the progeny of T 7-11. Finally T 9-12 was the only translocation that yielded both the expected tertiaries but only one primary. In addition, unrelated primary trisomics appeared in the progenies of three of the translocations.

The rates of nondisjunction, hence the frequency of trisomics in the progenies of translocation heterozygotes, vary from species to species and for different translocations of the same species. In the progeny of one translocation of *Pisum sativum* studied by Sutton (1939), there were as many as 18% trisomics. In

TABLE 2.V
The Frequency of Tertiary and Primary Trisomics in the Backcrossed Progenies of Four Translocation Heterozygotes of Tomato[a]

Translocation	Total progeny	$2n$	Type and frequency of aneuploid progeny			
T 1-11 (1S·11S) (1L·11L)	3575	3574	$2n$ + 1S·11S 0	$2n$ + 1L·11L 1	$2n$ + 1S·1L 0	$2n$ + 11S·11L 0
T 5-7 (5S·7L) (5L·7S)	2278	2271	$2n$ + 5S·7L 1	$2n$ + 5L·7S 0	$2n$ + 5S·5L 3	$2n$ + 7S·7L 3
T 7-11 (7S·11L) (7L·11S)	1931	1923	$2n$ + 7S·11L 1	$2n$ + 7L·11S 0	$2n$ + 7S·7L 4	$2n$ + 11S·11L 3
T 9-12 (9S·12S) (9L·12L)	2018	2002	$2n$ + 9S·12S 13	$2n$ + 9L·12L 2	$2n$ + 9S·9L 1	$2n$ + 12S·12L 0

[a]From Khush and Rick, 1967b.

barley, on the average 1% of the progeny of translocation heterozygotes is trisomic (Ramage, 1960 and personal communication). The rates of nondisjunction in the translocation heterozygotes of tomato seem to be much lower. For example, only one of 3575 tested gametes produced by T 1-11 was aneuploid, whereas T 5-7, T 7-11, and T 9-12 produced 7/2278, 8/1931, and 16/2018 aneuploid gametes, respectively (Table 2.V). The rates of nondisjunction probably depend upon the type of orientation of the translocation multivalent which is, in turn, influenced by such factors as the length of chromosomes and translocated segments, position of chiasma formation, and the degree of chiasma terminalization.

From the results of progeny tests of translocation heterozygotes of tomato discussed above, it may be noted that the two expected tertiary trisomics appeared in the progeny of only one translocation heterozygote while the other three yielded tertiary trisomics of one type only but not the other, although large populations were grown. *Datura* workers encountered similar difficulties in obtaining certain tertiary trisomics. For example, in order to obtain tertiary trisomics $2n$ + 4·22 and $2n$ + 3·21, prime type 4, which has the chromosomes 3·21 and 4·22 in place of 3·4 and 21·22, was crossed with the standard line 1A. The tertiary trisomic $2n$ + 4·22 frequently appeared in the offspring of such heterozygotes but $2n$ + 3·21 did not appear under similar conditions. This type was finally obtained (Blakeslee *et al.*, 1936) by the following procedure.

A secondary trisomic $2n$ + 3·3 was rendered homozygous for prime type 4 by continuous backcrossing to the latter as the male parent. Secondaries regularly throw primaries in their progenies (Chapter 4). Therefore, among the offsprings of the secondary $2n$ + 3·3 homozygous for the 3·21 and 4·22 chromosomes, a $2n$ + 3·21 appeared through segregation. This was a primary trisomic of prime type 4. By backcrossing to the standard line 1A, it was made homozygous for 3·4 and 21·22 chromosomes and thus became a desired tertiary trisomic of the standard line.

The primary trisomics may similarly be employed for inducing trisomy from the translocation heterozygotes. By crossing the primary trisomic as female with a translocation heterozygote, a primary trisomic rearrangement heterozygote can be obtained. Such an individual forms a chain of five chromosomes during prophase of meiosis in a majority of the cells. One type of disjunction from the chain of five would yield spores with an extra tertiary chromosome. For example, a primary trisomic for chromosome 1·2, heterozygous for a translocation between chromosomes 1·2 and 17·18, would form a chain of five chromosomes as follows:

The three middle chromosomes would occasionally go to the same pole and produce a spore having the 1·18 tertiary chromosome in addition to the full haploid complement. When fertilized by a haploid gamete such a gamete would yield the tertiary trisomic, $2n + 1 \cdot 18$. The tertiary trisomic $2n + 2 \cdot 17$ may be produced by crossing the primary trisomic for chromosome 17·18 with the same translocation.

Sometimes only one product of reciprocal translocation is recovered. For example, during radiation treatment of tomato pollen, two chromosomes are broken in the centromere regions. Two arms of different chromosomes join together to constitute a tertiary chromosome with a functional centromere. The other two arms either fail to join or cannot constitute a functional centromere. They are thus eliminated during either the mitotic divisions of the pollen grain or the fertilized zygote, and an individual with $2n - 1$ chromosomes (tertiary monosomic) is produced (Khush and Rick, 1966b). Such individuals form a trivalent at diakinesis-metaphase I, and occasionally the three chromosomes pass to the same pole at anaphase I yielding $n + 1$ gametes with the extra tertiary chromosome. When fertilized by pollen from normal disomics, such gametes produce tertiary trisomic individuals. Khush and Rick (1966b) raised progenies of nine such tertiary monosomics of tomato, and two of them yielded tertiary trisomics. Out of 595 plants of a backcross family of mono-4S·10S, three were tertiary trisomics for 4L·10L, three were other aneuploids, and the remaining plants were disomics. Similarly, mono-2S·10S yielded two $2n +$ 2L·10L in a progeny of 383 plants.

The tertiary trisomics isolated from tertiary monosomics have the extra chromosome composed of two complete arms of two different chromosomes because the breakage and reunion occurs in the centromere region of the chromosomes. The extra chromosome of the tertiary trisomics obtained from the tertiary monosomics is easily identified. Identification of the tertiary trisomics from the translocation heterozygotes is facilitated by determining the new arrangement of the translocated chromosomes. For example, a translocation between chromosomes 1 and 11 may produce chromosomes 1S·11L and 1L·11S or 1L·11L and 1S·11S. The heterozygotes for the former rearrangement yield $2n$ + 1S·11L and $2n$ + 1L·11S tertiaries while the latter produce $2n$ + 1L·11L and $2n$ + 1S·11S. It is helpful if the new arrangement of chromosome arms can be determined from the translocation configurations. Moreover, if the points of rearrangement are known, the exact amount of duplication caused by the tertiary chromosome can be ascertained. These analyses can best be made at pachytene stage of meiosis. However, the chromosomes of only a few species are amenable to pachytene analysis.

In most of the tertiary trisomics of *Datura,* as pointed out by Avery *et al.* (1959), the exact arm composition was not known. For example, it was not known whether the tertiary chromosome consisted of two complete arms of the interchanged chromosomes or more than one arm of one chromosome and less than one arm of the other. Such designations as $2n$ + 1·18, which implies that the extra chromosome comprised the ·1 arm of the 1·2 chromosome and ·18 arm of the 17·18 chromosome, were therefore arbitrary. Without precise knowledge of the points of breakage and reunion of the parental translocation and in the absence of pachytene analysis, the content of the tertiary chromosome could not be known exactly. If the chromosomes had not been broken and united in their centromeres, the supposed 1·18 chromosome might have consisted of more than the arm of one and less than the arm of the other interchanged chromosome.

All reported tertiary trisomics of *Datura, Pisum, Oenothera, Crepis, Secale,* and mouse, and probably of barley, suffer this uncertainty. Thanks to the feasibility of pachytene analysis in the tomato, the points of rearrangement can be ascertained and an exact identification of the tertiary chromosome made. The exact arm content of the seven tertiary trisomics of tomato reported by Khush and Rick (1967b) was determined by pachytene analysis of the translocation heterozygotes and of the tertiary trisomics, and these trisomics were employed in the cytogenetic analysis of the tomato genome (Chapter 5).

Sources of Secondary Trisomics

The origin of secondary trisomics is dependent upon four events: (1) origin of the isochromosome, (2) production of n + isochromosome spores, (3) viability of gametophytes with extra isochromosome, and (4) viability of $2n$ +

isochromosome zygotes. Of these, events (1) and (3) seem to be of crucial importance. The production of isochromosomes is mainly a chance event, although certain experimental techniques can be employed to produce them. However, if the isochromosome is present, n + isochromosome spores are produced by nondisjunction. The imbalance caused by certain isochromosomes may not be tolerated at the gametophytic level, and the secondary trisomics for these chromosome arms are not obtained.

Belling (1927) suggested that isochromosomes (secondary chromosomes) might originate from crossing over between sister strands that lie side by side but with the ends of one reversed. He proposed another explanation that unequal crossing over takes place between sister strands retaining the centromeres of the new secondary chromosomes formed by the joining together of the broken ends of the two similar arms affected. However, these hypotheses are of doubtful validity for lack of supporting evidence. On the other hand, the misdivision of univalents is known to give rise to isochromosomes (Darlington, 1939, 1940). A univalent chromosome may divide equationally at anaphase I or it may misdivide (Chapter 3). The products of misdivision may be two telocentrics, two isochromosomes, one for each arm, one telocentric and one isochromosome, or four telocentrics. Logical sources, then, of secondary trisomics are the genotypes with univalent chromosomes.

Monosomic individuals always have one univalent chromosome at metaphase I, and it frequently misdivides at anaphase I. Sears (1952a) observed misdivision in 39.7% of the microsporocytes of mono-5A of hexaploid wheat, and over half of these misdivisions were of the type that produced isochromosomes. The frequency of plants with an isochromosome for the long arm of 5A in the progeny of mono-5A was determined genetically to be 7.0%. Presumably the frequency of plants with an isochromosome for the short arm of 5A was similar. The misdivision of the univalents of other monsomics of wheat was also observed by Sears (1952a). When wheat monosomics are selfed, monoisosomics, e.g., plants with one isochromosome in place of a pair of chromosomes, are produced, and, rarely, a monoisodisomic ($2n - 1$ + isochromosome). By nondisjunction these monoisodisomics produce gametes with n + isochromosome, which produce monoisotrisomics (secondary trisomics) when fertilized by pollen from a normal disomic. The frequency of monoisodisomics may be increased by backcrossing the monosomics to normal disomics instead of selfing. Occasionally a secondary trisomic appears directly in the progeny of a backcrossed monosomic. Sears (1952a) reported such a plant in the backcrossed progeny of mono-5A.

The availability of monosomics for all the chromosomes of wheat and the frequent misdivision of univalents in this species, should make it relatively easy to obtain the secondary trisomics for all the chromosome arms. However, the secondary trisomics offer no particular advantage over monosomics or monoisosomics for the cytogenetic analysis of the genome, so no effort has been made to establish the secondary trisomic series in wheat. Only two secondaries have been reported, the one mentioned above and another reported by Sears (1954).

The frequency of misdivision in the monosomics of species such as oats, cotton, and tobacco is much lower. The diploid species in which the secondary trisomics are especially useful tolerate very little monosomy. Hence this source of secondary trisomics is not applicable to these species. However, the progenies of primary trisomics should be carefully watched for the occurrence of secondary trisomics. The extra chromosome may be present as a univalent in some sporocytes and may misdivide to produce an isochromosome which may yield an n + isochromosome gamete when included in one of the meiotic products. According to Blakeslee and Avery (1938) secondary trisomics of *Datura* arise in the progenies of related primary trisomics at a frequency of 0.07% which is much higher than the frequency of their occurrence in the progenies of unrelated primary, secondary, and tertiary trisomics, and in disomics. Of a total of fourteen secondary trisomics of *Datura* isolated by Blakeslee and co-workers, nine appeared in the progenies of related primary trisomics and four originated more than once from the same source. Thus $2n + 19 \cdot 19$ and $2n + 2 \cdot 2$ appeared on seven and five independent occasions, respectively, in the progenies of their related primaries. Nine secondary trisomics appeared in the progenies of unrelated primaries, but each of them only once, and five originated in the progenies of unrelated secondary trisomics. From all the sources put together, the $2n + 1 \cdot 1$ secondary appeared six times, the $2n + 2 \cdot 2$ thirteen times, the $2n + 5 \cdot 5$ nine times, and the $2n + 19 \cdot 19$ eleven times. One secondary trisomic of tomato, $2n$ + 2S·2S, was isolated by Moens (1965) from a progeny of a related primary trisomic, $2n$ + 2S·2L. Similarly, Tsuchiya (1960) reported occurrence of secondary trisomics in the F_2 progeny of the slender trisomic of barley segregating for a genetic marker. A secondary trisomic for the short arm of chromosome 5 of maize was isolated by Rhoades (1933b) from a progeny of an unrelated primary trisomic (triplo-6).

Asynaptics, desynaptics, triploids, and haploids which have many univalents may also produce isochromosomes by misdivision. Goodspeed and Avery (1939) obtained three secondaries from the progenies of asynaptics and triploids of *Nicotiana sylvestris*. Koller (1938) observed misdivision of univalents in asynaptic *Pisum sativum*, and some of the trisomics isloated by him in the progeny may have been secondary trisomics. Rajhathy and Fedak (1970) obtained one secondary trisomic in the progeny of a desynaptic *Avena strigosa*.

Monotelodisomics and telotrisomics may also yield secondary trisomics by conversion of the telocentric chromosome into the isochromosome. Thus, Lesley and Lesley (1941) obtained a secondary trisomic of tomato in the progeny of a plant deficient for one arm of a chromosome; and a secondary trisomic of maize ($2n$ + 5S·5S) was generated at a high frequency in the progeny of a telotrisomic for 5S (Rhoades, 1938, 1940). When $2n$ + ·5S was used as a male, the secondary trisomic appeared with a frequency of 0.42%. When used as a female, the frequency of the secondary trisomic in the progeny was 0.22%. The secondary chromosome 5S·5S did not transmit through the male but the telocentric ·5S did. Rhoades, therefore, concluded that the isochromosome was produced during the first pollen mitosis by doubling of the telocentric and reunion of the terminal centromeres of the two daughter telocentrics.

Secondary trisomics may originate spontaneously from almost any other chromosomal variant or from disomics. According to Blakeslee and Avery (1938), $2n + 10 \cdot 10$ of *Datura* appeared in the progeny of a disomic plant. Similarly, Khush and Rick (1969) obtained $2n + 9S \cdot 9S$, $2n + 6L \cdot 6L$, and $2n + 12L \cdot 12L$ secondaries of tomato in the progenies of mono-6S·7S, mono-4S·10S, and mono-10S·12S, respectively. The secondary $2n + 12L \cdot 12L$ also appeared in the progeny of a plant deficient for 4S and in the progeny of a monoisodisomic ($2n - 3S \cdot 3L + 3L \cdot 3L$).

A plant carrying an isochromosome in its chromosome complement, throws secondary trisomics by nondisjunction. For example, diiso compensating trisomics of tomato, in which a normal chromosome is replaced by two isochromosomes, one for each arm, regularly produce secondary trisomics in their progeny (Chapter 4, Table 4.XI).

From the foregoing paragraphs it may appear that the occurrence of secondary trisomics is a chance event. Avery *et al.* (1959) stated that no technique is known for producing the secondary trisomics experimentally. However, Sen (1952) and Khush and Rick (1967d) showed that certain chemical and physical agents can be employed for the experimental production of isochromosomes in tomato. Sen (1952) obtained two monoisodisomics, $2n - 8S \cdot 8L + 8L \cdot 8L$ and $2n - 9S \cdot 9L + 9L \cdot 9L$, in the progenies sired by pollen treated with formaldehyde and ammonia vapor. Khush and Rick (1967d) obtained five monoisodisomics from radiation treatment of pollen, two of which were the same ones obtained by Sen. It appears that chemical and radiation treatments cause breaks in the centromere region. One of the broken arms is lost, and the other is converted into an isochromosome by reunion of the terminal centromeres of two daughter telocentrics during pollen mitosis. If such a pollen grain deficient for one arm and duplicated for the other arm of the same chromosome fertilizes a normal haploid egg, a monoisodisomic zygote is produced. It may be noted that functional pollen grains of similar chromosomal constitution are produced by misdivision of the univalent chromosome in the monosomics of polyploid wheat. However, pollen grains of this type in the diploid species are inviable due to the deficiency of one arm. In these species, only those spores are functional which receive the full chromosome complement. If a mutagen is used after the full development of the pollen and chromosomal deficiencies are produced, the viability of the pollen is not impaired, and it can take part in fertilization (Khush and Rick, 1966b, 1967d). In a similar

TABLE 2.VI
The Frequency of Related Secondary Trisomics in the Progenies of Monoisodisomics of Tomato[a]

Monoisodisomics	Total	2*n*	Related secondary trisomics No.	Related secondary trisomics %	Other aneuploids
2*n* − 10S·10L + 10L·10L	1545	1462	78	5.04	5
2*n* − 9S·9L + 9L·9L	590	573	15	2.54	2
2*n* − 2S·2L + 2L·2L	1416	1412	0	0.00	4
2*n* − 3S·3L + 3L·3L	822	821	0	0.00	1
2*n* − 8S·8L + 8L·8L	284	275	8	2.81	1

[a]From Khush and Rick, 1967d.

manner Morris (1955) obtained one monoisodisomic of maize (2n − 4S·4L + 4S·4S) from irradiated seed.

When backcrossed as a female to a tester stock one of the monoisodisomics (2*n*− 9S·9L + 9L·9L) of tomato reported by Sen (1952) produced three secondary trisomics (2*n* + 9L·9L) in a progeny of 45 plants. In a selfed progeny of this monoisodisomic, 5% of the individuals were secondary trisomics. When two monoisodisomics, 2*n* − 9S·9L + 9L·9L and 2*n* − 10S·10L + 10L·10L, were discovered by Khush and Rick (1967d) in the progenies sired by irradiated pollen, they were also fecunded with pollen from a normal disomic, and both of the progenies yielded secondary trisomics (Table 2.VI). Encouraged by this success in obtaining secondary trisomics, a systematic search of tomato progenies was made and five more monoisodisomics were discovered. Progenies were obtained from three of them, and one yielded secondary trisomics (Table 2.VI).

Because of their usefulness as a source of secondary trisomics, monoisodisomics were sought using marker genes whose arm locations were known (Khush and Rick, 1967d). According to this method, stocks homozygous for recessive markers of those arms whose deficiencies are tolerated were pollinated with irradiated pollen carrying normal alleles. The F_1 progenies were grown, and plants showing the recessive phenotype were examined cytologically. Monoisodisomics were included in these pseudodominant plants. In a progeny of a cross in which the female parent was homozygous for *ℓ* a recessive marker of 8S, five *ℓ* plants appeared. Of the two examined, one proved to be monoisodisomic, 2*n* − 8S·8L + 8L·8L. A progeny of 284 plants of this monoisodisomic yielded eight secondary trisomics. A 2*n* − 7S·7L + 7L·7L plant was similarly obtained with the aid of *var,* a recessive marker of 7S, but this plant was so weak that it failed to yield a progeny.

Caldecott and Smith (1952) obtained one secondary trisomic of barley by X-ray treatment of the dormant seed. In addition, twenty-seven X_1 spikes (the plants were probably chimaeras) were doublemonoisosomics, i.e., had two isochromosomes, one for each arm, which replaced two normal homologs of the chromosome complement. These isochromosomes evidently originated as a result of reciprocal translocations between the two homologs with breaks occurring in or near the centromere regions and the joining together of homologous arms. In one spike there were four isochromosomes in place of two homologous pairs. In maize, Morris (1955) obtained seventeen doublemonoisosomic plants grown from X-rayed seed. Maguire (1962) also obtained a maize chimaera from irradiated seed in which two isochromosomes, one for the short arm of chromosome 6 and the other for the long arm, replaced both the normal chromosomes.

In the progenies of such plants obtained upon backcrossing to a normal disomic, most of the offspring would be diiso compensating trisomics, each yielding two secondary trisomics in the progeny of a second backcross.

It is evident that secondary trisomics can be produced experimentally by the techniques enumerated above. The statement of Avery *et al.* (1959) that no technique is available for producing secondary trisomics is unwarranted.

It may be noted from Table 2.VI that two of the monoisodisomics of tomato, $2n - 2S \cdot 2L + 2L \cdot 2L$ and $2n - 3S \cdot 3L + 3L \cdot 3L$, failed to yield secondary trisomics in their progenies. As discussed by Khush and Rick (1967d, 1969), the imbalance caused by isochromosomes for the long arms is not tolerated by the gametophytes of tomato. The gametophytes with extra $2L \cdot 2L$ or $3L \cdot 3L$ do not survive, hence the secondary trisomics for these two arms were not obtained. On the basis of arm length alone, Khush and Rick (1969) hypothesized that transmissible secondary trisomics for 1L, 2L, 3L, 4L, and 6L of tomato may never be obtained. One secondary trisomic for 6L was obtained, but it failed to transmit the isochromosome to the next progeny. Clearly the tolerance limits of gametophytes for the duplications are of crucial importance in obtaining transmissible secondary trisomics.

Of the possible twenty-four secondary trisomics of *Datura*, fourteen were isolated between 1921 and 1929. No new secondary appeared between 1929 and 1955, although several of the fourteen reappeared in *Datura* cultures a number of times during this period (Avery *et al.*, 1959). It seems reasonable to conclude that the imbalance caused by extra isochromosomes for the ten arms for which secondaries have not appeared is not tolerated by *Datura* gametophytes.

Sources of Telotrisomics

The telotrisomics appear spontaneously in the same sources as the secondary trisomics. Like secondary trisomics they can be induced by treatment with various mutagenic chemicals and radiation agents, and they may arise by misdivision of the univalents in trisomics. Rhoades (1936) isolated one telotrisomic ($2n + \cdot 5S$) in the progeny of the primary trisomic for chromosome 5 of maize. Blakeslee and Avery (1938) obtained a $2n + \cdot 11$ telotrisomic in the progeny of the $2n + 11 \cdot 12$ trisomic of *Datura*. Similarly, telotrisomics were obtained in the progenies of primary trisomics of rye (Kamanoi and Jenkins, 1962), barley (Frost and Ising, 1964; Tsuchiya, 1969; Fedak *et al.*, 1971), and *Lotus pedunculatus* (Chen and Grant, 1968b).

Telotrisomics may also be obtained from the progenies of triploids, haploids, and asynaptics. Since many univalents are present in such individuals during the first division of meiosis, the telocentrics are produced by misdivision. Of the resulting spores, those with complete chromosome complement and a telocentric upon fertilization by a haploid male gamete give rise to telotrisomics. Thus telotrisomics were obtained in the progenies of asynaptics and triploids of *Nicotiana sylvestris* (Goodspeed and Avery, 1939) and in triploids of *Hordeum spontaneum* (Tsuchiya, 1960), *Secale cereale* (Kamanoi and Jenkins, 1962), and *Lotus pedunculatus* (Chen and Grant, 1968a).

The telotrisomics appear regularly in the progenies of compensating trisomics in which the missing chromosome is compensated by the telocentric and a secondary or a tertiary chromosome. Smith (1947) and Moseman and Smith (1954) isolated a telotrisomic of *Triticum monococcum*, and Khush and Rick (1967a) obtained a telotrisomic of tomato in the progenies of compensating trisomics in which the missing chromosome was compensated by a telocentric and an isochromosome. Khush and Rick (1968b) isolated two telotrisomics in the progenies of tertiary monosomics—$2n + \cdot 3L$ appeared in the progeny of mono-$3S \cdot 11L$, and $2n + \cdot 8L$, in the progeny of mono-$8S \cdot 5L$; another telotrisomic of tomato, $2n + \cdot 10S$, was detected in the progeny of a monoisodisomic ($2n - 2S \cdot 2L + 2L \cdot 2L$).

Monotelodisomics which arise by the misdivision of univalents in the monosomics of wheat (Sears, 1952a, 1954), oats (McGinnis *et al.*, 1963), and cotton (White and Endrizzi, 1965) are potent sources of telotrisomics. The monotelodisomics produce gametes in which a telocentric is present in addition to the haploid chromosome complement, and such gametes yield telotrisomics when fertilized by a haploid gamete. However, in the diploid species where the telotrisomics are most useful for cytogenetic analysis, monosomics in general are not viable, and telocentrics cannot be obtained as in the polyploid species. Nevertheless, monotelodisomics for at least those arms of the chromosome complement whose deficiency is tolerated at the sporophytic level may be induced by radiation or chemical mutagenesis. Thus whole arm deficiencies for fifteen of the twenty-four arms of the tomato chromosome complement are tolerated at the diploid level (Khush and Rick, 1966b, 1968a), and terminal deficiencies for at least eleven of these arms have been induced by radiation treatment. In some terminal deficiencies the break occurs near the centromere so that the entire arm, except one or two small chromomeres, is lost. Such a chromosome is essentially telocentric, as the remaining chromomeres of the missing arm are genetically inert. Deficiencies of this kind were also sought with the help of marker genes as was explained in the previous section.

Although several whole arm deficiencies of this kind were obtained in tomato, only one deficiency was employed for obtaining the telotrisomics. A plant deficient for 4S was crossed with a normal disomic, and a progeny of 1601 plants yielded two desired $2n + \cdot 4L$ telotrisomics (Khush and Rick, 1968b).

Sources of Compensating Trisomics

The compensating trisomics can be synthesized from existing trisomic and translocation stocks or may be induced by radiation treatment or chemical mutagenesis. The first compensating trisomic of any species, a ditertiary of *Datura,* appeared in 1921 among the offspring of a plant exposed to radium emanation. The missing chromosome 1·2 of this trisomic is compensated by two tertiary chromosomes 1·9 and 2·5 so that the chromosomal formula of this trisomic is $2n - 1\cdot2 + 1\cdot9 + 2\cdot5$ (Blakeslee, 1927a). The exact mode of origin of this trisomic is not known. Presumably, two translocations involving three nonhomologous chromosomes were induced by radium treatment of the parent plant and three translocated chromosomes, 1·9, 2·5, and 6·10 were produced. In this plant, a chain or ring of six chromosomes was probably formed at metaphase I of meiosis as follows:

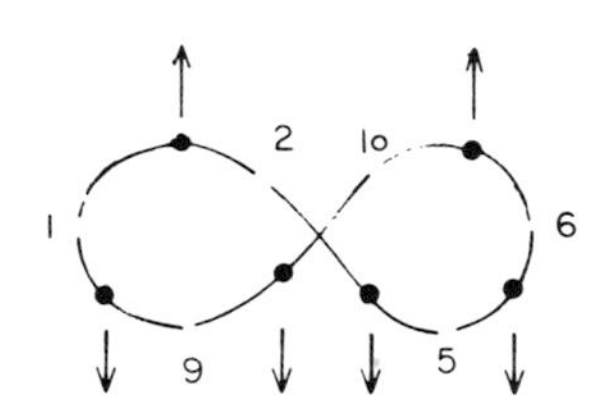

The type of disjunction from the translocation multivalent indicated by arrows above probably yielded gametes with $n - 1\cdot2 + 1\cdot9 + 2\cdot5$ chromosomes which produced the trisomic when fertilized by the haploid gamete.

Blakeslee and co-workers induced rings of six or more chromosomes in *Datura* by radiation treatment and produced several compensating trisomics experimentally (Avery *et al.*, 1959). For example, a plant with a ring of six chromosomes having 12·17, 11·13, and 14·18 translocated chromosomes was obtained. The progeny of this plant yielded a ditertiary compensating trisomic in which the missing 11·12 chromosome was compensated by the 11·13 and 12·17 chromosomes. The *Datura* workers produced plants with rings of six or more chromosomes by crossing two or more translocation stocks. They were able to obtain plants with rings of ten chromosomes. The larger the number of chromosomes involved in the translocations, the larger would be the number of compensating trisomics expected in the progeny. Thus, three, four, and five compensating trisomics are expected in the progeny of plants with rings of six, eight, and ten chromosomes.

The isotertiary compensating trisomics may be produced by crossing a translocation homozygote or a heterozygote with an appropriate secondary trisomic. Thus, when the $2n + 3 \cdot 3$ secondary of *Datura* was crossed with a translocation homozygote having the $3 \cdot 21$ and $4 \cdot 22$ instead of $3 \cdot 4$ and $21 \cdot 22$ chromosomes, the trisomic F_1 was heterozygous for the translocation and formed a ring of five chromosomes. Among the progeny of 76 plants of this trisomic, two plants of the desired compensating type appeared and had $2n - 3 \cdot 4 + 3 \cdot 3 + 4 \cdot 22$ chromosomal constitution (Avery *et al.*, 1959). The gametes which gave rise to these two trisomics probably originated by orientation and disjunction of five chromosomes from the ring of five as follows:

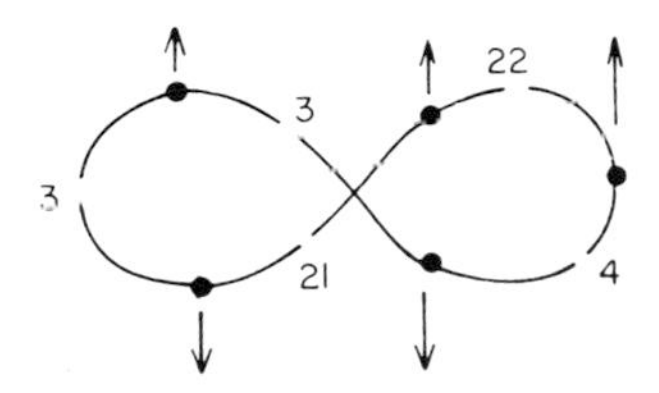

One isotelo compensating trisomic of *Triticum monococcum* reported by Smith (1947) appeared in the X_2 progeny of dormant seed treated with X rays. Similarly, one isotelo compensating trisomic of tomato (Khush and Rick, 1967a) occurred in the X_2 progeny of X-ray treated pollen. In both cases, the breaks probably occurred in the centromere region of one chromosome in such a way that one arm with a functional centromere remained telocentric and the other was converted into an isochromosome. Isotelo compensating trisomics may also be produced by crossing a telotrisomic for one arm of a chromosome with a secondary trisomic for the other arm of the same chromosome. In the F_1 generation of such a cross, some double trisomics having one extra telocentric chromosome and one isochromosome would appear. The selfed or backcross progenies of these double trisomics would yield the isotelo compensating trisomics. This procedure is outlined in Scheme 2.1 with the telotrisomic for the short arm of chromosome 10 and the secondary trisomic for long arm. Both of these have been isolated in tomatoes.

$2n + 10L \cdot 10L \times 2n + \cdot 10S$

↓

$F_1 = 2n + 10L \cdot 10L + \cdot 10S$

From the union of $n + 10L \cdot 10L$ and $n + \cdot 10S$ gametes

upon selfing ↓

$2n - 10S \cdot 10L + 10L \cdot 10L + \cdot 10S$

From the union of $n - 10S \cdot 10L + 10L \cdot 10L + \cdot 10S$ and n gametes

Scheme 2.1

It should be noted that one of the two trisomics used in this procedure must transmit the extra chromosome through the male and that the double trisomic must be viable and fruitful enough to yield a large progeny. Isotertiary, ditertiary, and diiso compensating trisomics may also be produced by this procedure.

The diiso compensating trisomics are also produced by radiation treatment. If one chromosome is broken in the centromere region and both the telocentric arms are converted into isochromosomes, a diiso compensating trisomic will be produced. The diiso compensating trisomic for chromosome 7 of tomato ($2n - 7S \cdot 7L + 7S \cdot 7S + 7L \cdot 7L$), which appeared in the X_2 of neutron-treated pollen, probably resulted from such a break (Khush and Rick, 1967a). Another diiso compensating trisomic of tomato ($2n - 3S \cdot 3L + 3S \cdot 3S + 3L \cdot 3L$) appeared spontaneously in the progeny of an isotelo compensating trisomic ($2n - 3S \cdot 3L + \cdot 3S + 3L \cdot 3L$), probably as a result of conversion of telo $\cdot$ 3S into iso $3S \cdot 3S$ (loc. cit.). Diiso compensating trisomics may also be obtained from doublemonoisosomic plants such as those of barley (Caldecott and Smith, 1952) and maize (Morris, 1955; Maguire, 1962) mentioned in an earlier section in this chapter.

3

Cytology of Trisomics

The cytological behavior of the extra chromosome in trisomics has an important bearing on the transmission rates of the extra chromosome to the next generation and on the genetic segregation of marker genes. Moreover, the different types of pairing associations which result at diakinesis and metaphase I due to the presence of the extra chromosome give clues to the nature of the trisomic. Pachytene analysis of the trisomic, if feasible, reveals the true identity of the extra chromosome. In view of its importance, this subject is treated in detail in this chapter.

Cytology of Primary Trisomics

The three homologous chromosomes of the primary trisomic generally pair to form a trivalent at the pachytene stage of meiosis. Only two of these chromosomes pair at a given point along the length of the chromosome but due to the change in pairing partners, the third member may pair with the other two at different sites (Fig. 3.1). In species such as maize and tomato, it is possible to identify each member of the chromosome complement at pachytene, and the extra chromosome of each of the trisomics can be identified by examining the trivalent associations at this stage of meiosis. The extra chromosomes of all of the primary trisomics of maize (Rhoades and McClintock, 1935) and tomato (Rick and Barton, 1954; Rick *et al.*, 1964) and of seven of the ten trisomics of *Sorghum* (Poon and Wu, 1967) were identified at the pachytene stage. Attempts to identify the extra chromosomes of trisomics of other species at pachytene have been less successful because the individual chromosomes are difficult to identify at this stage. Another source of uncertainty in the identification of the extra chromosome at this stage is the occurrence of nonhomologous pairing. Two nonhomologous arms of two members of the trivalent may pair (Fig. 3.2) and complicate the process of identification. This type of nonhomologous pairing occurs rarely in the primary trisomics of tomato but very frequently in the primary trisomics of *Sorghum* (Poon and Wu, 1967; Venkateswarlu and Reddi, 1968) and *Solanum* (Vogt and Rowe, 1968). In metacentric chromosomes, the two nonhomologously paired arms are equal in length, and the trivalent may be mistaken for the trivalent of the secondary trisomic. This type of nonhomologous pairing was observed frequently in what were obviously the primary trisomics of *Solanum* by Vogt and Rowe (1968). They were led to believe that the extra chromosomes in these trisomics might be secondary chromosomes.

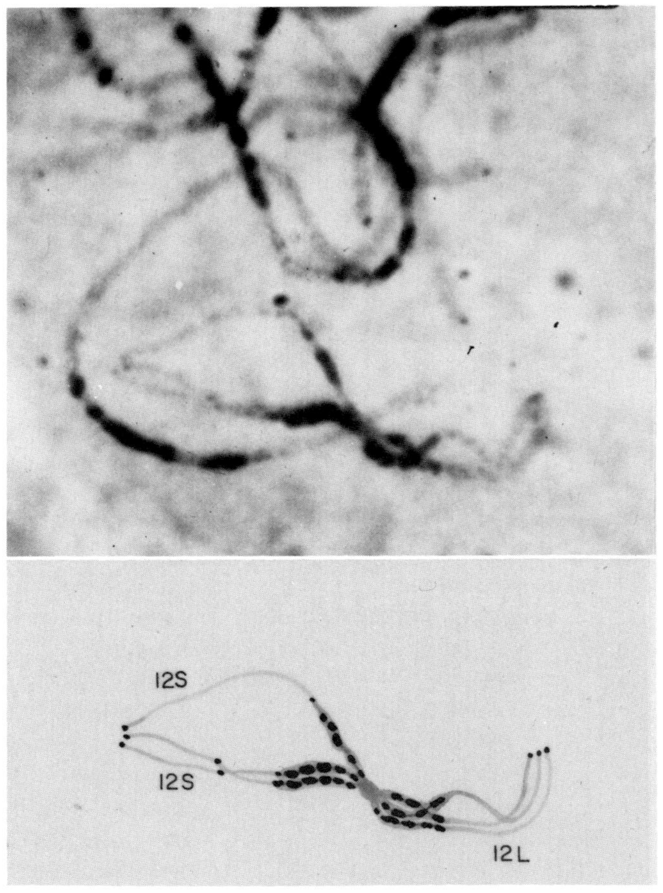

Fig. 3.1. Trivalent association at pachytene in triplo-12 of tomato. Note that only two chromosomes are paired at a given point. (From Rick *et al.*, 1964.)

In primary trisomics, pachytene cells with a univalent are extremely rare. Einset (1943) found only 2–4% of pachytene cells with univalents in the primary trisomics of maize. However, the frequency of cells with univalents increases due to failure of chiasma formation at pachytene and chiasma terminalization at later stages of meiosis, so that the three members of the pachytene trivalent may resolve themselves into a bivalent and univalent or, rarely, into three univalents. The following types of chromosome associations may be observed at diakinesis or metaphase I in the primary trisomics: (1) $n - 1$ bivalents + 1 trivalent; (2) n bivalents + 1 univalent; (3) $n - 1$ bivalents + 3 univalents.

The frequency of various types of associations may be variable for the trisomics of different species and is generally different for the different trisomics of the same species. For example, the longer chromosomes of maize and tomato have a higher chiasma frequency than the shorter ones, and the trisomics for these chromosomes form trivalents at a higher frequency (Table 3.I). The different members of the chromosome complements of barley (Tsuchiya, 1960) and *Nicotiana sylvestris* (Goodspeed and Avery, 1939) are almost of the same size, and no significant differences in the frequency of microsporocytes with trivalents were observed.

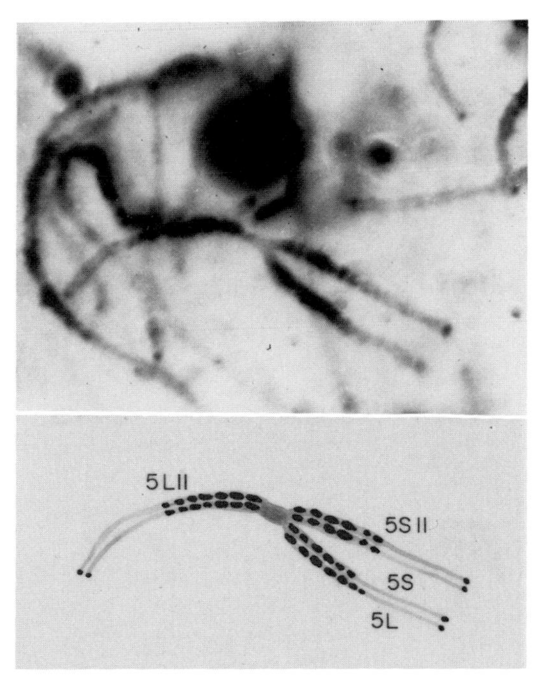

Fig. 3.2. Trivalent association at pachytene in triplo-5 of tomato. 5S and 5L are paired nonhomologously.

TABLE 3.I
The Frequency of Microsporocytes with Trivalents and Bivalents plus a Univalent in the Trisomics of Maize,[a] Tomato,[b] and Barley[c]

Tomato (diakinesis) % cells with				Maize (metaphase I) % cells with			Barley (diakinesis) % cells with			
Trisomic	11 II + 1 III	12 II + 1 I	11 II + 3 I	Trisomic	9 II + 1 III	10 II + 1 I	Trisomic	6 II + 1 III	7 II + 1 I	6 II + 3 I
1	64	34	2	2	80	20	Bush	82.9	17.1	0.0
2	68	28	4	3	72	28	Slender	89.3	10.7	0.0
3	70	26	4	5	86	14	Pale	94.3	5.7	0.0
4	56	44	0	6	72	28	Robust	92.9	7.1	0.0
5	50	44	6	7	74	26	Pseudo-normal	79.5	20.5	0.0
7	36	48	16	8	60	40	Purple	89.0	11.0	0.0
8	24	68	8	9	56	44	Semi-erect	93.9	6.1	0.0
9	42	50	8	10	57	43				
10	46	46	8							
12	24	70	12							

[a]From Einset, 1943.
[b]From Rick and Barton, 1954.
[c]Tsuchiya, 1960.

The type of configuration a trivalent may assume depends upon the extent of pachytene pairing and number and position of chiasmata. Generally, six types of trivalent configurations have been observed at diakinesis in the primary trisomics of different species. These configurations and the type of pachytene pairing and number and position of chiasmata which yield them are shown in Fig. 3.3.

It may be noted that there is no fundamental difference between types a and b and c and d. Types d and e may occasionally be mistaken for one another, although they result from different kinds of chiasma distribution. Types a and b require the least number of chiasmata and should form a frequent class in species with low chiasma frequency. In a sample of 109 trivalents from ten primary trisomics of *Datura* (Belling and Blakeslee, 1924), 57 were of type a or b. These two types together formed the most frequent class of trivalents in the PMC's of four trisomics of *Nicotiana sylvestris* (Goodspeed and Avery, 1939). However, in a large sample of trivalents of different trisomics of barley (Tsuchiya, 1960), type c formed the most frequent class, indicating that the

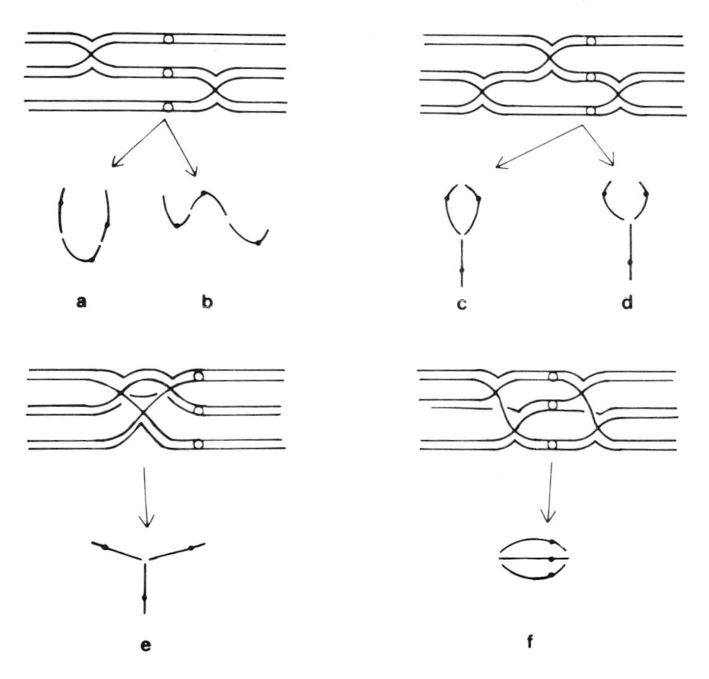

Fig. 3.3. The pachytene pairing, position of chiasmata, and the trivalent configurations expected at diakinesis in the primary trisomic.

TABLE 3.II
The Frequency of Microsporocytes with Different Kinds of Trivalents Observed at Diakinesis in Six Primary Trisomics of Tomato[a]

Trisomic	No. of cells scored	No. with trivalent	No. of cells with different types of trivalents[b]					
			a	b	c	d	e	f
$2n$ + 1S·1L	50	39	6	6	14	10	2	1
$2n$ + 2S·2L	50	34	0	9	0	21	4	0
$2n$ + 3S·3L	50	39	8	10	14	3	3	1
$2n$ + 6S·6L	50	26	13	8	5	0	0	0
$2n$ + 8S·8L	50	28	10	11	3	4	0	0
$2n$ + 11S·11L	50	26	13	8	5	0	0	0

[a]From G. S. Khush, unpublished.
[b]See Fig. 3.3.

chiasma frequency of barley per bivalent may be higher than that of *Datura* or *Nicotiana sylvestris*.

Since the trivalent types c and d require a minimum of three chiasmata, the frequency of these types should be higher in the trisomics for the longer chromosomes. Thus 24/39, 21/34, and 17/39 trivalents in the trisomics for chromosomes 1, 2, and 3 (long chromosomes) of tomato were either of type c or d compared with 7/28 and 5/26 in trisomics for chromosomes 8 and 11 (short chromosomes) (Table 3.II). The trivalent of type e requires three chiasmata in one arm and none in the other. Therefore, this type may be expected in trisomics for chromosomes with subterminal centromeres. The frequency of type e was extremely low in barley (Tsuchiya, 1960) as the chromosomes of this species are mostly metacentric, while about 17% of the microsporocytes of *Datura* trisomics formed this type of configuration, indicating the hetero-brachial nature of its chromosomes. The trivalent of type f requires two chiasmata in each arm, and as expected, its occurrence is rather rare. Less that 1% of the trivalents of *Datura* (Belling and Blakeslee, 1924) and tomato (Table 3.II) were of this type. However, barley has higher chiasma frequency per bivalent and the chromosomes are metacentric. These two conditions favor the formation of trivalents of type f in the trisomics of this species (Tsuchiya, 1960). It may be noted that trivalents of type f can be formed by primary trisomics only, and not by any other kind of trisomic.

At anaphase I, two members of the trivalent generally go to one pole and the third to the other. Rarely, all three go to one pole and result in the production of $n - 1$ and $n + 2$ spores. When present as a univalent the extra chromosome behaves irregularly: (1) it may pass to one of the poles at anaphase I undivided and divide normally at the second division; (2) it may lag and fail to be included in any of the telophase I nuclei; (3) it may divide equationally, and sister chromatids pass to opposite poles or one or both may lag and fail to be included in the telophase I nuclei; or (4) it may misdivide and give rise to telocentrics and isochromosomes which may also lag or become included in one of the sister halves.

At anaphase II, the univalent may divide normally if it remained undivided during anaphase I. If, however, it underwent division earlier during anaphase I, it may misdivide at anaphase II. Similarly, the telocentrics and isochromosomes produced by misdivision during anaphase I may misdivide at anaphase II and lag behind.

As a result of this irregular behavior of the univalent chromosome, the proportion of $n + 1$ spores will be lower than the expected value of 50%. The deviation from the 1:1 ratio for n and $n + 1$ spores is governed by the degree of irregularity. The trisomics for the longer chromosomes of maize form trivalents more frequently than those for the shorter ones and there is little chance for the extra chromosome to lag and misdivide. Consequently, $n + 1$ spores are produced at a frequency which is only slightly lower than the expected. For example, the average frequency of $n + 1$ microspores produced by the trisomics for the longer chromosomes of maize was 46%. However, the trisomics for the shorter chromosomes of maize had a higher univalent frequency, and there was a greater amount of lagging and misdivision. Consequently, the proportion of $n + 1$ spores was lower, the average for two shorter chromosomes being 29% (Einset, 1943). As a result of the irregular behavior of the univalent, the spores with n + telocentric and n + isochromosome are also produced at a low frequency.

Cytology of Tertiary Trisomics

The two arms of an extra tertiary chromosome have homologies with two members of the chromosome complement. Therefore, the extra chromosome may take part in pairing with either of the two chromosome pairs or with both, or rarely, may be left out as a univalent at pachytene stage of meiosis. In species where pachytene analysis is possible, an examination of such multivalents is helpful in identification of arms or arm segments which make up the extra chromosome, and the exact length of the duplicated segments can also be ascertained at this stage. To date only the tertiary trisomics of tomato have been identified by pachytene analysis. Khush and Rick (1967b) examined the pentavalent associations of seven tertiary trisomics and were able to determine the constituent arms of the extra tertiary chromosomes. Two such pentavalents are shown in Figs. 3.4 and 3.5.

Failure of chiasma formation or terminalization may reduce the pachytene

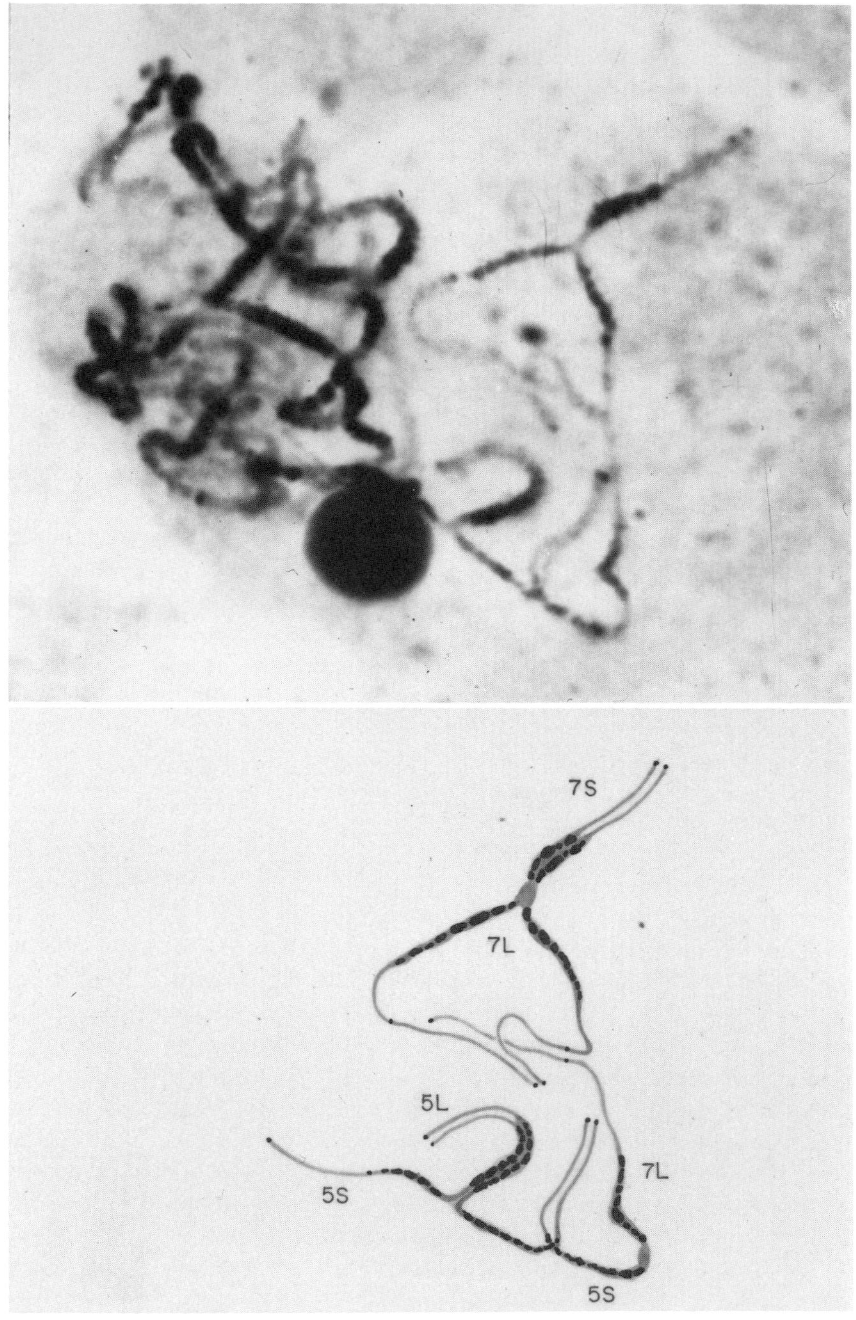

Fig. 3.4. Pentavalent association at pachytene in $2n + 5S \cdot 7L$ tertiary trisomic of tomato. (From Khush and Rick, 1967b.)

pentavalent associations into several other configurations at diakinesis and metaphase I. The following chromosomal associations might be observed at these stages in a tertiary trisomic: (1) $n - 2$ II + 1 V; (2) $n - 2$ II + 1 IV + 1 I; (3) $n - 1$ II + 1 III; (4) $n - 2$ II + 1 III + 2 I; (5) n II + 1 I; (6) $n - 1$ II + 3 I.

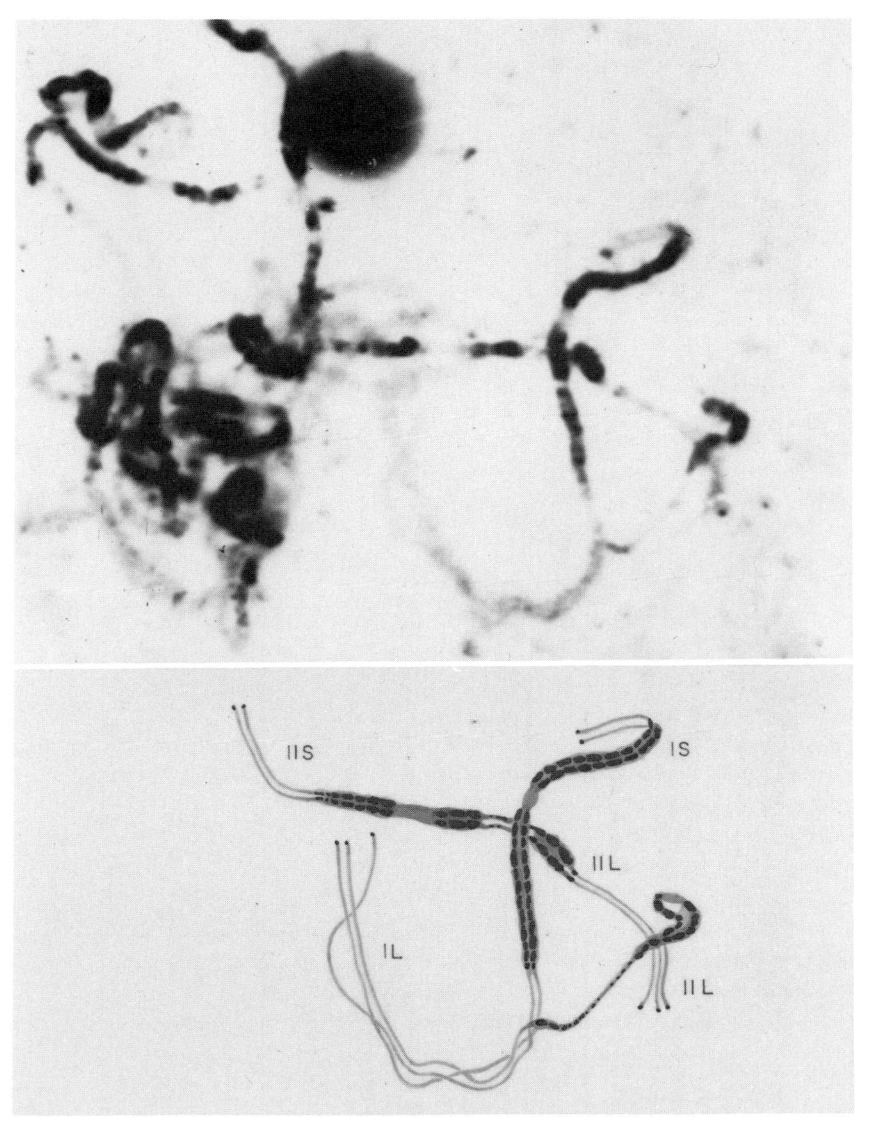

Fig. 3.5. Pentavalent association at pachytene in $2n$ + 1L·11L tertiary trisomic of tomato. (From Khush and Rick, 1967b.)

TABLE 3.III
The Frequency of Microsporocytes with Different Chromosomal Associations at Diakinesis in the Tertiary Trisomics of Tomato[a]

		No. of cells with						
Tertiary trisomic	No. of cells scored	10 II + 1 V	11 II + 1 III	12 II + 1 I	10 II + 1 IV + I	10 II + 1 III + 2 I	11 II + 3 I	10 II + 5 I
2*n* + 1L·11L	50	15	20	13	1	1	0	0
2*n* + 2L·10L	50	12	17	12	1	6	2	0
2*n* + 4L·10L	50	22	20	6	1	1	0	0
2*n* + 5S·7L	50	13	20	15	0	2	0	0
2*n* + 7S·11L	50	5	22	21	1	0	1	0
2*n* + 9L·12L	50	14	16	19	0	0	0	1
2*n* + 9S·12S	50	11	17	17	2	2	0	1

[a]From Khush and Rick, 1967b.

The frequencies of the different associations obviously depend upon the length of the chromosomes involved, the length of the duplicated segments, and the chiasma frequency. The longer chromosome arms of the complement have a greater chance to form chiasmata at pachytene than the shorter arms. Similarly, if the whole arms of the two chromosomes are involved in the formation of a tertiary chromosome, the frequency of multivalents with a higher number of chromosomes is higher. The frequencies of different associations observed in the seven tertiary trisomics of tomato (Khush and Rick, 1967b) are given in Table 3.III. It may be noted that the extra chromosome in all of these tertiaries was composed of two complete arms of nonhomologous chromosomes.

From Table 3.III it is clear that the tertiaries differ from each other in their ability to form pentavalents since some of the differences shown are statistically significant. In general, the extra chromosomes with two long arms tended to form pentavalents at a higher frequency than the ones with two short arms, while those with one long and one short arm tended to have intermediate frequencies. The modal configuration among these tertiaries was 11 II + 1 III; the next most frequent class was 12 II + 1 I. The former resulted from pairing and chiasma formation between the tertiary chromosome and one of the related homologous pairs, the latter, probably from failure of such events. Sporocytes with the other four types of association were rarely observed (Table 3.III, columns 6–9).

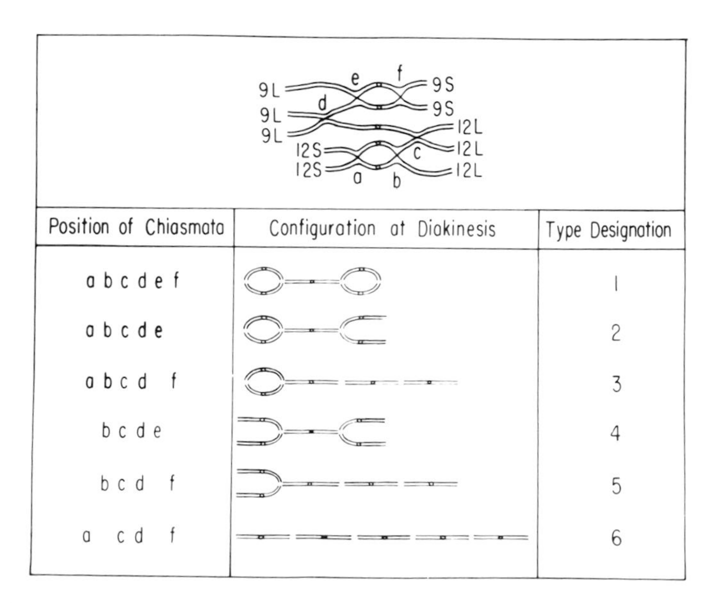

Fig. 3.6. Pentavalent configurations expected in a tertiary trisomic according to various positions of chiasmata. (From Khush and Rick, 1967b.)

TABLE 3.IV
The Frequency of Various Types of Pentavalents Observed in the Microsporocytes of Tertiary Trisomics of Tomato at Diakinesis[a]

Trisomic	No. of cells scored	No. with pentavalents	No. of cells with each type of pentavalent[b]					
			1	2	3	4	5	6
2n + 1L·11L	50	15	2	1	5	–	1	6
2n + 2L·10L	50	12	–	1	4	1	2	4
2n + 4L·10L	50	22	2	–	5	3	–	12
2n + 5S·7L	50	13	–	–	6	–	1	6
2n + 7S·11L	50	5	–	–	1	–	–	4
2n + 9L·12L	50	14	–	–	5	–	–	9
2n + 9S·12S	50	11	1	–	–	–	3	7

[a]From Khush and Rick, 1967b.
[b]See Fig. 3.6.

The pentavalent in a tertiary trisomic may assume various configurations, depending upon the number and position of chiasmata. For example, in the $2n$ + 9L·12L tertiary of tomato, six different types of pentavalents may be expected (Fig. 3.6). The configuration of type 1 requires at least six chiasmata; both sets of three homologous arms must form at least two chiasmata apiece, which accounts for the rarity of this type (Table 3.IV). The low frequency of types 2 and 4 is similarly explained. Type 6, which requires the least number of chiasmata and no more than one per pairing arm, was most frequent in all the tertiaries. The next most frequent class, type 3, requires a minimum of two chiasmata in one set of three homologous arms, but only one in the other. Such longer arms in the tomato tertiaries as 1L, 2L, and 4L evidently provide a better opportunity for the fulfillment of this requirement.

At anaphase I, if the different members of the multivalents segregate at random and the univalents also go to any of the poles at random, a tertiary trisomic would produce four types of gametes. For example, the tertiary $2n$ + 9L·12L would produce gametes with n, n + 9L·12L, n + 9S·9L, and n + 12S·12L chromosomes. The proportion of different kinds of gametes would be determined by the frequency of different kinds of associations at metaphase I and the type of disjunction at anaphase I. The segregation of chromosomes from the pentavalents of types 1, 2, and 4 would be mainly alternate and result in n and n + 9L·12L gametes only, while adjacent segregations from pentavalents of types 3, 5, and 6 would give rise to n + 9S·9L and n + 12S·12L gametes as well. Similarly, adjacent segregation from the trivalent would result in n + 9S·9L and n + 12S·12L gametes. However, the orientation of the trivalents of the tertiary trisomics is generally such that the two fully homologous chromosomes separate to the opposite poles while the tertiary chromosome goes to one pole. Thus a great majority of the gametes produced by the microsporocytes with $n - 1$ II + 1 III would be either n or n + 9L · 12L. The n + 9S · 9L or n + 12S · 12L gametes would be produced only at a very low frequency.

In addition to the four types of gametes enumerated above, a tertiary trisomic may rarely produce a gamete in which the extra chromosome is either a telocentric or an isochromosome, produced as a result of misdivision of the univalent as in the primary trisomics.

Cytology of Secondary Trisomics

In a secondary trisomic, in contrast to a primary or a tertiary, the two arms of the extra chromosome are completely homologous and may pair internally (Fig. 3.7) resulting in the formation of n bivalents and a small ring univalent at diakinesis. The isochromosome may also pair fraternally with two homologous chromosomes to form a trivalent and $n - 1$ bivalents. Figures 3.8, 3.9, and 3.10 show trivalents resulting from fraternal pairing of the isochromosomes in the secondary trisomics of tomato. In rare instances, the three chromosomes may fail to pair or form a chiasmata; when this happens the sporocyte has three univalents at diakinesis and metaphase I. The extra isochromosome arm can be identified by observing the internally paired isochromosome (Fig. 3.7) or by examination of trivalents (Figs. 3.8, 3.9 and 3.10).

The proportion of sporocytes with n bivalents and a univalent depends upon the length of the isochromosome arm and the chiasma frequency. In a random

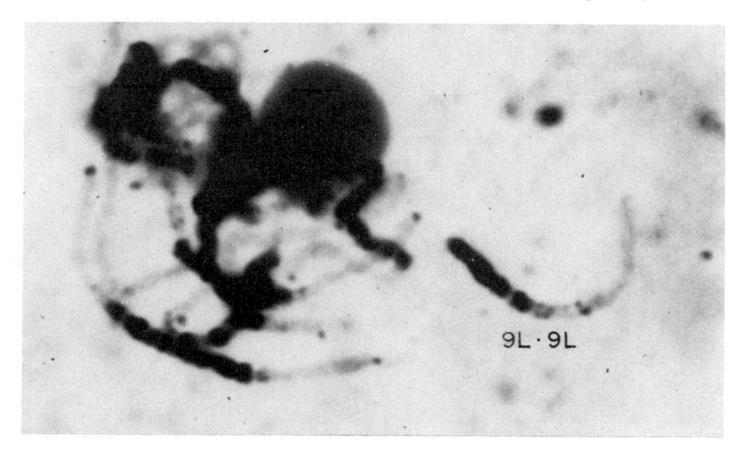

Fig. 3.7. Internally paired isochromosome 9L · 9L in $2n$ + 9L · 9L secondary trisomic of tomato. (From Khush and Rick, 1969.)

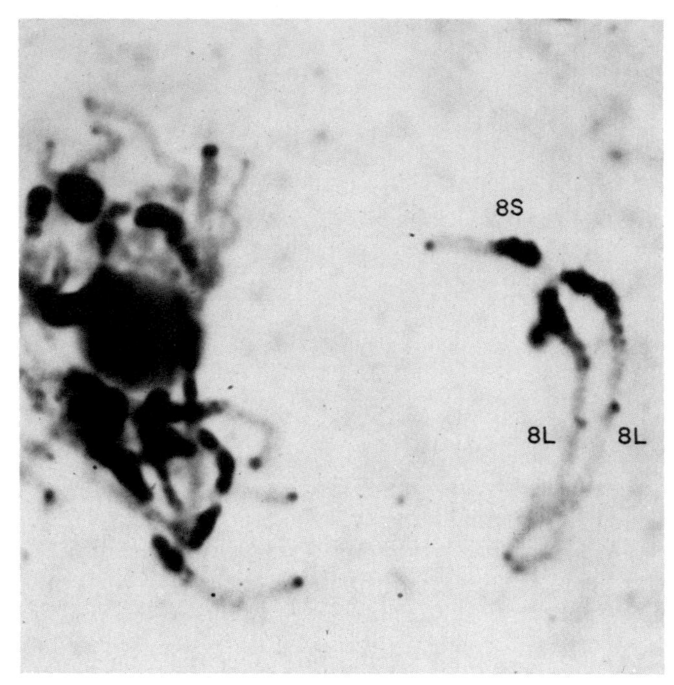

Fig. 3.8. Trivalent association in $2n$ + 8L·8L secondary trisomic of tomato. One chiasma can be easily seen in the 8L's. (From Khush and Rick, 1969.)

sample of microsporocytes of eight secondary trisomics of *Datura,* Belling and Blakeslee (1924) observed 17.1% with n bivalents and a univalent. Rhoades (1933b) reported that 41.2% of the sporocytes of the secondary trisomic for the short arm of chromosome 5 of maize had a univalent and n bivalents. In the secondary trisomics for the three short arms of the tomato chromosome complement, the frequency of sporocytes with univalents was considerably higher than it was in the secondary trisomics for the long arms of the complement (Table 3.V). A univalent in sporocytes of the secondary trisomic, is almost always the isochromosome. Thus all the univalents observed by Belling and Blakeslee (1924) in the secondary trisomics of *Datura* were ring univalents which can be formed only by the isochromosome. Rhoades (1933b) recorded twenty sporocytes with a ring univalent and five in which the univalent was present in the form of a rod. However, all the ten bivalents in these cells formed closed rings. He concluded that the univalent in these sporocytes must have been the isochromosome. Had the isochromosome paired with one of the normal chromosomes, it could not have formed the ring bivalent. It is therefore evident that the two normal chromosomes pair preferentially and the

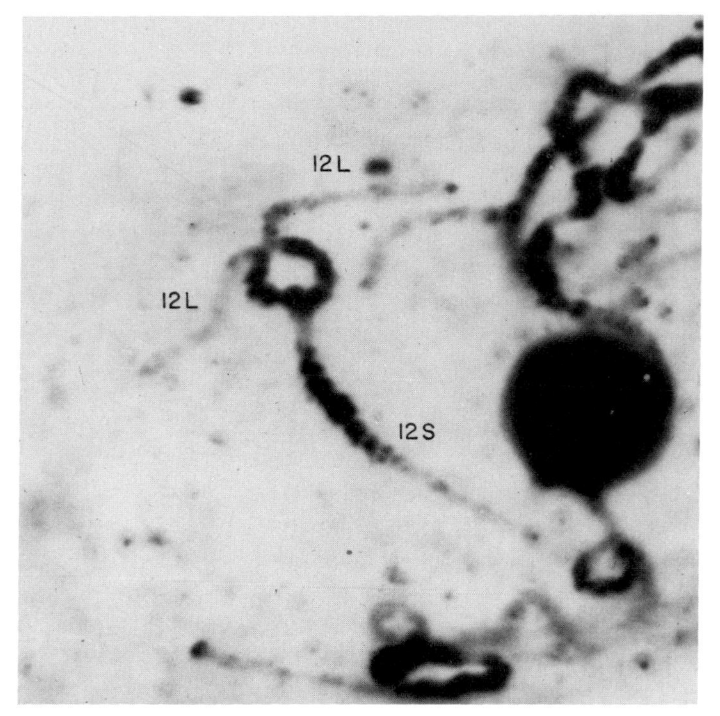

Fig. 3.9. Trivalent association in $2n$ + 12L·12L secondary trisomic of tomato. One chiasma can be easily seen in the 12L's. (From Khush and Rick, 1969.)

isochromosome is left out as a univalent. No sporocytes with three univalents were observed in the secondary trisomics of *Datura* and maize, and in all but one of tomato (Table 3.V).

The trivalents of the secondary trisomics may assume eight different kinds of configurations resulting from a variable number and position of chiasmata. The pachytene associations and the number and position of chiasmata that give rise to these configurations at diakinesis are shown diagramatically in Fig. 3.11. It may be noted that configurations a, b, c, and d can appear in both primary and tertiary trisomics, but configurations e, f, g, and h occur only in secondary trisomics. The trivalents of type e appear most frequently in the majority of the secondary trisomics examined to date. Thus, in eight secondary trisomics of *Datura* (Belling and Blakeslee, 1924), 53 trivalents out of a random sample of 98 were of the e type. Similarly 27 out of 37 trivalents in $2n$ + 5S·5S of maize were of the e type (Rhoades, 1933b). The e type of trivalent was formed with the highest frequency in the tomato secondaries for the longer arms of the complement (Table 3.V), while in the secondaries for the shorter

TABLE 3.V
The Frequency of Different Types of Trivalents in Eight Secondary Trisomics of Tomato[a] from a Sample of Fifty Cells Each

Trisomic	Type of trivalent[b]								No. of cells with		
	a	b	c	d	e	f	g	h	11 II + 3 I	12 II + 1 I	11 II + 1 III
$2n$ + 3S·3S	8	5	3	3	3	0	0	0	0	28	22
$2n$ + 7S·7S	8	2	1	0	3	0	0	2	0	34	16
$2n$ + 9S·9S	3	6	1	4	5	0	0	1	2	28	20
$2n$ + 6L·6L	4	5	4	2	12	5	0	2	0	16	34
$2n$ + 8L·8L	5	6	3	2	9	2	0	2	0	21	29
$2n$ + 9L·9L	5	5	4	3	12	1	0	1	0	19	31
$2n$ + 10L·10L	8	5	3	3	8	0	0	3	0	20	30
$2n$ + 12L·12L	6	6	3	1	16	1	2	1	0	14	36

[a]From Khush and Rick, 1969.
[b]See Fig. 3.11.

arms of the complement, the frequency of this type was quite low. The type g trivalent is formed most infrequently. Only two sporocytes with a trivalent of this type were observed in a random sample of 218 from eight secondary trisomics of tomato (Khush and Rick, 1969); none were observed in the secondaries of *Datura* and maize.

At anaphase I, the segregation of chromosomes from sporocytes with n bivalents and a univalent would result in the formation of n and n + isochromosome spores. However, the segregation of three chromosomes from the trivalent would yield two additional kinds of spores. The three chromosomes of the trivalent may segregate in two different ways. For example, from the trivalent 10S·10L-10L·10L-10S·10L of $2n$ + 10L·10L secondary of tomato, the 10S·10L and 10L·10L chromosomes may go to one pole and the 10S·10L chromosome to the other. This type of segregation would produce n and n + 10L·10L spores similar to the ones produced by the sporocytes with a univalent isochromosome. However, the segregation in which the 10L·10L chromosome went to one pole and two 10S·10L chromosomes to the other would produce n − 10S·10L + 10L·10L spores. The n − 10S·10L + 10L·10L spore would most certainly abort in the diploid species due to the deficiency of one arm. The n + 10S·10L spore would be functional and have an extra primary chromosome. Therefore, three types of functional spores, and hence gametes, produced by a secondary trisomic would have: (1) n, (2) n + secondary chromosome, and (3) n + related primary chromosome.

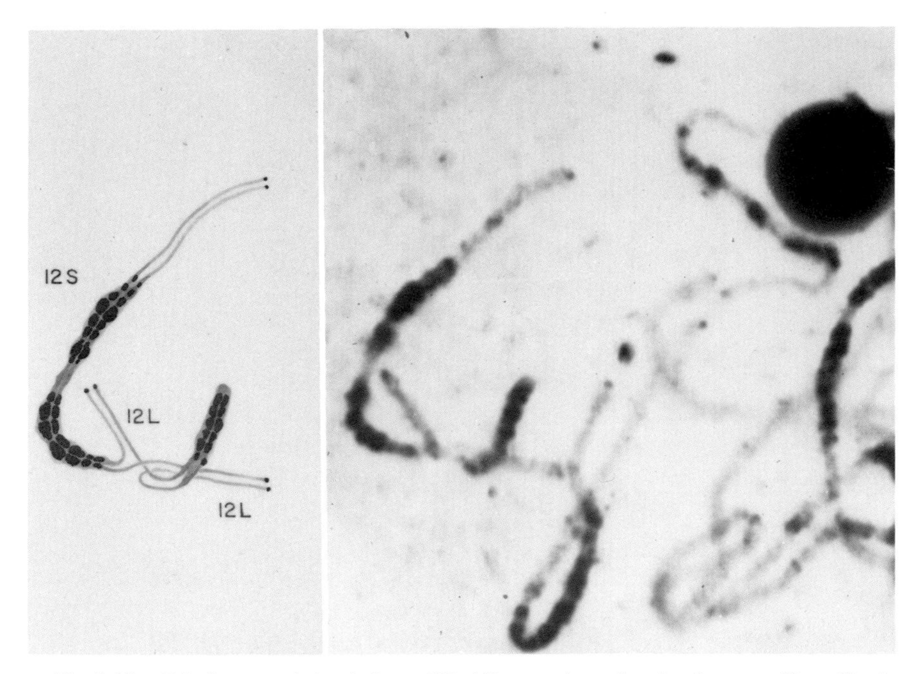

Fig. 3.10. Trivalent association in $2n + 12L \cdot 12L$ secondary trisomic of tomato. (From Khush and Rick, 1969.)

As was pointed out by Khush and Rick (1969), the frequency of $n + 1$ gametes in which the extra chromosome is primary instead of secondary depends upon the frequency of trivalent formation and the type of segregation it undergoes at anaphase I. The secondary trisomics for short arms of the tomato chromosome complement form trivalents less frequently than those for the long arms. The isochromosome in these trisomics is shorter than the two normal chromosomes, and the orientation of the trivalent when formed tends to result in the two normal chromosomes passing to different poles and the isochromosome segregating at random. Conversely, the trivalent formation in the secondary trisomics for long arms of the complement is more frequent than those for the short arms, and the relatively larger size of the isochromosome tends to favor its orientation toward one pole while the two normal chromosomes are oriented toward the other pole. Thus, on the basis of cytological observations, the secondary trisomics for the short arms of the chromosome complement should produce gametes with the extra primary chromosome at a very low frequency. On the other hand, the frequency of such gametes in the secondary trisomics for the long arms should be higher. This subject is considered further in Chapter 4.

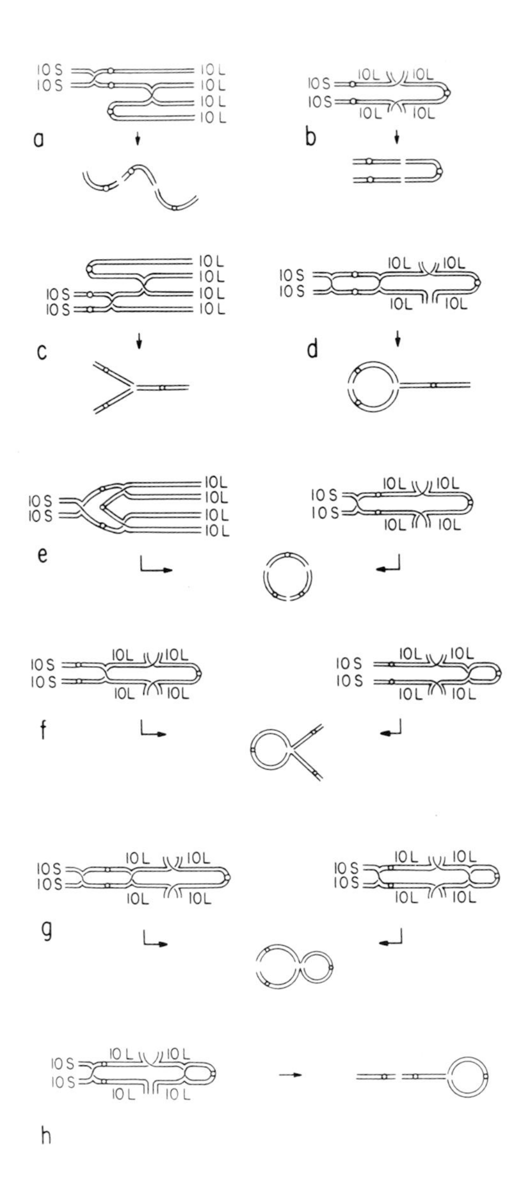

Fig. 3.11. Patterns of chromosomal association and chiasma formation at pachytene in the trivalents of the secondary trisomic $2n$ + 10L·10L. Arrows point to the respective configurations anticipated at diakinesis. (Khush and Rick, 1969.)

Cytology of Telotrisomics

The extra telocentric chromosome in the telotrisomics may pair with the homologous arms of two normal chromosomes to form a trivalent or may remain unpaired as a univalent. The relative proportion of the sporocytes with trivalents and univalents depends upon the chiasma frequency of the telocentric arm which in turn may be dependent upon its length. The telocentrics for the long arms have a greater chance to pair and form chiasmata (Fig. 3.12) compared with telocentrics for the short arms (Fig. 3.13). Therefore, the frequency of trivalents at diakinesis is higher in the telotrisomics for the long arms compared with that of telotrisomics for short arms. Thus the proportion of sporocytes with trivalents in $2n + \cdot 3L$, $2n + \cdot 4L$, $2n + \cdot 7L$, and $2n + \cdot 8L$ of tomato was 42, 60, 58, and 58%, respectively, while it was only 26% in $2n + \cdot 3S$ and 24% in $2n + \cdot 10S$ (Khush and Rick, 1968b). In the rest of the sporocytes, the telocentric was present as a univalent.

The difference in the size of the telocentric and the two normal chromosomes influences the orientation of the trivalent in favor of the segregation of two

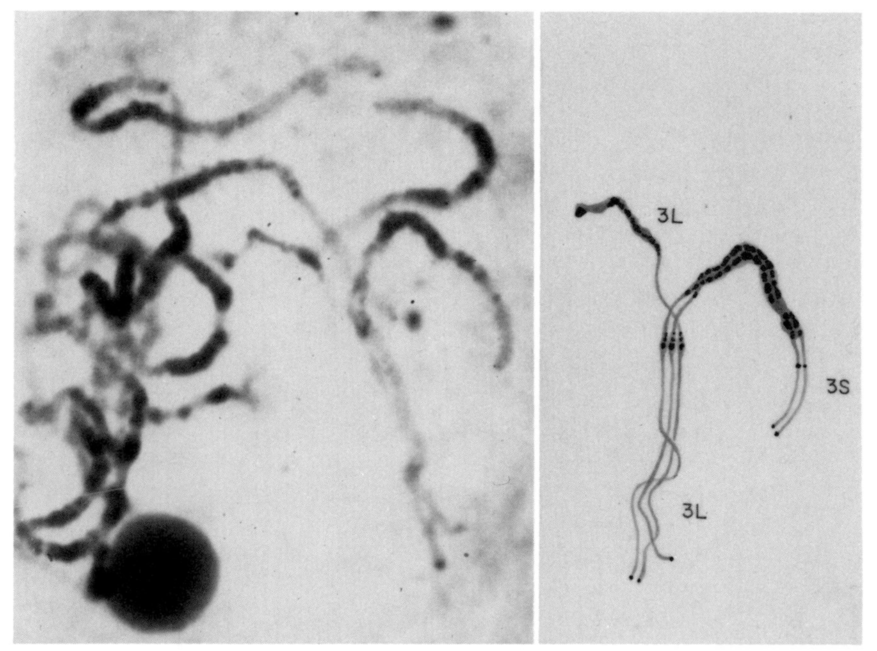

Fig. 3.12. Trivalent association in $2n + \cdot 3L$ telotrisomic of tomato. (From Khush and Rick, 1968b.)

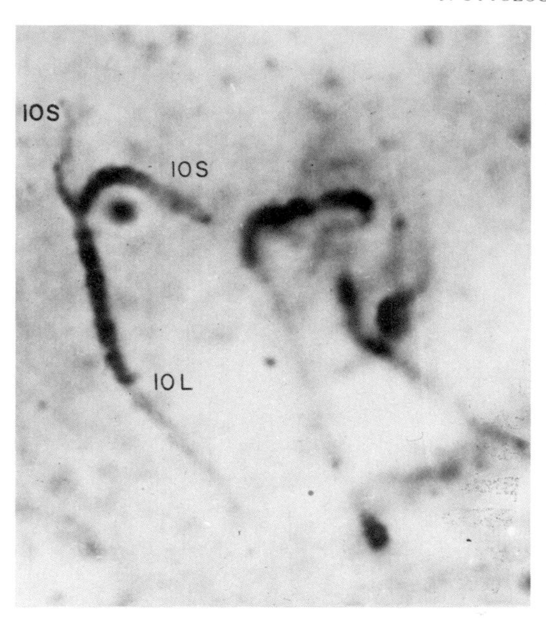

Fig. 3.13. Trivalent association in $2n + \cdot 10S$ telotrisomic of tomato. (From Khush and Rick, 1968b.)

normal chromosomes to opposite poles and the segregation at random of the telocentric to either of the poles. For the most part, two kinds of spores are produced by the telotrisomics, n and n + telocentric. In rare instances, the telotrisomic is converted into an isochromosome and thus yields a spore with an extra secondary chromosome.

Occasionally the segregation from the trivalent may yield a spore with an extra primary chromosome. In diploid species the other product of this type of segregation would abort.

Cytology of Compensating Trisomics

A thorough cytological examination of compensating trisomics has not yet been attempted. The pachytene analysis of compensating trisomics of tomato was made by Khush and Rick (1967a). In the diiso compensating trisomics ($2n - 3S \cdot 3L + 3S \cdot 3L + 3L \cdot 3L$ and $2n - 7S \cdot 7L + 7S \cdot 7S + 7L \cdot 7L$), the isochromosomes tended to pair internally, leaving the normal chromosome as a univalent (Fig. 3.14). On the other hand, in the teloiso compensating trisomic ($2n - 3S \cdot 3L + \cdot 3S + 3L \cdot 3L$), "trivalent" pairing (Fig. 3.15) was somewhat more frequent. The bivalent associations, consisting of the normal chromosome and the isochromosome in the diiso compensating trisomics and of the normal

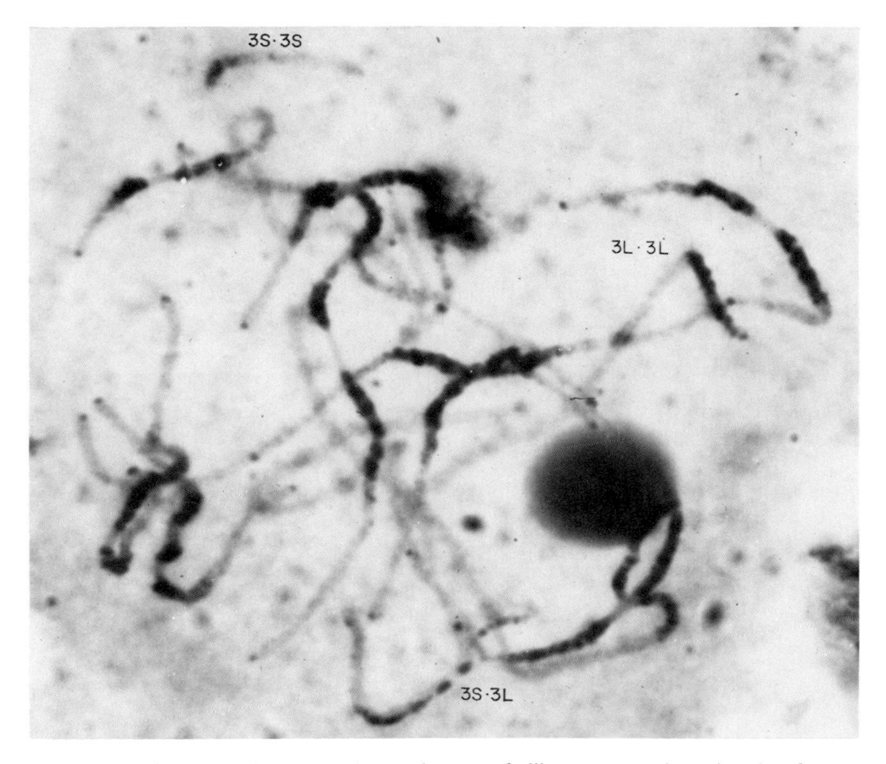

Fig. 3.14. Complete chromosomal complement of diiso compensating trisomic of tomato ($2n$ − 3S·3L + 3S·3S + 3L·3L). All the elements of three complex are univalents. The long arm (bottom of figure) of the normal (3S·3L) chromosome is entirely paired nonhomologously with itself in foldback fashion. Both isochromosomes are paired internally. (From Khush and Rick, 1967a).

chromosome and either the isochromosome or the telocentric in the teloiso compensating trisomic, were also observed. The maximum association at pachytene in a ditertiary compensating trisomic would be of seven chromosomes, while in a telotertiary or an isotertiary compensating trisomic, the maximum association at pachytene would be of five chromosomes.

In the teloiso compensating trisomics of wheat (Smith, 1947) and tomato (Khush and Rick, 1967a), the following configurations were observed at diakinesis: (1) a highly heteromorphic trivalent; (2) a distinctly heteromorphic bivalent and a ring-shaped univalent; or, rarely, (3) a heteromorphic bivalent and a rod univalent. In the diiso compensating trisomics, sporocytes with a rod univalent and two ring univalents were more frequent, although heteromorphic trivalents and bivalents were also observed.

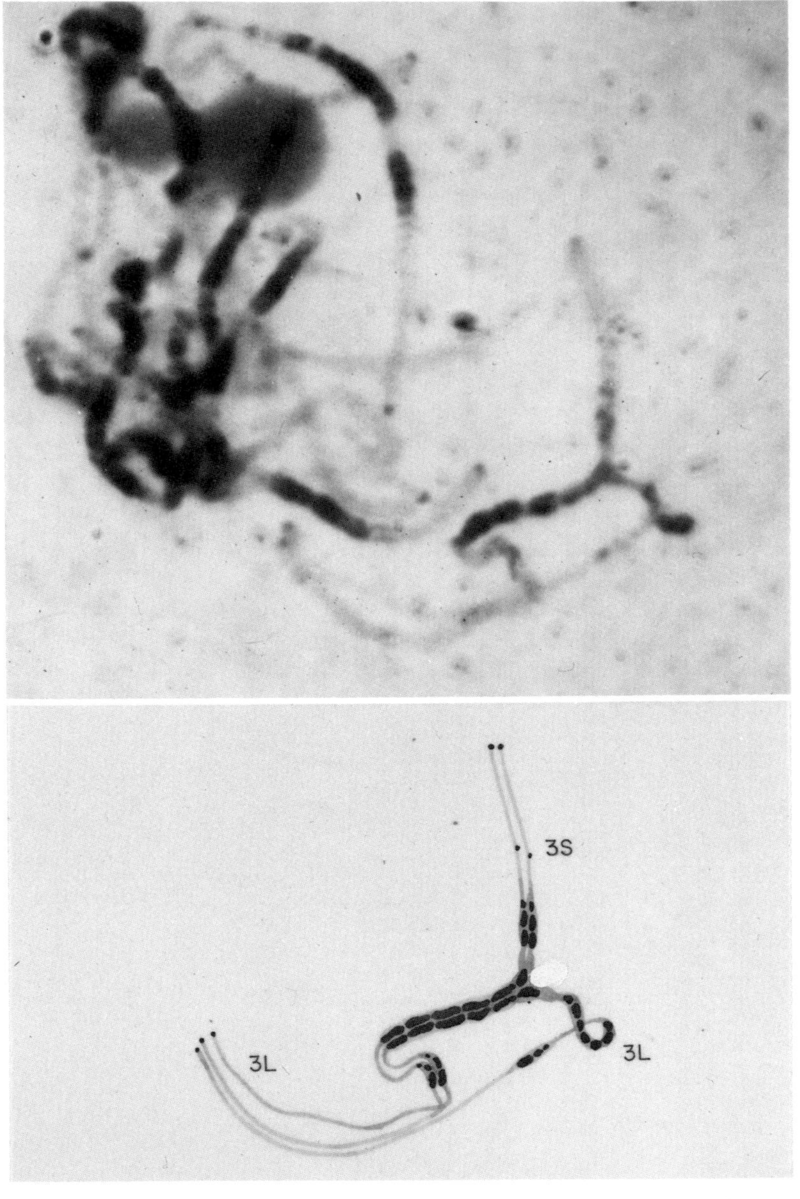

Fig. 3.15. Trivalent association at pachytene in teloiso compensating trisomic of tomato ($2n - 3S \cdot 3L + \cdot 3S + 3L \cdot 3L$). $\cdot 3S$ telocentric and $3L \cdot 3L$ isochromosome are paired, respectively, with 3S and 3L of normal $3S \cdot 3L$ chromosome. (From Khush and Rick, 1967a.)

At anaphase I, the segregation of the compensating elements as well as the other associated chromosomes would yield additional aneuploid spores besides the parental combinations. For example, a diiso compensating trisomic would produce two kinds of n + iso spores and a teloiso compensating trisomic would yield n + telo and n + iso spores. The ditertiary compensating trisomic, on the other hand, would produce four kinds of additional aneuploid spores. In two of them the extra chromosome would be tertiary while in the other two it would be primary. The new aneuploid spores produced by the isotertiary compensating trisomic would be (1) n + iso, (2) n + tertiary, and (3) n + primary (related to tertiary).

4

Transmission of the Extra Chromosome in Trisomics

Theoretically, one expects an individual with $2n + 1$ chromosomes to produce n and $n + 1$ gametes in equal frequency on the female and the male sides. Similarly, one should be able to obtain $2n + 1$ and $2n$ individuals in equal frequency from $(2n + 1) \times 2n$ as well as $2n \times (2n + 1)$ crosses. Furthermore, $2n + 1$ individuals when selfed should yield $2n$, $2n + 1$, and $2n + 2$ progeny in the ratio of 1:2:1. However, as the data of Table 4.I show, these expectations are never realized, and the percentage of $2n + 1$ individuals in the progeny of a $(2n + 1) \times 2n$ cross is considerably lower than 50%. It is even lower in the $2n \times (2n + 1)$ crosses, and in most cases one fails to recover any $2n + 1$ individuals from such a cross. The tetrasomics rarely appear in the progenies of selfed trisomics. Another unique feature of the progenies of the trisomics is the appearance of nonparental but related, and a few unrelated, $2n + 1$ types in addition to parental trisomics.

In this chapter, the following three aspects of the breeding behavior of the trisomics will be discussed separately. In the first part the causes of lowered yield of $2n + 1$ progeny from $(2n + 1) \times 2n$ crosses (transmission of the extra chromosome through the female) will be presented. The second part will deal with the problems of extremely low or no transmission of trisomics from $2n \times (2n + 1)$ crosses (transmission through the male). In the third part the causes of the appearance of nonparental trisomics in the progenies of trisomics of various kinds will be explored.

Transmission of the Extra Chromosome through the Female

The reduction in the number of $2n + 1$ individuals to less than the expected 50% has been attributed to several causes (1) elimination of the extra chromosome

TABLE 4.I

Transmission Frequency of the Extra Chromosome in $(2n + 1) \times 2n$ and $2n \times (2n + 1)$ Crosses of *Nicotiana sylvestris*,[a] *Hordeum spontaneum*,[b] and *Secale cereale*[c]

Nicotiana sylvestris			*Hordeum spontaneum*			*Secale cereale*		
Trisomic	$(2n + 1) \times 2n$	$2n \times (2n + 1)$	Trisomic	$(2n + 1) \times 2n$	$2n \times (2n + 1)$	Trisomic	$(2n + 1) \times 2n$	$2n \times (2n + 1)$
Recurved	16.3	0.0	Bush	24.9	0.0	Slender	16.0	0.0
Enlarged	24.0	6.9	Slender	27.5	0.0	Feeble	19.2	0.0
Narrow	28.6	9.5	Pale	27.2	0.0	Stout	35.7	0.0
Pointed	28.8	34.1	Robust	11.3	0.0	Pseudonormal	39.3	2.4
Late	25.0	24.1	Pseudonormal	9.8	0.0	Semisterile	17.3	2.6
Puckered	27.9	1.8	Purple	17.4	0.0	Dwarf	32.8	0.0
Bent	20.0	0.0	Semierect	24.5	0.0	Bush	25.4	1.0
Stubby	20.5	19.6						
Compact	18.2	1.9						
Inflated	26.7	5.8						
Stiff	38.0	–						

[a]From Goodspeed and Avery (1939, 1941).

[b]From Tsuchiya (1960).

[c]From Kamanoi and Jenkins (1962).

during meiosis, due to lagging or misdivision or both; (2) megaspore replacement; (3) reduced viability of $n + 1$ spores or gametes; (4) subnormal development of $2n + 1$ zygotes, endosperm, and embryos; (5) poor and delayed germination of $2n + 1$ seeds; (6) reduced vigor of $2n + 1$ seedlings; and (7) the effect of genetic background.

Elimination of the Extra Chromosome during Meiosis

This probably occurs to some extent in all species. As discussed in Chapter 3, the extra chromosome may take part in metaphase I pairing or be left out as a univalent. If it pairs with the other homologs and segregates regularly to one of the poles, spores with n and $n + 1$ chromosomes will be produced in equal frequency. However, when the extra chromosome is present as a univalent, it tends to lag and misdivide, and if it is not included in any of the telophase I nuclei, all the spores produced by the sporocytes will have n chromosomes. The net result of lagging and misdivision is production of n spores in excess of 50% and a corresponding decrease in the frequency of $n + 1$ spores. As discussed in Chapter 3, the presence of the extra chromosome as a univalent is a regular feature of the cytology of trisomics. If the reduced transmission of the extra chromosome is due to univalent formation and elimination, there should be a positive correlation between the frequency of univalent formation and transmission frequency. In maize, in which an extensive study of this nature has been carried out, such a correlation does exist (Einset, 1943). Einset found that those trisomics having the lowest percentage of univalents at metaphase I, e.g., chromosomes 2, 3, and 5, produced the highest percentage of $n + 1$ spores and $2n + 1$ plants. The trisomics with the highest percentage of univalents, e.g., triplo-9 and -10, produced the lowest proportion of $n + 1$ spores and $2n + 1$ plants (Table 4.II). Triplo-6, -7, and -8 had the intermediate proportion of univalents, $n + 1$ spores, and $2n + 1$ plants.

Einset (1943) also reported that three homologs in all the trisomics were paired in almost all the cells observed at pachytene, but as meiosis progressed into later stages, the frequency of univalents increased. He found a positive correlation between the chromosome length and the univalent frequency at metaphase I. The trisomics for long chromosomes 2, 3, and 5 had a lower univalent frequency compared to those of short ones 9 and 10. Einset attributed these differences in the frequency of univalent formation to chiasma frequency of a particular chromosome. In the trisomic stocks of maize involving the long chromosomes, the opportunity for any two chromosomes to synapse over relatively long regions is greater than it is among the short chromosomes and, for this reason, the opportunity for chiasma formation is greater, resulting in higher trivalent frequency at metaphase I. In the trisomics of the short chromo-

TABLE 4.II
Data on the Trisomics of Maize Showing Relation between Chromosome Length and Frequency of Trisomics in the Progeny, Frequency of Microsporocytes with Univalents, and Frequency of $n + 1$ Microspores[a]

		Progeny of $(2n + 1) \times 2n$		Microsporocytes		Microspores	
Chromosome	Relative length	Total no.	% $2n + 1$	Total no.	% with univalents	Total no.	% $n + 1$
2	80	320	47	602	20	454	48
3	74	91	45	342	28	167	41
5	73	89	52	209	14	198	50
Average	76		48		21		46
6	60	155	38	109	28	109	34
7	56	80	41	246	26	193	50
8	57	191	31	267	40	245	36
Average	58		37		31		40
9	52	113	22	214	44	218	23
10	45	198	28	672	43	299	34
Average	49		25		44		29

[a]From Einset, 1943.

somes, the opportunity for chiasma formation is lower thus fewer trivalents were observed. Einset's observations on the trisomics of maize agree with the data presented by Belling (1925) on triploid *Hyacinths,* by Newton and Darlington (1929), and Darlington and Mather (1932) on triploid *Tulipa,* which indicate that the longer chromosomes form trivalents at metaphase I more frequently than the shorter ones do. Darlington (1934) and Mather (1939) also found the linear correlation between the chromosome length and the number of chiasma formed in $2n$ and $3n$ maize.

Einset (1943) reported an average transmission frequency of 48% for the three longer chromosomes, 37% for the three medium chromosomes, and 25% for the three shorter chromosomes. The proportion of $n + 1$ spores for these three categories was 46, 40, and 29%, respectively, and the frequency of microsporocytes with univalents was 21, 31, and 44%, respectively (Table 4.II). It is evident that in maize a strong correlation exists between chromosome length, univalent frequency at metaphase I, and transmission frequency of the extra chromosome. Einset reported that the spikelet (ear) fertility of all the trisomics of maize was normal and that the kernels (seeds) of trisomic maize showed normal germination. He concluded that the variable reduction in the transmission frequency of maize trisomics must be due only to univalent formation and subsequent lagging and elimination of the extra chromosome.

The transmission frequency of trisomics of hexaploid wheat (Sears, 1954) is high (40–45%). As in maize, the female fertility is normal. Therefore, it appears that a minor reduction in the transmission frequency of the extra chromosome in wheat trisomics is caused only by chromosome lagging and elimination.

Megaspore Replacement

McClintock and Hill (1931) suggested another theory to account for the reduced transmission of the extra chromosome through the female. In maize, the four megaspores produced during meiosis lie in a row. The three upper ones degenerate, and the fourth, at the chalazal end, forms the embryo sac. McClintock and Hill reasoned that, if the basal megaspore produced by the $2n + 1$ plant contained the $n + 1$ chromosomes, it may be replaced by one of the megaspores with n chromosomes in a certain percentage of the ovules and thus reduce the frequency of $2n + 1$ plants. Theoretically, this mechanism could operate in all the species which have a monosporic or bisporic embryo sac. To test this possibility, a study of embryo sac development in a large number of ovules in the trisomics of the monosporic and bisporic species is needed, but so far no such study has been undertaken. However, indirect evidence rules out such a possibility. In a study of gametic lethals on the fourth chromosome of maize, Singleton and Mangelsdorf (1940) found no indication of megaspore substitution. Similarly, while investigating an abnormal type of chromosome 10 in maize, Rhoades (1942) followed the development of 200 embryo sacs and found no replacement of chalazal megaspores. In a study of corn plants carrying an extra chromosome derived from *Tripsacum,* which was transmitted to 91% of the progeny through the female side, Maguire (1963) found no megaspore replacement.

Reduced Viability of $n + 1$ Spores and Gametes

The positive correlation between length of the extra chromosome, univalent frequency, and transmission rates observed for the maize trisomics does not hold true for the trisomics of tomato. In the trisomics for the longer chromosomes of the tomato (chromosomes 1, 2, and 3), the univalent frequency is lowest (Rick and Barton, 1954) as expected. However, the trisomics of these chromosomes also show the lowest rates of transmission. The trisomics for the shorter chromosomes of the complement (chromosomes 7 to 12) have the higher frequency of univalents, but their transmission rates are also higher (Table 4.III). In other species in which the rates of transmission of different trisomics have

TABLE 4.III
Transmission Rates of Tomato Trisomics in the Genotypical Background of Three Varieties, Dwarf Aristocrat,[a] San Marzano,[b] and VF 36[c]

Chromosome	Chromosome length (μm)	Percentage transmission in Dwarf Aristocrat $2n + 1$ (self)	San Marzano $(2n + 1) \times 2n$	VF 36 $(2n + 1) \times 2n$
1	52.0	–	4.6	7.6
2	42.1	23.0	4.4	15.2
3	40.3	–	1.1	6.3
4	35.0	45.0	24.7	32.2
5	31.4	26.7	22.2	31.3
6	31.3	–	0.4	8.7
7	27.5	32.0	14.8	38.4
8	27.5	37.0	21.7	33.4
9	26.8	14.0	16.6	20.0
10	25.7	–	20.0	37.2
11	24.8	–	–	22.0
12	22.5	20.0	15.7	29.2

[a]From Lesley, 1928, 1932.
[b]From Rick and Barton, 1954.
[c]From G. S. Khush, unpublished.

been investigated, such as *Nicotiana sylvestris* (Goodspeed and Avery, 1939), *Spinacea oleracea* (Tabushi, 1958; Janick *et al.*, 1959), *Datura stramonium* (Avery *et al.*, 1959), *Hordeum spontaneum* (Tsuchiya, 1960), *Secale cereale* (Kamanoi and Jenkins, 1962), and *Lotus pedunculatus* (Chen and Grant, 1968b), the pachytene analysis of the trisomics has not been made. Consequently, the relative lengths of the different chromosomes are not known, and any speculation regarding the correlations between chromosome length and the transmission frequency is unjustified. The minor differences in the lengths of the members of the chromosome complement detected in the somatic tissues really do not give the true picture of their comparative lengths. It is known, for example, that the euchromatic and heterochromatic regions contract differentially, and in the condensed stage, the heterochromatic regions occupy a greater bulk of the length of the chromosome (Brown, 1949). A short pachytene chromosome with a relatively long heterochromatic segment may measure longer in the somatic tissues than a longer pachytene chromosome with little heterochromatin. For example, chromosome 10 of tomato, which is one of the shortest at pachytene, measures one of the longest in the somatic tissues because of its large heterochromatic segment, but chromosome 6, which has a medium length at pachytene, appears as the shortest of the complement in the somatic tissues by virtue of its extremely small heterochromatic regions.

To be sure, the univalent formation occurs in the trisomics of all of the species discussed above, but the transmission rates are generally so low that the reduction cannot be wholly accounted for on the basis of lagging and elimination of the extra chromosome during meiosis. Also, there is the obvious dilemma of explaining the inverse relationship which exists between the frequency of univalent formation and the transmission rates of tomato trisomics. To account for the reduced transmission of the extra chromosome in these species, the explanation must be sought in the postmeiotic events. The first postmeiotic stage which needs to be explored is the maturation of gametophytes.

It is generally agreed that the haploid or gametophytic stage is more sensitive to losses or duplications of chromosomal material and has narrower tolerance limits for deficiencies and duplications than the sporophytic or diploid stage. In tomato and *Datura,* for example, the loss of one entire chromosome may be tolerated by the sporophyte, but none of the gametophytes with the missing chromosome will function. Similarly, segmental deficiencies of different kinds and length are tolerated by the sporophytes of tomato, but none is transmitted by the gametophytes. (For a detailed discussion, see Khush and Rick, 1966b, 1967c, 1968a, 1969). The evidence for differential tolerance of duplications by the gametophytes and sporophytes comes from a study of the secondary trisomic for the long arm of chromosome 6 of tomato (Khush and Rick, 1969). This trisomic ($2n$ + 6L·6L) originated spontaneously in the progeny of mono-4S·10S. The plant had good vigor, flowered profusely, and set a moderate fruit load. When a large progeny was raised, the parental trisomic did not appear. It was concluded that gametophytes receiving an extra 6L·6L isochromosome do not tolerate the imbalance. This is a clear-cut example of different tolerance of duplications by the sporocytes and gametophytes. The sporophytes with extra 6L·6L are viable and vigorous, but gametophytes with an extra 6L·6L are inviable and nonfunctional.

On the basis of this evidence it can be argued that $n + 1$ gametophytes have lower viability, and some of them abort at different stages of development. A systematic study of the development of spores and gametophytes in the trisomic of one or more diploid species would be diagnostic, but no such investigation has been carried out. However, the extensive and painstaking researches of Satina and Blakeslee (1937a,b) on the male and female gametophytes of triploid *Datura* are illuminating. Male and female spores of *Datura* with chromosome numbers varying from n to $n + 11$ developed normally in the initial stages. The first mitotic division of all the microspores was normal and two nuclei were produced. However, after the first division some of the pollen grains started to disintegrate. As the pollen grains grew to maturity, the percentage of nonstainable grains increased. Satina and Blakeslee (1937a) concluded that the disintegrating pollen grains contained extra chromosomes. On the female side, all the megaspores went through three mitotic divisions to form perfectly normal embryo sacs. Satina and Blakeslee (1937b) surmised that the female gametophytes with extra chromosomes were able to develop normally, at least morphologically. To study their functional viability, Satina *et al.* (1938) examined the ovules 6–8 days after pollination. They found a very large number with unfertilized female gametes and central nuclei, and some of these ovules were disintegrating. They also found many fertilized female gametes, evident by the presence of pollen tubes near the micropylar end, but the eggs did not divide and were doomed to disintegrate. It is clear that the female gametophytes with extra chromosome, although morphologically normal, did not function. It should be borne in mind that a majority of the nonfunctional gametophytes in this study may have carried more than one or two extra chromosomes. However, it is conceivable that some of the $n + 1$ female gametophytes produced by the trisomics are also unable to function because of the imbalance caused by the extra chromosome. In a study of *Datura* capsules produced from hand pollinations of the trisomic Globe (trisomic × normal), Buchholz and Blakeslee (1922) found 29.5% aborted ovules, some apparently unfertilized.

If the functional inviability of the $n + 1$ gametophytes was due to the imbalance caused by the extra chromosome, then the gametophytes with longer extra chromosomes should abort more frequently, as the imbalance caused by them must be greater than the imbalance caused by shorter chromosomes. Transmission rates of tomato trisomics are in line with this reasoning. The transmission rates of long chromosomes (1, 2, and 3) are the lowest of the complement, while those of the short chromosomes (7–12) are much higher (Table 4.III). It may be noted that the transmission rate of triplo-6 is as low as that of the other three long chromosomes. This can easily be explained on the basis of a proportionately greater amount of euchromatin in this chromosome than in others. It is known, in the tomato, that deficiencies of euchromatin produce more imbalance than heterochromatic deficiencies (Khush *et al.*, 1964; Khush and Rick, 1967c, 1968a). The amount of imbalance caused by the duplications of euchromatin may be similarly higher than that caused by heterochromatin. It is interesting to note that in *Datura* no correlation exists between the length of the extra chromosome and the transmission rates of the trisomics. As can be seen from Table 4.I, the trisomic for the third longest chromosome gives the highest rate of transmission, while the trisomic for the third shortest chromosome gives the lowest rate of transmission. However, the chromosomes of *Datura* were measured at diakinesis, a stage at which euchromatic regions show extreme contraction, and heterochromatic regions comprise the bulk of the volume of the paired bivalents. It may be conjectured that, if the chromosomes of *Datura* are measured at pachytene, the 19·20 chromosome, which shows the lowest transmission rate when extra, may turn out to be one of the longest of the complement. Similarly, the chromosomes which give the highest rates of transmission when extra, such as 5·6, 9·10, and 23·24, may be found to be the shortest of the complement.

In the progenies of triploids of *Datura* (Satina *et al.*, 1938) and of tomato (Rick and Barton, 1954; Rick and Notani, 1961), the majority of the aneuploids are of $2n + 1$ chromosomal constitution. Only a few $2n + 2$ and still fewer $2n + 3$ individuals are recovered, and plants with higher aneuploid numbers, expected on the basis of random segregation of chromosomes in the $3n$ parent, are not obtained. Moreover, double and triple trisomics and tetrasomics in tomato are obtained for the shorter members of the complement only. These results also indicate that the tolerance limits for the extra chromosomes in these two species, and others which yield similar progenies from their triploids (cf. Chapter 6), are extremely narrow. These limits can be determined by tolerance at the gametophytic or the zygotic stage, or at both, but, as already discussed, the tolerance limits at the gametophytic stage are more critical. Therefore, it is likely that the low gametophytic tolerance accounts for these results.

Subnormal Development of $2n + 1$ *Zygotes, Endosperm, and Embryos*

The reduced transmission of the trisomics has been attributed by several authors to the reduced viability of $2n + 1$ zygotes and embryos. Buchholz and Blakeslee (1922) found many aborted but enlarged ovules in the Globe trisomic of *Datura*. They believed these ovules had been fertilized, but the zygotes ceased to develop at various stages of development. Similarly, the presence of many small inviable seeds in trisomic types of *Matthiola* (Frost, 1927) was considered due to the lethality of $2n + 1$ zygotes in earlier stages of development. In an extensive study of the developing zygotes and embryos of the $3n \times 2n$ cross of *Datura*, Satina *et al.* (1938) found many abnormalities caused by extra chromosomes. Some of the zygotes disintegrated after going through a few cell divisions, while some seemed to be developing normally but were doomed because of abnormalities in the development of the endosperm. A further study of the small seeds revealed many defective embryos. In some, a small undifferentiated body was present below the seed coat; in others, a decayed condition of cotyledons, hypocotyl, or radical was evident. In still others the embryos were misshapen and rudimentary. There is no doubt that a large proportion of defective zygotes and embryos carried more than one extra chromosome, but conceivably some of the $2n + 1$ zygotes and embryos of the trisomic individuals meet the same fate. This contention is supported by the lowered seed fertility of the trisomics of *Datura* (Buchholz and Blakeslee, 1922), tomato (Rick and Barton, 1954), and barley (Tsuchiya, 1960). A systematic study of female gametophytes, fertilized ovules, developing zygotes, and embryos of $2n + 1$ plants should reveal whether the reduced viability of the $n + 1$ gametophytes plays a more important role in lowering the transmission of the trisomics than the subnormal development of the $2n + 1$ zygotes and embryos. However, as already discussed, the developmental and functional abnormalities of the $n + 1$ gametophytes probably are the main cause of lowered transmission of the extra chromosomes.

Reduced and Delayed Germination of the 2n + 1 Seeds

It has been known for some time that the seeds obtained from $2n + 1$ plants of tomato do not give as good germination rates as the seeds obtained from $2n$ plants. One reasonable explanation is that some of the $2n + 1$ seeds either do not germinate at all, due to malformed embryos, endosperms, or seed coats, or take much longer to germinate than $2n$ seeds. These two factors apparently influence the transmission rates of the extra chromosome in the trisomics of many species. Tsuchiya (1960) divided the seeds of two barley trisomics, Bush and Slender, into small and large classes and studied the germination percentage, as well as the frequency of the $2n + 1$ individuals (Table 4.IV). The larger seeds of the two trisomics showed 94.5 and 84.4% germination, respectively, while the smaller seeds showed 81.5 and 43.5% germination. The proportion of trisomics among the progenies of the large seeds was 17.8 and 15.7%, while the smaller seeds yielded 80.7 and 90.0% trisomics. These results show that the smaller seeds of barley are more often trisomics, and that their lowered germination must lower the frequency of $2n + 1$ individuals in the progeny.

According to Blakeslee and Avery (1938), the trisomic progenies of *Datura* which germinate later have a much higher frequency of $2n + 1$ individuals. The pooled data of the trisomic families for which records of first transplanting and second transplanting from the same seed pans were kept separately are

Table 4.IV
Data Showing the Differential Germination and Differential Frequency of Trisomics in the Large and Small Seeds of Two Trisomics of Barley[a]

Trisomic	Seed size	No. of seeds: Sown	No. of seeds: Germinated	% germination	Trisomics obtained: No.	Trisomics obtained: %
Bush	Large	261	247	94.6	44	17.80
	Small	108	88	81.5	71	80.68
Slender	Large	128	108	84.4	17	15.74
	Small	46	20	43.5	18	90.00

[a]From Tsuchiya, 1960.

TABLE 4.V
Data Showing Differential Recovery of Primary and Secondary Trisomics in the First and Second Pottings[a]

Parental trisomics	First pottings		Later pottings	
	Total plants grown	% $2n + 1$	Total plants grown	% $2n + 1$
Primary	4499	19.76	388	39.95
Secondary	5033	13.39	666	29.73

[a]From Blakeslee and Avery, 1938.

presented in Table 4.V. The first transplanting of the primary trisomic families gave 19.7% parental trisomics, while the second transplanting yielded 39.9%. Similarly, the frequency of parental trisomics in the first transplanting of progenies of secondary trisomics was 13.4%, while in the second transplanting it was 29.7%. *Datura* workers discarded the seed pans 1 month after transplanting. The majority of those seeds which remained ungerminated by that time were probably $2n + 1$. Seeds of *Datura* are known to germinate as long as 6 months after planting, but it is not feasible to hold the seed pans until the last seed germinates. Therefore, the seeds which show delayed germination (a great majority are $2n + 1$) tend to cause a slight reduction in the transmission of $2n + 1$ individuals.

The reduction in the transmission frequency of trisomics caused by delayed germination may vary for different trisomics of the same species. Satina *et al.* (1938) reported that the presence of one particular chromosome as an extra greatly delays germination of the seed. Seeds of the trisomic Inflated of *Nicotiana sylvestris* (Goodspeed and Avery, 1939) germinate very poorly compared to those of other trisomics, and the trisomic plants from the progeny of the Super Cup trisomic of *Gossypium hirsutum* germinate later than the disomic sibs (Kohel, 1966).

Reduced Vigor of $2n + 1$ *Seedlings*

Some reduction in the transmission frequency of the trisomics is obviously caused by the reduced vigor of the seedlings. Some of the trisomics, especially for the long chromosomes of tomato, are very weak in the initial stages of growth, and many $2n + 1$ seedlings die under adverse conditions. Many seedlings of triplo-9 of tomato cease growth after producing two to three leaves, thus causing a severe reduction in the transmission frequency of the trisomic (Rick and Barton, 1954).

The Effect of Genetic Background

Certain genotypes of a species are able to tolerate the imbalance caused by the extra chromosomes better than others. Rick and Notani (1961) found a statistically significant difference for tolerance of the extra chromosomes by the tomato variety Red Cherry compared to variety San Marzano. Similarly, the transmission rates of the different trisomics of tomato vary according to the genetic background. As Table 4.III shows, the trisomics of the variety San Marzano (Rick and Barton, 1954) gave much lower transmission than the trisomics of Dwarf Aristocrat (Lesley, 1932) and VF 36 (G. S. Khush, unpublished). The transmission of the extra chromosome is much higher in heterozygous backgrounds compared to the homozygous background of the parental variety. Thus, the frequency of $2n + 1$ individuals in the trisomic stocks of homozygous varieties such as San Marzano and VF 36 is consistently lower compared to their frequency in the F_2 or backcross progenies of trisomics and genetic marker stocks with different genetic backgrounds.

Transmission of the Extra Chromosome through the Male

As stated in the beginning of this chapter, the transmission of an extra chromosome through the male, if any, is extremely low compared to that through the female. The data available on male transmission in various species are summarized in Table 4.VI. At one extreme are species such as barley (Tsuchiya, 1960) in which the extra chromosome is not transmitted. In other species, as *Datura, Pennisetum typhoides,* tomato, rye, maize, and spinach, only a few trisomics of the complement transmit the extra chromosome at low frequency. In *Nicotiana sylvestris,* eight of the ten trisomics on which data have been published are transmitted through the male, and three of them give as good transmission through the male as through the female. All the trisomics of common wheat are transmitted through the male, although at a lower frequency than through the female. By virtue of this transmission, it has been possible to establish all the twenty-one tetrasomics in this species. Finally, the transmission rates of four of the six trisomics of *Lotus pedunculatus* have been investigated (Chen and Grant, 1968b), and this is the only species studied to date whose trisomics show equal transmission through male and female.

The complete trisomic series have not been established in any other polyploid species except common wheat. It is possible that in *Gossypium hirsutum, Nicotiana tabacum, Triticum durum,* and *Avena sativa* all the trisomics might transmit through the male as in wheat. The transmission rate through the male of a single trisomic of *Nicotiana tabacum* reported by Clausen and Goodspeed (1924) was 3.4%, and those of two *Gossypium hirsutum* studied by Endrizzi *et al.* (1963) and Kohel (1966) were 9.5 and 12.3%. Six trisomics of *Triticum*

TABLE 4.VI
Transmission of the Extra Chromosome from $2n \times (2n + 1)$ Crosses in Different Species

Species	n	Transmission through male	Reference
Lycopersicon esculentum	12	Only three trisomics, triplo-7, -8, and -10, are transmitted with a frequency of less than 5.0%	Lesley (1928, 1932); G. S. Khush (unpublished)
Zea mays	10	Male transmission of only two trisomics, triplo-5 and triplo-10, has been studied. Triplo-5 did not transmit, but triplo-10 transmitted to 1.4% of progeny	McClintock and Hill (1931); Rhoades (1933b)
Antirrhinum majus	8	None of the trisomics transmitted through male in study reported by Rudorf-Lauritzen, but Stubbe reported transmission through male at a low frequency of 0.13 to 0.22%	Stubbe (1934); Rudorf-Lauritzen (1958)
Datura stramonium	12	Four primary trisomics, $2n + 11 \cdot 12$, $2n + 13 \cdot 14$, $2n + 15 \cdot 16$, and $2n + 21 \cdot 22$, transmit the extra chromosome at low frequency, generally less than 10.0%. Secondary trisomic $2n + 2 \cdot 2$ is transmitted to an ocassional progeny	Blakeslee and Avery (1938)
Nicotiana sylvestris	12	Eight of the ten trisomics for which data are available transmit the extra chromosome with a frequency varying from 1.8% for Puckered to 34.1% for Pointed (Table 4.I)	Goodspeed and Avery (1939, 1941)
Triticum aestivum	21	All the trisomics transmitted at a low frequency, the average being about 7.0%. This permitted the establishment of all 21 tetrasomics	Sears (1954)
Spinacea oleracea	6	None of six trisomics studied by Janick *et al.* transmitted through male, but three out of four studied by Tabushi transmitted at frequencies of 1.8, 3.5, and 7.5%	Janick *et al.* (1959); Tabushi (1958)
Hordeum spontaneum	7	None of the seven primary trisomics transmits through male (Table 4.I)	Tsuchiya (1960)
Clarkia unguiculata	9	Only one of the trisomic lines transmitted the extra chromosome through the male	Vasek (1961)

TABLE 4.VI (Continued)

Species	n	Transmission through male	Reference
Secale cereale	7	Four of the seven primary trisomics transmit through male with a maximum frequency of 2.6% (Table 4.I).	Kamanoi and Jenkins (1962)
Collinsia heterophylla	7	Some transmission through male. (No quantitative data presented.)	Garber (1964)
Collinsia tinctoria	7	Only one of the trisomic lines transmitted the extra chromosome through the male.	Chomchalow and Garber (1964)
Lotus pedunculatus	6	Transmission through male of four trisomics investigated is as good as through the female.	Chen and Grant (1968b)
Pennisetum typhoides	7	Only two trisomics, Dark Green and Spindle, are transmitted through the male at frequencies of 3.66 and 1.00%, respectively.	Virmani (1969)
Arabidopsis thaliana	5	Four trisomics show very good transmission through male. Only one does not transmit.	Sears and Lee-Chen (1970)

durum studied by Tsunewaki (1964b) were transmitted through the male at an average rate of 2% as compared to 23% through female.

All the factors which lower the transmission of the extra chromosome through the female presumably play a similar role in reducing transmission through the male. For example, lagging and elimination of the univalent chromosome during meiosis, reduced viability of gametophytes, subnormal development of the $2n + 1$ zygotes endosperm and embryos, and poor and delayed germination of the $2n + 1$ seeds should have similar effects on male as on female transmission. The reasons for the differential transmission rates of the extra chromosome through the male and female should, therefore, be sought in the differential morphology and development of the male and female gametophytes. The female gametophytes are known to tolerate genetic and chromosomal imbalance, as well as various types of environmental stresses, to a greater extent than male gametophytes. Satina and Blakeslee (1937a), as discussed in the previous section, showed that in *Datura* some of the male gametophytes or pollen grains with extra chromosomes started to disintegrate after the first mitotic division, and that there was an increase in the number of sterile grains as the pollen matured and grew in size. Pollen sterility has been noted in all the trisomics of different species: in *Datura* (Blakeslee and Cartledge, 1926), in *Nicotiana sylvestris* (Goodspeed and Avery, 1939), in barley (Tsuchiya, 1952), in spinach (Tabushi, 1958), and in *Secale cereale* (Kamanoi and Jenkins, 1962). Generally, pollen sterility (abortion) in trisomics is not very high, the average for all the species probably is less than 15%. The real cause of the differential transmission of the extra chromosome through the male and female seems to lie in the differential ability of the apparently normal n and $n + 1$ pollen grains to take part in fertilization. The ability of the $n + 1$ pollen to compete with n pollen in effecting fertilization may be impaired at the following stages: (1) the $n + 1$ pollen may mature later than n pollen; (2) the $n + 1$ pollen grains may produce slow growing pollen tubes; (3) the $n + 1$ pollen may produce defective pollen tubes; or (4) the $n + 1$ pollen may fail to germinate.

Ramage (1965) reported that the $n + 1$ pollen grains of barley trisomics are immature when the anthers shed the pollen, but that the n grains are fully mature by that time and able to effect fertilization. This observation of Ramage is supported by the studies of Tsuchiya (1960), who found that the pollen grains of barley trisomics can be divided into two groups according to size. Presumably, in the small-sized group are the $n + 1$ grains, which are not full sized at anthesis due to immaturity.

Our understanding of the subject of differential behavior of n and $n + 1$ grains in germination and pollen tube growth comes from the pioneering studies of Blakeslee and his associates. These workers carried out extensive experiments with the primary and secondary trisomics of *Datura stramonium*. In order to find out the stage or stages at which the normal course of development of $n + 1$ grains was inhibited, they followed the development of the pollen tubes from the time of applying the pollen to the receptive stigma, through its germination, and the growth of the pollen tubes through the styles into the ovary.

The style of *Datura,* as in most other angiosperms, contains a central core of conducting tissue, soft and fibrous in nature, which is made up of narrow, linear-shaped cells. These cells terminate in the stigma, where they become papillate and exude the stigmatic fluid which induces the pollen grains to germinate. The pollen tubes aided by the process of digestion grow down the style toward the ovary. Buchholz and Blakeslee (1922) pollinated the stigmas of $2n$ plants of *Datura* with pollen from $2n$ plants and removed the styles after a given period of time. They scalded the styles in boiling water for 2 minutes, slit them lengthwise, and removed the cortical tissues. The central strands of conducting tissues were stained with magenta (acid red), washed in water, and mounted whole on slides. By applying pressure to the coverglass, they spread this tissue in a thin layer and were then able to count the pollen tubes, which were stained dark red in contrast to the pink-stained elongated cells of the conducting tissue. The remains of germinated pollen grains appeared transparent and were recognized as empty shells; the ungerminated pollen stained deep red. Thus they were able to get reliable counts of the number of ungerminated grains and of the number of pollen tubes in various portions of the style at any given time after pollination.

Buchholz and Blakeslee (1922) counted the ungerminated pollen grains on the stigmas of selfed $2n$ and Globe trisomic plants and found that the germination of pollen was 95.6 and 94.9%, respectively. This evidence was sufficient to show that n and $n + 1$ pollen of the Globe trisomic germinated equally well. They then counted the pollen tubes of a dozen or more styles in both treatments and found that 14 hours after pollination the foremost pollen tubes had penetrated about 42 mm. The counts of pollen tubes in both treatments for 2-mm intervals showed a striking difference. The distribution of pollen tubes in the styles of selfed $2n$ was unimodal, while that of the styles of the selfed Globes was bimodal. The investigators concluded that the pollen tubes with $n + 1$ chromosomes grew at a slightly slower rate than those with n chromosomes, and soon after germination the population of gametophytes became resolved into two groups. This bimodal character increased with time. Some of the slow growing pollen tubes failed to fertilize because they failed to reach the ovary before abscission of the style, and others, which managed to reach the ovary, could not take part in fertilization because all the ovules had already been fertilized by the n pollen tubes.

Datura workers (Davenport, 1924) found that the fertilization of ovules proceeded downward from the upper part of the ovary, that most of the ovules in the upper half of the ovary were fertilized by the first arriving pollen tubes, and that most of the seeds in the lower portion of the seed capsule resulted from fertilization by the slower or later arriving pollen tubes. When an excessive amount of pollen from the $2n + 1$ plant was applied to the stigma, the slow growing $n + 1$ tubes had small chance of accomplishing fertilization, as the n pollen tubes had already fertilized all the ovules. In making routine pollinations in *Datura,* abundant quantities of pollen were used. However, when a medium or small amount of pollen was used, there was a chance for fertilization by $n + 1$ pollen tubes, since some ovules were left unfertilized after all the faster growing tubes had taken part in fertilization.

The Cocklebur trisomic of *Datura* transmits the extra chromosome to about 9.2% of its progeny through the male in routine crosses. Buchholz and Blakeslee (1930a) could increase the male transmission of this extra chromosome threefold to 27.6% by applying a restricted quantity of pollen of the Cocklebur to the stigmas of $2n$ plants. In one capsule, the proportion of $2n + 1$ plants was 34.1%. This experiment proved beyond doubt that the slow growing tubes were, in fact, of $n + 1$ constitution and, given a chance, could fertilize the ovules and produce $2n + 1$ zygotes.

Buchholz and Blakeslee (1930a) carried out another experiment with the Cocklebur trisomic to verify further that the slower group of pollen tubes were those carrying $n + 1$ chromosomes. They compared the proportion of $2n$ and $2n + 1$ seeds in the upper and lower portions of the seed capsules obtained from restricted pollination. The average proportion of $2n + 1$ individuals in the lower halves of the capsules was 56.4%, while in the upper halves it was only 1.4%. One capsule yielded as many as 65.4% trisomics from the lower half. This, again, proved that the vast majority of the slow growing tubes were from $n + 1$ pollen which had arrived late and fertilized the ovules only in the lower part of the ovary. To verify further that the rapidly growing tubes were of n constitution, Buchholz and Blakeslee (1930a) pollinated $2n$ plants with pollen from Cocklebur trisomic and cut off the styles after the rapidly growing tubes had entered the ovary and the tubes of the slower group were still in the style. All of the seeds from this experiment were $2n$, and no trisomics were recovered.

In still another experiment, Buchholz *et al.* (1932) cut off the styles at predetermined times after pollination, discarded the portion of the style containing a specific group of pollen tubes, and retained, by splicing, the part with the other group. By this method any midportion of a style could be removed and the remaining tip replaced on the cut stump of the same style or placed at any point desired on the style of a different flower. The portion to be removed was determined from data on the position of pollen tubes at specific times after pollination. When the portion of the style with fast growing n tubes was removed and the slow growing $n + 1$ tubes allowed to fertilize the ovules, Buchholz *et al.* (1932) obtained a transmission rate of 31% for the Globe trisomic. This was fifteen times the normal rate of transmission through pollen for this trisomic. In similar experiments involving the Cocklebur trisomic, it was possible to obtain 75% or more of $2n + 1$ offspring, an eightfold increase over the transmission rate obtained through the male in routine pollinations.

These experiments prove conclusively that the $n + 1$ pollen tubes, if able to grow, do so at a slower rate than n pollen tubes. The $n + 1$ tubes are unable to take part in fertilization because n tubes reach the ovules first and fertilize all of them. However, through various manipulations $n + 1$ tubes can be enabled to take part in fertilization and increase the male transmission of the extra chromosome. In some trisomics the $n + 1$ pollen grains do not germinate, those of others swell and burst without producing pollen tubes, while in others a few germinate, but produce only defective pollen tubes which swell and burst after growing a short distance from the stigma. In still other trisomics, a combination of poor germination and defective pollen tubes prevents the $n + 1$ pollen from functioning, and finally, combinations of defective and slow growing pollen tubes or poor germination and defective and slow growing tubes characterize other trisomics.

The results of studies of Buchholz and Blakeslee (1930a,b; 1932) on pollen tube growth of $n + 1$ pollen of twelve primary and fourteen secondary trisomics of *Datura* are summarized in Table 4.VII. As indicated in column 4 of the table only two primaries, Cocklebur and Globe, transmit through the pollen at a low but predictable frequency. The $n + 1$ pollen of these trisomics germinates normally and normal pollen tubes are produced, but they grow slowly, and only a few are able to take part in fertilization. The pollen tube behavior of Microcarpic and Reduced is similar and these trisomics are transmitted occasionally through the pollen. The pollen tube behavior of Glossy, Elongate, and Poinsettia is variable. The $n + 1$ pollen grains germinate poorly. Sometimes a few pollen grains produce defective pollen tubes, and at other times a few normal ones. By restricted pollinations or by other experimental techniques, the pollen tubes of these trisomics may be manipulated to take part in fertilization and male transmission obtained, but under routine breeding these trisomics are not transmitted through the male. The $n + 1$ pollen grains of Ilex and Buckling do not germinate at all, while those of Rolled, Echinus, and Spinach germinate and produce defective pollen tubes which burst after growing a short distance. Consequently, these five trisomics do not transmit through the pollen, and no experimental technique has been devised to overcome these defects and obtain male transmission of these trisomics.

The pollen tube behavior of the secondary trisomics, as expected, was found to be similar to that of the related primary trisomics, and sometimes the behavior of the primary trisomic was intermediate between that of its two secondaries. Thus, Buchholz and Blakeslee (1930b) studied the pollen tube growth of primary Rolled and its secondaries, Polycarpic and Sugarloaf. The $n + 1$ pollen grains of the Polycarpic do not germinate, but those of Sugarloaf germinate normally and produce normal pollen tubes which grow at about two-thirds of the normal rate of n pollen tubes. The behavior of $n + 1$ pollen grains of Rolled is intermediate between these two extremes. These pollen grains germinate normally, but the pollen tubes burst after growing a short distance. Predictably, Polycarpic and Rolled give no transmission through the pollen, but Sugarloaf is transmitted occasionally through the male. Buchholz and Blakeslee (1930b) were able to obtain 20% transmission of Sugarloaf through the pollen by means of restricted pollinations.

As discussed in Chapter 3, the secondary trisomics produce four types of spores. For example, the $n + 2\cdot2$ (Sugarloaf) secondary of *Datura stramonium* produces spores of n, $n + 2\cdot2$, $n + 1\cdot2$, and $n - 1\cdot2 + 2\cdot2$ constitution. During their studies of pollen tube growth, Buchholz and Blakeslee (1930b) identified these four categories of gametophytes. The n pollen tubes formed a forward mode in the style, while the $n + 2\cdot2$ pollen tubes, growing at a lower rate but normally, formed a mode at an intermediate position. The $n + 1\cdot2$ pollen tubes grew a short distance, swelled, and burst not far from the stigma. The fourth group of gametophytes, $n - 1\cdot2 + 2\cdot2$, was represented by the

TABLE 4.VII

Primary and Secondary Trisomics of *Datura*: Data on Pollen Sterility, Transmission through the Male, and Causes of Functional Inviability of the $n + 1$ Pollen[a]

Trisomic				
Extra chromosome	Laboratory name	% pollen sterility	% transmission through pollen	Causes of functional inviability of the $n + 1$ pollen
1·2	Rolled	2.6	0.0	Pollen tubes grow a short distance, then swell and burst
3·4	Glossy	2.2	Possible	Germination poor. Some pollen tubes burst; a few grow at half the normal rate
5·6	Buckling	2.6	0.0	Pollen swells and bursts, but no germination. Very few may produce burst pollen tubes
7·8	Elongate	7.1	Possible	Germination poor. Some pollen tubes burst; a few grow at half the normal rate
9·10	Echinus	3.7	0.0	Germination normal. Pollen tubes burst near stigma; very few may grow a short distance
11·12	Cocklebur	6.7	9.7	Germination normal. Pollen tubes normal, but grow at half the normal rate
13·14	Microcarpic	10.2	Some	Germination normal. Pollen tubes normal, but grow at half the normal rate
15·16	Reduced	3.8	Some	Germination normal. Some pollen tubes burst; others grow at half the normal rate
17·18	Poinsettia	5.2	Possible	Pollen tubes slow growing. Behavior variable
19·20	Spinach	27.1	0.0	Poor germination; burst pollen tubes
21·22	Globe	4.0	2.5	Germination normal. Some pollen tubes burst; others grow at two thirds the normal rate.
23·24	Ilex	5.2	0.0	No germination
1·1	Polycarpic	20.8	0.0	No germination
2·2	Sugarloaf	12.4	Possible	Germination and pollen tubes normal. Rate of growth of pollen tubes two thirds of normal
3·3	Smooth	9.7	0.0	Germination poor; those which germinate burst near stigma.
5·5	Strawberry	10.7	0.0	No germination
6·6	Aerolate	–	Possible	Pollen tube growth one third of normal rate

TABLE 4.VII (Continued)

Trisomic				
Extra chromosome	Laboratory name	% pollen sterility	% transmission through pollen	Causes of functional inviability of the $n + 1$ pollen
7·7	Undulate	10.2	0.0	Pollen tubes burst near the stigma; a few may grow a short distance
9·9	Mutilated	12.5	0.0	No germination
10·10	Thistle	–	0.0	No germination
11·11	Wedge	8.4	0.0	Germination normal, but pollen tubes burst
13·13	Marbled	–	0.0	Germination very poor, a few produce burst pollen tubes
14·14	Mealey	–	0.0	Germination poor; pollen tubes slow and burst
15·15	Scalloped	7.3	Possible	Germination normal; some pollen tubes burst, others grow at half the normal rate
17·17	Dwarf	14.3	0.0	No germination
19·19	Divergent	4.8	0.0	Poor germination; burst pollen tubes

[a]From Blakeslee and Cartledge, 1926; Buchholz and Blakeslee, 1930a, b, 1932.

aborted pollen grains lying unstained on the stigma (Fig. 4.1D). This type of analysis was carried out with all the secondary trisomics of *Datura* (Buchholz and Blakeslee, 1932). Figure 4.1 graphically shows the pollen tube behavior of the disomic, two primary trisomics, and the secondary trisomic. Studies of pollen tube growth in the trisomics of other species have not been made except for one trisomic of castor bean (*Ricinus communis* L.). In this trisomic the length of the $n + 1$ pollen tubes was much shorter than that of n pollen tubes (Jakob, 1963).

The Appearance of Nonparental Trisomics in the Progenies of Different Trisomics

One of the unique and useful features of the breeding behavior of all kinds of trisomics is that they throw nonparental but related trisomics in their progenies. A primary trisomic occasionally may yield a related secondary or a telotrisomic among its progeny, but all the secondaries and tertiaries throw related primaries at a consistent frequency. Similarly, the compensating trisomics throw related secondaries, tertiaries, or telotrisomics at consistent frequencies. This feature

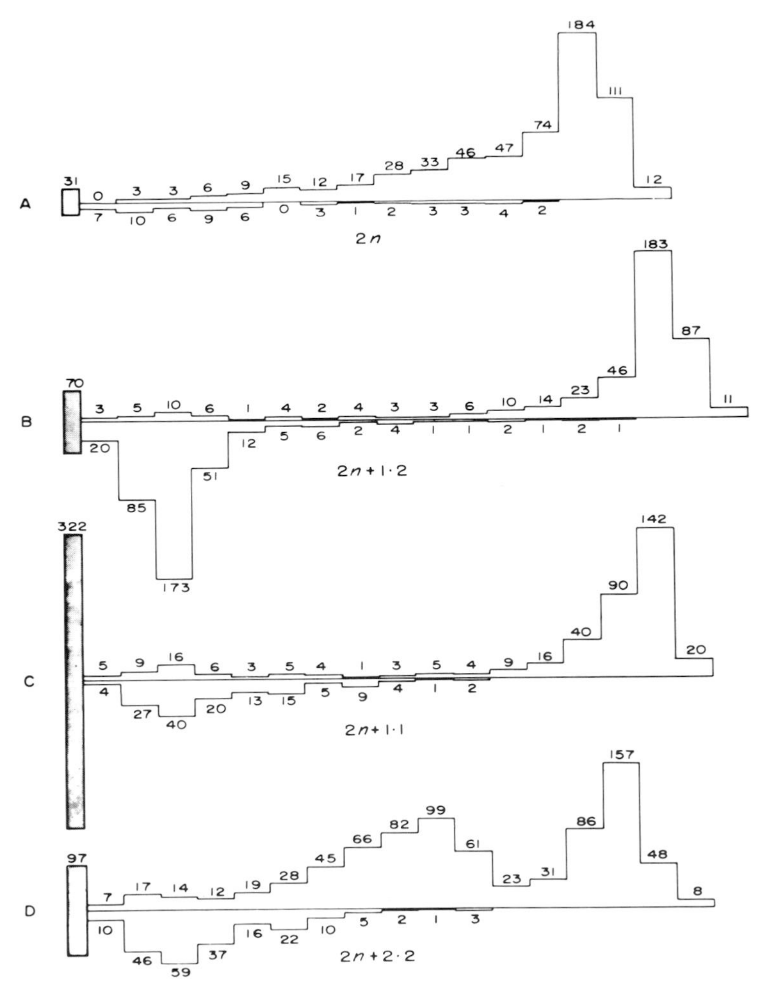

Fig. 4.1. Diagram of pollen tube distributions in the tests of the pollen of $2n$, $2n + 1 \cdot 2$, $2n + 1 \cdot 1$ and $2n + 2 \cdot 2$. The normal appearing pollen tubes are plotted above the datum line at 2-mm interval corresponding to their distance from the stigma, while the abnormal appearing pollen tubes are plotted below the datum line. The shaded vertical bar at the left in each diagram indicates the number of pollen grains remaining ungerminated on the stigma. (A) The pollen tube distribution of pollen of $2n$. The distribution is unimodal. (B) The pollen tube distribution of pollen of $2n + 1 \cdot 2$. The distribution is bimodal. The pollen tubes at the forward mode are from n pollen while the burst pollen tubes near the stigma are from $n + 1 \cdot 2$ pollen. (C) The pollen tube distribution of pollen of $2n + 1 \cdot 1$. The n pollen has formed the pollen tubes of the forward mode. The $n + 1 \cdot 2$ pollen is recognized by its behavior in that the pollen tubes burst near the stigma and $n + 1 \cdot 1$ group of gametophytes is recognizable as the ungerminated pollen. (D) The pollen tube distribution of pollen of $2n + 2 \cdot 2$. The n pollen tubes are at the forward mode. The $n + 2 \cdot 2$ pollen tubes grow normally but at a slower rate and have formed the second mode. The $n + 1 \cdot 2$ pollen tubes are recognized as having burst near the stigma. (From Buchholz and Blakeslee, 1932.)

TABLE 4.VIII
Transmission Rates of Primary Trisomics of *Datura* and the Occurrence of Related and Unrelated Secondary Trisomics among Their Selfed Progenies[a]

Trisomic	No. plants scored	2n	2n + 1					
			Parental type		Related secondary		Unrelated primary and secondary	
			No.	%	No.	%	No.	%
$2n + 1 \cdot 2$	3401	2947	421	12.37	6	0.18	27	0.79
$2n + 3 \cdot 4$	2616	2085	530	20.26	0	0.00	1	0.04
$2n + 5 \cdot 6$	2869	1975	860	30.15	2	0.07	27	1.12
$2n + 7 \cdot 8$	2432	2165	254	10.44	0	0.00	13	0.54
$2n + 9 \cdot 10$	2437	1689	731	30.12	1	0.04	13	0.66
$2n + 11 \cdot 12$	2798	2234	544	19.51	2	0.07	16	0.64
$2n + 13 \cdot 14$	3671	2884	741	20.21	0	0.00	42	1.25
$2n + 15 \cdot 16$	2762	2193	546	19.20	1	0.04	21	0.80
$2n + 17 \cdot 18$	2641	2013	614	23.32	1	0.04	11	0.45
$2n + 19 \cdot 20$	5561	5294	151	2.73	7	0.13	106	1.97
$2n + 21 \cdot 22$	2894	2070	806	28.29	0	0.00	4	0.21
$2n + 23 \cdot 24$	2379	1635	742	31.23	0	0.00	1	0.08

[a]From Avery, A. G., Satina, S., and Rietsema, J. (1959). "Blakeslee: The Genus *Datura*," 289 pp. © 1959, The Ronald Press Company, New York.

of their breeding behavior provides excellent sources of related trisomics. Another desirable aspect is that, by virtue of a high frequency of related types among their progenies, the secondary and compensating trisomics can be employed in studying more than one kind of genetic ratios from the same segregating progeny. This topic will be discussed in more detail in Chapter 5.

In addition to the related $2n + 1$ types, all the trisomics occasionally produce unrelated trisomics (Table 4.VIII). This is true also of disomics. However, the rate of spontaneous appearance of unrelated trisomics among the progenies of trisomics seems to be much greater than the rate at which trisomics appear spontaneously in the progenies of disomics. In *Datura,* for which extensive records on the spontaneous occurrence of different trisomics were kept, Blakeslee and Avery (1938) reported that disomics produced $2n + 1$ types at the rate of 0.16% while primary and secondary trisomics produced unrelated $2n + 1$ types at the rate of 0.86 and 0.62%, respectively. Chen and Grant (1968b) reported the occurrence of unrelated trisomics in the progenies of four trisomics of *Lotus pedunculatus* at a frequency of 0.6%. Why the presence of an extra chromosome increases the rate of spontaneous occurrence of other trisomics is not clear. The increase probably is brought about by the extra chromosome causing interference in the regular mechanics of meiotic division to increase the rate of nondisjunction.

In contrast, however, the occurrence of related $2n + 1$ types among the trisomic progenies can easily be explained and will be discussed in the following paragraphs.

Related 2n + 1 Types among the Progenies of Primary Trisomics

The two types of related trisomics thrown by the primary trisomics are secondary and telotrisomics. In Chapter 3 we discussed the behavior of the extra chromosome when present as a univalent. Because of its instability, it tends to misdivide and give rise to telocentrics or isochromosomes, which are sometimes included in the meiotic products, and give rise to $n + 1$ telocentric and $n + 1$ isochromosome gametophytes and gametes. When such gametes are fertilized by an n gamete they give rise to secondary and telotrisomics. These related $2n + 1$ types have been reported among the progenies of primary trisomics of tomato (Lesley, 1928, 1932), maize (Rhoades, 1933b, Einset, 1943), *Datura* (Blakeslée and Avery, 1938), *Nicotiana sylvestris* (Goodspeed and Avery, 1939), barley (Tsuchiya, 1960), *Secale cereale* (Kamanoi and Jenkins, 1962) and *Lotus pedunculatus* (Chen and Grant, 1968b). As can be seen from Table 4.VIII, seven of the primary trisomics of *Datura* yielded secondary trisomics among their progenies, and one of them, $2n + 19 \cdot 20$, gave rise to the $19 \cdot 19$ secondary on seven independent occasions. Five of the primary trisomics did not yield any secondary trisomics among their progenies. It is possible that the univalent extra chromosomes of these primaries do not misdivide. Alternatively, imbalance produced by the isochromosomes of these chromosomes may be so great as to be lethal to the $n + 1$ spores or $2n + 1$ zygotes. The evidence for the latter hypothesis comes from the studies of Khush and Rick (1967d), who showed that isochromosomes of the long arms of chromosomes 2 and 3 produce imbalance to such an extent that the gametophytes carrying these extra isochromosomes are unable to survive.

The primary trisomics (discussed in Chapter 3) occasionally produce $n - 1$ gametes in which none of the three members of the trisome is included. In the diploid species, such gametes abort due to the deficiency of the chromosome concerned. Consequently, $2n - 1$ plants are not recovered among the progenies of trisomics of the diploid species. In the polyploid species,however, the $n - 1$ gametes are functional, and $2n - 1$ plants are recovered among the progenies of $2n + 1$ plants (Sears, 1954). The case of maize trisomics is of great interest. Einset (1943) obtained five monosomics and one plant in which one arm of a chromosome was missing among 1916 plants of trisomic progenies. Whether these monosomics were for the same chromosome or different chromosomes of the complement is not known. In this respect, however, maize seems to resemble the polyploid species more than the diploids.

Related 2n + 1 Types in the Progenies of Secondary Trisomics

As discussed in the previous paragraphs, the primary trisomics occasionally throw related secondary trisomics as well as telotrisomics in their progenies. The secondary trisomics, however, regularly produce related primary trisomics and at a predictable frequency. We know from Chapter 3 that secondary trisomics produce four types of gametes compared to two types produced by the primary trisomics. A secondary trisomic of $2n + 1 \cdot 1$ constitution would produce $n + 1 \cdot 1$, $n + 1 \cdot 2$, $n - 1 \cdot 2 + 1 \cdot 1$, and n gametes. In the diploid species, the gametes, which do not receive the full haploid chromosome complement, abort due to the deficiency. Therefore, $n - 1 \cdot 2 + 1 \cdot 1$ gametes produced by the secondary trisomics of diploid species are abortive because they are deficient for one arm and duplicated for the other arm of the same chromosome. All the secondary trisomics of *Datura* (Blakeslee and Cartledge, 1926) and tomato (Khush and Rick, 1969) produce a certain amount of pollen abortion. The type of disjunction which gives rise to these abortive gametes also produces $n + 1$ gametes in which the extra member is the primary chromosome, and these gametes, when fertilized by n gametes, produce the related primary trisomics. All the secondary trisomics of *Datura* and all but one of those of tomato, whose breeding behavior has been studied, regularly produce related primary trisomics in their progenies (Table 4.IX). The frequency of related primaries in the progenies of secondaries varies rather widely. Of the *Datura* secondaries $2n + 9 \cdot 9$ may produce as many as 10.01% of $2n + 9 \cdot 10$ individuals, while $2n + 19 \cdot 19$ produces only 0.06% of $2n + 19 \cdot 20$ primaries. The $2n + 10L \cdot 10L$ secondary of tomato produces 14.9% of related primaries in its progeny, while $2n + 7S \cdot 7S$ failed to yield any related primary in a family of 259 plants. Since the progeny of $2n + 7S \cdot 7S$ was rather small, it is likely that in a larger population triplo-7 will be recovered among the offspring of this secondary trisomic. The frequency of the related primaries in the progenies of these and other secondaries is given in Table 4.IX. The ratio of related primaries to the parental secondaries of tomato was also interestingly variable (Khush and Rick, 1969). In the progeny of $2n + 6L \cdot 6L$, for example, only primaries (no secondaries) appeared; for $2n + 7S \cdot 7S$ the opposite was obtained. The proportion of primaries was more than double the proportion of secondaries in the progeny of $2n + 8L \cdot 8L$, while it was lower in the progenies of other secondaries. The ratio of related primaries to the parental secondaries was similarly variable in *Datura*.

The cytological requirements which give rise to $n + 1$ gametes having a primary extra chromosome instead of a secondary were discussed in Chapter 3. The correlation between low frequency of trivalent formation in $2n + 7S \cdot 7S$

TABLE 4.IX
Transmission Frequency of Secondary Trisomics of *Datura*[a] and Tomato[b] and the Frequency of Related Primary Trisomics in Their Progenies

Species	Trisomic	Total	$2n$	Parental secondary		Related primary	
				No.	%	No.	%
Datura	$2n + 1 \cdot 1$	1965	1870	48	2.43	47	2.39
	$2n + 2 \cdot 2$	2124	1661	364	17.08	99	4.64
	$2n + 3 \cdot 3$	2932	2681	154	5.22	97	3.29
	$2n + 5 \cdot 5$	2715	1767	796	29.17	152	5.59
	$2n + 6 \cdot 6$	2859	2192	482	16.84	185	6.46
	$2n + 7 \cdot 7$	3164	2269	839	26.51	56	1.77
	$2n + 9 \cdot 9$	2938	2247	396	13.34	295	10.01
	$2n + 10 \cdot 10$	2410	1721	631	26.01	58	2.38
	$2n + 11 \cdot 11$	2702	2225	447	16.55	30	1.11
	$2n + 13 \cdot 13$	2303	1920	255	11.03	128	5.82
	$2n + 14 \cdot 14$	2759	2141	537	19.43	81	2.96
	$2n + 15 \cdot 15$	2701	2071	541	20.02	89	3.24
	$2n + 17 \cdot 17$	3218	2711	321	9.95	186	5.76
	$2n + 19 \cdot 19$	3329	2347	980	29.25	2	0.06
Tomato	$2n$ + 6L·6L	681	510	0	0.00	71	10.40
	$2n$ + 7S·7S	259	170	87	33.60	0	0.00
	$2n$ + 8L·8L	268	216	15	5.60	34	12.70
	$2n$ + 9S·9S	298	234	61	20.40	3	1.00
	$2n$ + 9L·9L	397	300	67	15.30	29	7.30
	$2n$ + 10L·10L	681	447	130	19.10	102	14.90
	$2n$ + 12L·12L	594	496	81	15.30	17	2.80

[a]From Avery *et al.*, 1959.
[b]From Khush and Rick, 1969.

and $2n$ + 9S·9S of tomato, on one hand (Table 3.V), and the low frequency of the primaries in their progenies, on the other (Table 4.IX), is clear.

Related 2n + 1 Types in the Progenies of Tertiary Trisomics

As was discussed in Chapter 3, the segregations from the chains of 5, 4, or 3 chromosomes, or the random segregation of univalents when more than one is present, can give rise to $n + 1$ gametes in which the extra chromosome is not the tertiary but one of the two related primary chromosomes. When such gametes are fertilized by n gametes, they produce primary trisomics. Therefore, the frequency of each of the related primary trisomics in the progeny

TABLE 4.X
Transmission Frequency of Tertiary Trisomics of *Datura* and the Frequency of Related Primary Trisomics in Their Progenies[a]

Trisomic	Total no. of plants	2n	2n + 1: Parental type No.	%	Primary a[b] No.	%	Primary b[b] No.	%	Related secondary
2n + 1·9	2723	2289	393	14.39	14	0.51	27	0.99	3
2n + 1·18	2091	1773	295	14.14	5	0.23	10	0.48	
2n + 2·5	2057	1360	665	32.05	16	0.77	21	1.01	1
2n + 2·17	2779	2210	553	19.88	12	0.43	2	0.07	
2n + 4·6	2805	1966	821	29.26	1	0.04	3	0.10	1
2n + 4·22	978	715	262	26.55	0	0.00	0	0.00	
2n + 6·19	596	433	159	26.58	5	0.84	0	0.00	
2n + 9·20	1422	1075	331	23.17	17	1.19	0	0.00	

[a]From Avery, A. G., Satina, S., and Rietsema, J. (1959). "Blakeslee: The Genus *Datura*," 289 pp. © 1959, The Ronald Company, New York.
[b]Related primary *a* is the one having an extra chromosome with lower end numbers. For example, the related primary *a* of 2n + 1·9 is 1.2, and the related primary *b* is 2n + 9·10.

of a tertiary trisomic depends upon the frequency of various associations of the chromosomes involved and orientation and anaphase separation of these associations. That the type of segregations which would yield primary trisomics do occur, but at a low frequency, is evident from Table 4.X. Of the eight tertiaries of *Datura* reported in Table 4.X, five yielded both the primaries, two yielded one related primary, and only one failed to throw any related primaries in its progeny. Similarly, four of the seven tertiary trisomics of tomato yielded both the related primary trisomics, while the remaining three produced only one type of related primary trisomic (Khush and Rick, 1967b). In addition to related primaries, the tertiaries may occasionally throw a related telotrisomic or secondary trisomic (Table 4.X) as a result of misdivision of the univalent during meiosis.

Related 2n + 1 Types in the Progenies of Telotrisomics

The extra telocentric pairs with the related homologous chromosomes at a low frequency in the telotrisomics. Even when the telocentric pairs with the normal homologs to form a trivalent, the segregation of the three members of the trivalent is not random, and the two normal homologs segregate to the opposite poles. Therefore, the related primary trisomics are found only rarely in the progenies of telotrisomics. Rhoades (1936) obtained primary triplo-5 at a frequency of 1.5% in the progeny of the telotrisomic for the short arm of chromosome 5 of maize. Blakeslee and Avery (1938) did not obtain any $2n + 11 \cdot 12$ plants in the progenies of $2n + \cdot 11$ of *Datura*. Only one of six telotrisomics of tomato yielded a single related primary trisomic (Khush and Rick, 1968b). The telocentric chromosomes may be converted into isochromosomes and throw corresponding secondary trisomics. Thus, Rhoades (1938, 1940) reported that the $2n + \cdot 5S$ telotrisomic of maize yielded the $2n + 5S \cdot 5S$ secondary at frequencies of 0.42 and 0.22% when used as male and female, respectively.

Occurrence of Related 2n + 1 Types in the Progenies of Compensating Trisomics

Since the compensating trisomics are of several types in regard to chromosomal constitution, the type of related trisomics produced by them are different for each kind. A ditertiary compensating trisomic yields two tertiary trisomics as well as two primary trisomics. Nubbin, a compensating trisomic of *Datura* with a chromosomal formula of $2n - 1 \cdot 2 + 1 \cdot 9 + 2 \cdot 5$ (Blakeslee, 1927a), regularly produced $2n + 1 \cdot 9$ and $2n + 2 \cdot 5$ tertiaries as well as $2n + 9 \cdot 10$ and $2n + 5 \cdot 6$ primaries (Table 4.XI). Avery *et al.* (1959) studied the breeding behavior of six other compensating trisomics of *Datura* of similar constitution. The average frequency of related tertiaries among the progenies of all of them was somewhat higher (2.5–5.3%) than that of related primaries (0.13–1.9%).

The isotertiary compensating trisomic would yield a tertiary trisomic, a secondary trisomic, and a primary trisomic. Similarly, a telotertiary compensating trisomic would produce a telotrisomic, a tertiary trisomic, and a primary trisomic. The compensating trisomic of tomato, in which the missing chromosome 3 was compensated by the $3L \cdot 3L$ isochromosome and $\cdot 3S$ telocentric (Khush and Rick, 1967a) produced telotrisomic $2n + \cdot 3S$ at a frequency of 2.7% (Table 4.XI). The secondary trisomic $2n + 3L \cdot 3L$ expected from this trisomic was not recovered, probably due to the inviability of $n + 3L \cdot 3L$ spores. Finally, the compensating trisomic, in which the missing chromosome 7 was replaced by its two isochromosomes, produced both the secondary trisomics among its progeny (Khush and Rick, 1969).

Conclusions

The present status of our knowledge of transmission of the extra chromosome in the trisomics can be summarized as follows:

1. Transmission of the extra chromosome in $(2n + 1) \times 2n$ crosses is lower than the expected 50% in most species. Transmission from $2n \times (2n + 1)$ crosses

TABLE 4.XI
Transmission Rates of Compensating Trisomics and the Occurrence of Related $2n + 1$ Types among Their Progenies

Trisomic	Total	$2n$	Parental trisomic	Related $2n + 1$ types			
				$2n + 1 \cdot 9$	$2n + 2 \cdot 5$	$2n + 9 \cdot 10$	$2n + 5 \cdot 6$
$2n - 1 \cdot 2 + 1 \cdot 9 + 2 \cdot 5$ (Blakeslee, 1927a)	1221	668 (54.5%)	424 (34.6%)	36 (2.9%)	58 (4.7%)	14 (1.1%)	9 (0.7%)
				$2n + \cdot 3S$	$2n + 3L \cdot 3L$		
$2n - 3S \cdot 3L + \cdot 3S + 3L \cdot 3L$ (Khush and Rick, 1967a)	182	103 (56.3%)	74 (40.5%)	5 (2.7%)	0 (0.0%)		
				$2n + 7S \cdot 7S$	$2n + 7L \cdot 7L$		
$2n - 7S \cdot 7L + 7S \cdot 7S + 7L \cdot 7L$ (Khush and Rick, 1969)	185	116 (62.7%)	48 (25.9%)	20 (10.8%)	1 (0.54%)		

is lower than in $(2n + 1) \times 2n$ crosses in all species except *Lotus pedunculatus*.

2. The causes of reduced transmission in $(2n + 1) \times 2n$ crosses in such diploid species as *Datura*, tomato, barley, *Nicotiana sylvestris*, *Secale cereale*, etc., may be manifold. The reduction may be caused by elimination of the extra chromosome during meiosis due to lagging, by reduced viability of $n + 1$ spores and gametophytes, by subnormal development of $2n + 1$ zygotes and embryos, by poor and delayed germination of $2n + 1$ seeds, by poor survival of $2n + 1$ seedlings, and by the effect of the genetic background. Although no existing data permit precise comparisons, it seems likely that the main cause of reduced transmission is the reduced viability of $n + 1$ gametophytes. A systematic study of developing female gametophytes, fertilized ovules, developing zygotes, and embryos of $2n + 1$ plants is needed to verify this conclusion.

The slight decrease in the transmission frequency of trisomics of polyploid species such as wheat (transmission rates 40–45%) can be wholly accounted for on the basis of elimination of the extra chromosome during meiosis. The extra chromosome causes no imbalance, or very little, in the spores, gametophytes, zygotes, and embryos of the polyploid species, as is revealed by the normal female fertility of the trisomics (Sears, 1954). It is noteworthy that maize resembles the polyploid species rather than the diploids with respect to tolerance of the extra chromosome. Einset (1943) has shown that the reduction in the transmission rates of maize trisomics is caused only by the elimination of the extra chromosome during meiosis. The spores, gametophytes, zygotes, and embryos with extra chromosomes are fully viable and, consequently, the female fertility of the trisomics of maize is normal.

3. All the factors which lower the transmission frequency of the extra chromosome through the female presumably play a similar role in reducing the transmission rate through the male. In addition, reasons for the much lower transmission rates through the male may be one of the following: (1) late maturity of $n + 1$ pollen, (2) slow growth of $n + 1$ pollen tubes, (3) abnormal development of $n + 1$ pollen tubes, or (4) failure of $n + 1$ pollen to germinate.

4. Trisomics of different types regularly throw nonparental but related trisomics in their progenies. Therefore, trisomic of one kind is a good source of another kind. The special genetic applications of this aspect of transmission are discussed in Chapter 5.

5

Genetic Segregation and Other Uses of Trisomics

Since various kinds of trisomics carry an extra chromosome, the genetic ratios for the genes which are located on that chromosome are modified in the segregating progenies. This modified ratio technique has been used in the cytogenetic analysis of several plant species. In fact, trisomic segregations have been studied extensively in all the genetically well investigated plant species, such as maize, tomato, barley, *Datura stramonium, Sorghum vulgare,* and *Arabidopsis thaliana*. In addition to determining gene–chromosome relationships by the modified ratio technique, trisomics have been used in several other cytogenetic investigations. This chapter deals with theoretical considerations and experimental results of genetic segregation in trisomics of different types. Other special uses of trisomics are also reviewed in this chapter.

Genetic Segregation in Primary Trisomics

Primary trisomics provide an excellent cytogenetic tool for testing the independence of linkage groups and for assigning linkage groups to particular chromosomes. Since there are three homologous chromosomes in a primary trisomic instead of two for a particular member of the complement, the genetic ratios in segregating progenies for genes that are located on this chromosome are very different from the 3:1 or 1:1 ratios obtained in the F_2 and backcross generations of a normal disomic heterozygous for a recessive gene. Consider, for example, a trisomic individual heterozygous for a recessive marker gene located on the trisomic chromosome, so that two homologs carry the normal alleles and the third carries the recessive allele. The genotype can be expressed as *AAa*. Such an individual would produce two types of gametes with regard to chromosomal constitution, n and $n + 1$, and four with respect to allelic constitution in the following frequencies: $1AA{:}2Aa{:}2A{:}1a$. The genotypic frequencies in the F_2 would be $1AAAA{:}4AAAa{:}4AAaa{:}4AAA{:}10AAa{:}4Aaa{:}4AA{:}4Aa{:}1aa$. In other words, there would be thirty-five normal individuals to one recessive. In the backcross there would be five normals to one recessive.

The above calculations are based on two assumptions: (1) the extra chromosome is transmitted by 50% of the gametes on the female as well as on the male side, so that there are 25% tetrasomics, 50% trisomics, and 25% disomics in the progeny; and (2) there is no crossing over between the locus of the marker gene and the centromere of the chromosome. In practice, however, the phenotypic frequencies calculated above are never realized, as has been shown in the trisomic inheritance studies in more than half a dozen species. The departures from the 35:1 ratio are always very strong, because the assumptions enumerated above seldom hold true. As we observed in Chapter 4, the $n + 1$ gametes are rarely transmitted through the male, so that the functional male gametes are produced in the ratio of $2A{:}1a$. Calculated on this basis, the genotypic frequencies of the selfed progeny should be $2AAA{:}5AAa{:}2Aaa{:}4AA{:}4Aa{:}1aa$. In other words, the ratio of normals to recessives should be 17:1 for the entire progeny, 8:1 for the disomic fraction, and 1:0 for the $2n + 1$ fraction.

As was discussed in the last chapter, $2n$ and $2n + 1$ individuals are rarely produced in equal frequency. Generally, the proportion of $2n + 1$ individuals is much lower than 50%. Since all the $2n + 1$ individuals in the above calculations are phenotypically normal, the net result of their lowered frequency is a decrease in the proportion of phenotypically normal progeny. If we assume that one-third of the viable gametes on the female side are $n + 1$ and two-thirds are n, the genotypic frequencies in the F_2 can be then adjusted and would be as follows: $1(2AAA{:}5AAa{:}2Aaa) + 2(4AA{:}4Aa{:}1aa)$. In other words, the normal and recessive individuals would be produced in a ratio of 12.5:1, but the 1:0 ratio for the $2n + 1$ and 8:1 for the $2n$ fractions would not change. Therefore, it is clear that the lower the transmission of the extra chromosome through the female, the lower would be the proportion of normal progeny in the F_2, and vice versa. With no transmission, the ratio among all progeny would be 8:1, and with 50% transmission, 17:1. The effect of reduced transmission on the female backcrosses is similar. With no transmission, the ratio of normal to recessive would be 2:1; with 33.3% transmission, 3.5:1; and with 50% transmission, 5:1. In the male backcrosses with no transmission of the extra chromosome, the ratio would be 2:1.

The above ratios have been calculated on the basis of chromosome segregation (Muller, 1914). According to this type of segregation, the chromatids derived from a particular multivalent belong to separate chromosomes in that multivalent. For example, from a trisomic of AAa genotype, only AA, Aa, A, and a gametes would be formed—aa gametes would never appear. In practice, however, trisomic individuals with the *aaa* genotype, which can only result from *aa* gametes, appear at a low frequency in some of the F_2 and backcross generations. Two other types of segregations have been proposed as alternatives to chromosome segregation to account for these individuals. These are random chromatid segregation (Haldane, 1930) and complete equational segregation (Mather, 1935; Sansome and Philp, 1939). Burnham (1962) proposed that the more descriptive term "maximum equational" be used instead of "complete equational." The random chromatid segregation assumes an infinite number of crossovers between the centromere and the locus, so that at least those parts of the chromatids which carry the locus assort at random to the gametes, independent of the centromere, where the first division is always reductional. Therefore, assuming equal production of the $n + 1$ and n gametes on the female side, *AAAAaa* chromatids of the *AAa* trisomic would produce 6*AA*:8*Aa*:1*aa*:10*A*:5*a* gametes. It should be noted that 3/15 of the $n + 1$ gametes are double reductional, or derived from sister chromatids, and 1/15 carry both the recessive alleles. On selfing, the genotypic frequencies would be 12*AAA*:22*AAa*:10*Aaa*:1*aaa*:20*AA*:20*Aa*:5*aa*. According to this segregation, 2.22% of the trisomics would have a recessive phenotype, while the ratio of the normals to the recessives among the $2n$ fraction would still be 8:1. The overall proportion of the normals to the recessives in the progeny would now be 14:1. With the transmission frequency of the $n + 1$ eggs at 33.3%, the genotypic frequencies on the basis of random chromatid segregation would be 12*AAA*:22*AAa*:10*Aaa*:1*aaa*:40*AA*:40*Aa*:10*aa*, and the overall ratio of normals to recessives would be 11.27:1. However, Mather (1935, 1936), Sansome and Philp (1939), Little (1945, 1958), and, more recently, Burnham (1962) have shown that random chromatid segregation does not lead to maximum value for double reduction in autopolyploids. On the basis of cytological considerations, they have shown that a proportion of double reductional gametes and recessive genotypes higher than expected on the basis of chromatid segregation can be obtained. Therefore, they have proposed the hypothesis of complete equational segregation. Their arguments applied to the trisomic of the *AAa* genotype are summarized in Fig. 5.1, in which the alleles *A* and *A'* are identical. If the three chromosomes carrying *A*, *A'*, and *a* always pair as a trivalent, and if the chiasmata are formed as shown in the figure, between the centromere and the locus of the gene and segregate 2:1 at random, then three types of dyads are produced in equal frequency after the first division. Each of these dyads can give two types of segregations at second division with regard to allelic constitution, as shown in the figure. Thus, 3 out of 12 or one-fourth of the $n + 1$ spores produced (*AA*, *A'A'*, and *aa*), are double reductional. The frequency of double reductional gametes in this case is clearly higher than that obtained on the basis of random chromatid segregation, where the frequency is one-fifth; and the frequency of *aa* gametes is also higher (1/12) in maximum equational segregation than in random chromatid segregation (1/15).

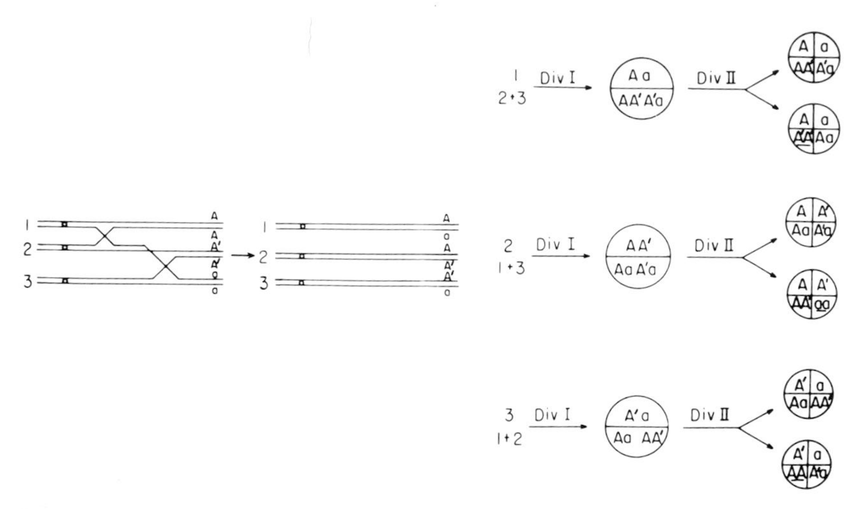

Fig. 5.1. Crossovers and segregations in a trisomic *(AAa)* which lead to the production of spores receiving two alleles derived from sister chromatids (double reduction), one of which is *aa*. Note that two more spores *AA* and *A'A'* are also produced by double reduction. Therefore, 3 (underlined in diagram) out of 12 $n+1$ spores are double reductional and 1/12 carry the two recessive alleles. *A* and *A'* are identical but have been so drawn for ease of identification.

The gametic frequencies in the trisomic of the *AAa* genotype, on the basis of maximum equational segregation, therefore, would be 5*AA*:6*Aa*:1*aa*:8*A*:4*a*, and genotypic frequencies in the F_2 would be 10*AAA*:17*AAa*:8*Aaa*:1*aaa*:16 *AA*:16*Aa*:4*aa*. Thus, 2.77% of the trisomics would be of the recessive phenotype, and the overall ratio of normals to recessives would be 67:5 or 13.4:1 compared to 17:1 and 14:1 obtained on the basis of chromosome and random chromatid segregations. If only 33.3% of the female gametes transmit the extra chromosome, the overall ratio would be modified to 11:1.

The frequencies of different types of gametes expected on the basis of various types of segregations from a trisomic of *AAa* (Simplex) genotype are given in Table 5.I, and the genotypic and phenotypic ratios expected among the F_2 progeny of this trisomic are given in Tables 5.II and 5.III. The gametic frequencies and genotypic and phenotypic ratios expected in the segregating generations of a trisomic of the duplex (*Aaa*) genotype can be similarly calculated.

To date, inheritance studies with primary trisomics have been carried out in ten species: *Datura* (Blakeslee and Farnham, 1923; Gager and Blakeslee, 1927; Blakeslee *et al.*, 1927; Blakeslee and Avery, 1934; Buchholz and Blakeslee, 1927), maize (McClintock, 1929a; McClintock and Hill, 1931; Rhoades, 1933a), tomato (Lesley, 1926, 1928, 1932, 1937; Rick and Barton, 1954; Rick *et al.* 1964; Hagemann, 1969), barley (Tsuchiya, 1956, 1958, 1959; Tsuchiya *et al.*, 1960), *Arabidopsis thaliana* (Lee-Chen and Steinitz-Sears, 1967; Sears

TABLE 5.1
Gametic Types and Their Frequencies Expected from Various Types of Segregations in a Trisomic of *AAa* Genotype

Type of segregation	50% Transmission of $n + 1$ gametes through the female			33.3% Transmission of $n + 1$ gametes through the female		
	$n + 1$	n	% $aa + a$	$n + 1$	n	% $aa + a$
Chromosome	1*AA* + 2*Aa*	2*A* + 1*a*	16.7	1*AA* + 2*Aa*	4*A* + 2*a*	22.2
Random chromatid	6*AA* + 8*Aa* + 1*aa*	10*A* + 5*a*	20.0	6*AA* + 8*Aa* + 1*aa*	20*A* + 10*a*	24.4
Maximum equational	5*AA* + 6*Aa* + 1*aa*	8*a* + 4*a*	20.8	5*AA* + 6*Aa* + 1*aa*	16. 16*A* + 8*a*	25.0

TABLE 5.II

Genotypical Frequencies Expected from Various Types of Segregation in an F_2 of Trisomic of *AAa* Constitution

Type of segregation	50% Transmission of $n + 1$ gametes through female		33.3% Transmission of $n + 1$ gametes through female	
	$2n + 1$	$2n$	$2n + 1$	$2n$
Chromosome	2*AAA* + 5*AAa* + 2*Aaa*	4*AA* + 4*Aa* + 1*aa*	2*AAA* + 5*AAa* + 2*Aaa*	8*AA* + 8*Aa* + 2*aa*
Random chromatid	12*AAA* + 22*AAa* + 10*Aaa* + 1*aaa*	20*AA* + 20*Aa* +5*aa*	12*AAA* + 22*AAa* + 10*Aaa* + 1*aaa*	40*AA* + 40*Aa* + 10*aa*
Maximum equational	10*AAA* + 17*AAa* + 8*Aaa* + 1*aaa*	16*AA* + 16*Aa* + 4*aa*	10*AAA* + 17*AAa* + 8*Aaa* + 1*aaa*	32*AA* + 32*Aa* + 8*aa*

TABLE 5.III

Phenotypic Ratios (Normal:Recessive) Expected from Various Types of Segregations in the F_2 and Female Backcross of a Trisomic of *AAa* Genotype

Type of segregation	50% Transmission of $n + 1$ gametes through female						33.3% Transmission of $n + 1$ gametes through female					
	$2n + 1$		$2n$		Total		$2n + 1$		$2n$		Total	
	F_2	BC[a]	F_2	BC	F_2	BC	F_2	BC	F_2	BC	F_2	BC
Chromosome	1:0	1:0	8:1	2:1	17:1	5:1	1:0	1:0	8:1	2:1	12.5:1	3.5:1
Random chromatid	44:1	14:0	8:1	2:1	14:1	4:1	44:1	14:1	8:1	2:1	11.27:1	3.1:1
Maximum equational	35:1	11:1	8:1	2:1	13.4:1	3.8:1	35:1	11:1	8:1	2:1	11:1	3:1

[a] BC, backcross.

and Lee-Chen, 1970), *Antirrhinum majus* (Rudorf-Lauritzen, 1958), *Oenothera blandina* (Catcheside, 1954), spinach (Janick *et al.*, 1959), *Collinsia heterophylla* (Rai and Garber, 1961), and *Sorghum vulgare* (Hanna and Schertz, 1970). All of these studies except those of Lesley (1937) and Rhoades (1933a) were aimed at associating the linkage groups or the genes with their respective chromosomes. Although sufficient for discriminating between disomic and trisomic inheritance, the populations studied by these authors were not large enough to permit any conclusions about the type of segregation obtained. However, the data of Lesley (1937) on the segregation of three markers of linkage group 2 of tomato are relevant. Triplo-2 plants heterozygous for three linked markers, *d, p,* and *s,* were obtained, so the trisomic plants carried two dominant alleles and one recessive of each. These were backcrossed as females to the plant that was homozygous recessive for all three markers. Combined data of four progenies from this experiment are given in Table 5.IV. There were 327 disomic plants (58%) and 234 trisomics (42%). Seventeen of the trisomic plants were homozygous for *d,* and seventeen for *p,* but only five were homozygous for *s*. This gives a ratio of 12.7:1 for *d* and *p,* and 45.8:1 for *s* in the trisomic fraction. The ratio of 12.7:1 obtained for *d* and *p* lies between the ratios expected on the basis of random chromatid segregation and maximum equational segregation. However, the number of homozygous trisomics is not large enough to decide which of the ratios is the better fit. Nevertheless, it is clear that both of these markers must be quite far from the centromere and that the gene *s* should be at least some crossover distance from the centromere. Actually, it is now known that *s, p,* and *d* are situated on the same arm of chromosome 2 at crossover distances of 30, 75, and 79 from the centromere (Khush and Rick, 1968a). Two aspects of Lesley's data are difficult to explain. In the first place, the proportion of recessives among the 2*n* fraction is much lower than the expected value of 2:1. At the same time, the proportion of

TABLE 5.IV
Progeny Tests of Triplo-2 of Tomato (*DDdPPpSSs* × *ddppss*) and Triplo-5 of Maize ($V_2V_2v_2 \times v_2v_2$)[a]

Species	Gene	2*n* Normal	2*n* Recessive	2*n* Ratio	2*n* + 1 Normal	2*n* + 1 Recessive	2*n* + 1 Ratio
Tomato	*d*	221	106	3:1	217	17	12.7:1
Lesley (1937)	*p*	191	136	2.4:1	217	17	12.7:1
	s	243	84	3.9:1	229	5	45.8:1
Maize	v_2	2210	1146	1.9:1	1439	61	23.6:1
Rhoades (1933a)							

[a]Data have been pooled from four progenies of tomato and fifteen of maize.

$2n + 1$ individuals (42%) is much higher than normally obtained for this trisomic (see Chapter 4).

The number of plants in the study involving the trisomic for chromosome 5 of maize and the marker v_2 (Rhoades, 1933a) was considerably larger, and the observed ratio for the $2n$ fraction fits the expected ratio (2:1) rather well. Only 4.1% of the trisomic individuals were recessive and gave a ratio of 23.6:1. This value is lower than expected on the basis of random chromatid segregation and maximum equational segregation, and thus indicates that the locus of v_2 is not far enough from the centromere to permit maximum double reduction.

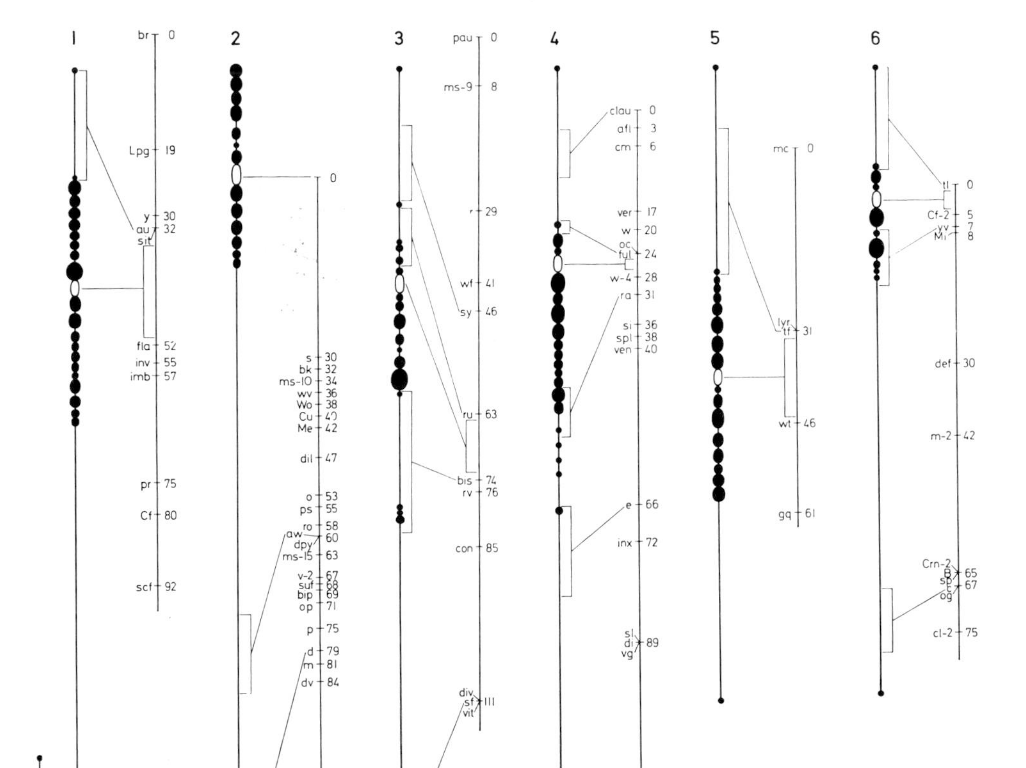

The frequency of trisomics with the recessive phenotype gives only an approximate indication of the crossover distance between the centromere and the locus of the marker. In most cases, the data from trisomic segregations cannot be used for exact mapping of loci. Similarly, if the map distance of the marker from the centromere is known, only an approximate prediction about the amount of double reduction can be made. The meiotic events which permit double reduction at a given locus—multivalent formation, crossing over between the gene and the centromere, random separation of three chromosomes at anaphase I, and random separation of chromatids at anaphase II—are hard to quantify.

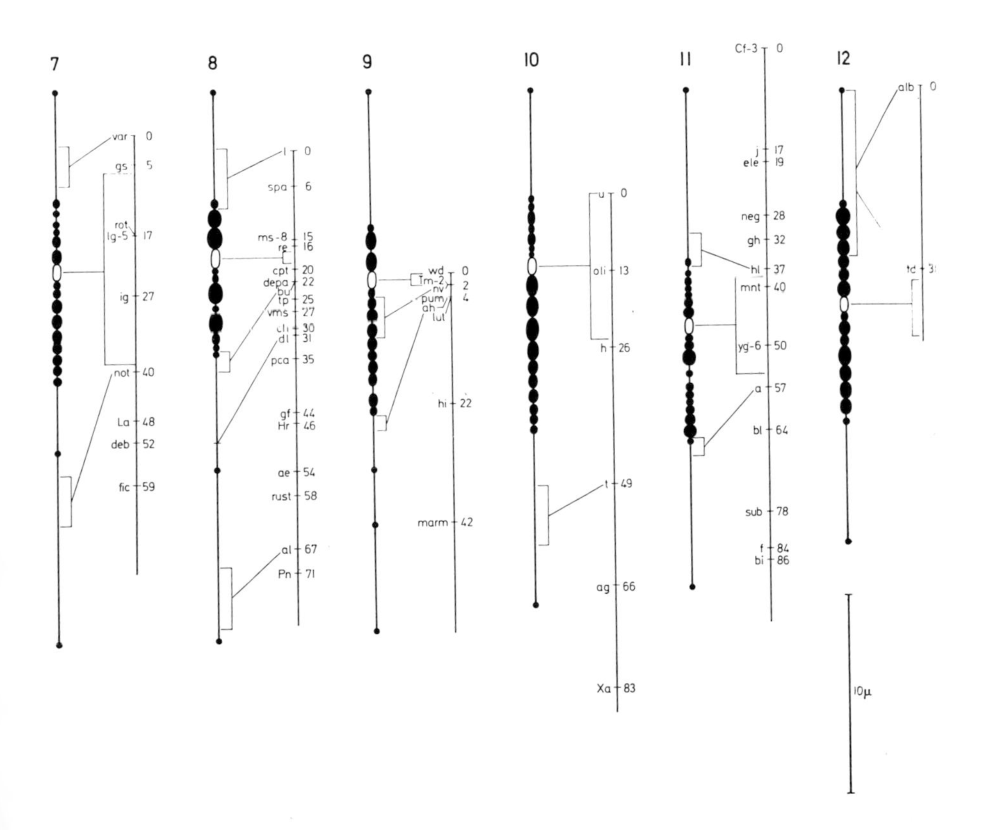

Fig. 5.2. Summary of cytological and genetic linkage maps of tomato chromosomes. Gene loci approximated on cytological maps and centromere approximated on genetic maps according to information gained from induced deficiencies and trisomic tests. (From Khush and Rick, 1968a.)

For example, the three homologous chromosomes of the trisomic may be present as a bivalent and a univalent, or as three univalents. Even if a trivalent is formed, one of the chromosomes of a trivalent may have formed a chiasma not in the arm carrying the marker, but in the other one. Conversely, the observed univalents may have taken part in pairing and crossing over and then separated from the trivalent due to early terminalization of the chiasmata. The separation of the three chromosomes at anaphase I and the chromatids at anaphase II may not be random.

The case of chromosome 2 of tomato, discussed above, is somewhat unusual. This chromosome has a completely heterochromatic short arm which does not form any chiasmata. All the crossovers occur in the very long 2L arm (second longest of the tomato complement), and the genes *d* and *p* are located in the terminal segment of this arm (Fig. 5.2). The frequency of trivalent formation in this trisomic is very high, and the minimum of two crossovers required for the trivalent formation must occur between the centromere and the loci of these markers. No wonder, then, that random chromatid segregation was obtained for these markers.

In the previous paragraphs the basis of the trisomic test and the various theoretical ratios expected in a segregating population of a trisomic were discussed. The actual procedures of the trisomic test and the experimental data obtained in different species can now be reviewed. When some or all of the linkage groups of a species have been discovered, it is desirable to assign these groups to the respective chromosomes of the complement and thereby test their independence. Primary trisomics provide excellent tools for these purposes. Theoretically, if all the linkage groups are known and all the possible primary trisomics are available, each of the primary trisomics can be made heterozygous for one representative marker of each of the linkage groups. Association between the linkage groups and the chromosomes can be made by studying the F_2 or backcross progenies. If the linkage groups are independent, one particular trisomic should give the trisomic ratio with only one of the markers, and each of the markers should segregate in a trisomic fashion in only one of the families. However, if the linkage groups are not independent, some of the trisomics will give trisomic ratios with more than one marker. The linkage groups to which these markers belong can then be combined. For an organism with $n = 12$, 144 segregating progenies need to be grown. The number can be reduced if, for instance, three or four markers of different linkage groups are combined into multiple marker stocks and the trisomics made heterozygous for these markers. Then only 48 or 36 segregating progenies are needed.

In practice, these requirements are rarely met. Either all of the linkage groups are not known or all the primary trisomics are not available. Furthermore, the multiple tester stocks carrying genes belonging to different chromosomes may not have been synthesized. When McClintock isolated all the primary trisomics of maize, the ten linkage groups of this species had been discovered. She was able to associate six of the ten with the respective chromosomes by means of trisomic tests within a period of 4 to 5 years. Three of the remaining four linkage groups were associated by means of translocations and the other one by means of a radiation-induced deficiency technique (Rhoades and McClintock, 1935). Two linkage groups previously considered independent were consolidated on one chromosome. Of the six associations in maize determined by means of trisomic tests, extensive data on only one of them, the *r–g* linkage group and chromosome 10, has been published (McClintock and Hill, 1931). The trisomic for chromosome 10 of maize cannot be distinguished morphologically from the disomic sibs. Therefore, the authors scored the progenies for the dominants and the recessives only and tested them for conformity to trisomic or disomic inheritance of the total progeny. The trisomic transmits through the male rarely, and about one-third of the gametes transmit the extra chromosome through the female side.

The data from various crosses are summarized in Table 5.V. The observed ratios fit the trisomic inheritance very well. The ratio of 9.6:1 obtained for the F_2 of a trisomic of the *RRr* genotype clearly is closer to the ratios expected on the basis of different types of trisomic segregations (Table 5.V). The observed ratio differs markedly from the expected 3:1 disomic inheritance. The small difference between the observed and the expected ratios can be easily explained by the transmission of $n + 1$ female gametes at a rate lower than the 33.3% on which the expected values have been calculated. The backcross ratio of 3.8:1 also fits the trisomic ratio rather than the disomic ratio of 1:1. In the control crosses involving the disomic individuals, only normal disomic ratios for *r* were observed. To test the independence of the *r–g* linkage group from other known linkage groups of maize, McClintock and Hill (1931) crossed triplo-10 with a representative marker of each of the other linkage groups. All of these markers segregated in a normal disomic fashion and established the independence of the *r–g* linkage group.

It was shown above how the trisomics of maize were used in associating the linkage groups with the respective chromosomes. In maize, most of the trisomics cannot be recognized phenotypically. However, the task of trisomic analysis becomes simpler if the $2n + 1$ individuals can be identified morphologically. The appearance of an appreciable proportion of homozygous recessive trisomics in the progeny of a trisomic of the *AAa* genotype proves that the segregating gene cannot be located on the extra chromosome of the given trisomic. Therefore, the scoring of a population can be discontinued as soon as an appreciable number of recessive trisomics are encountered. If the population size is too small to permit discrimination between the disomic and trisomic inheritance, an examination of the trisomic part of the progeny is often helpful. The appearance of a number of recessive trisomics rules out the possibility of trisomic inheritance.

TABLE 5.V

Data on Various Tests of Maize Plants Trisomic for Chromosome 10 and Heterozygous for *R* Versus *r*[a]

Cross	Normal	Recessive	Total	Observed ratio	Expected ratios on the basis of 33.3% transmission of $n + 1$ gametes through female		
					Chromosome seg.	Chromatid seg.	Max. equ. seg.
RRr selfed	396	41	437	9.6:1	12.5:1	11.27:1	11:1
RRr × *rr*	819	213	1032	3.8:1	3.5:1	3.1:1	3:1
rr × *RRr*	646	355	1001	1.8:1	2:1	2:1	2:1
Rrr × *rr*	679	836	1515	1:1.2	1:1	1:1.37	1:1.38
rr × *Rrr*	1282	2451	3733	1:1.9	1:2	1:2	1:2
Control							
Rr selfed	608	204	812	3:1			
Rr × *rr*	1161	1196	2357	1:1			

[a]From McClintock and Hill, 1931.

Trisomics of tomato can be distinguished easily from the disomic sibs, as well as from one another (Rick and Barton, 1954), and have been used extensively in trisomic analysis. Some of the results are discussed below.

Genetic investigations with the primary trisomics of tomato were started by Lesley (1926, 1928), who isolated from a spontaneous triploid, eleven of the twelve primary trisomics. Lesley (1928, 1932) was able to associate the genes *d, r,* and *e* with triplo-A, triplo-I, and triplo-B by studying the segregating progenies of the trisomics. Few genetic markers of tomato were available at that time, and Lesley did not continue his work with trisomics. Later, Barton (1950) and Gottschalk (1951) showed that each of the twelve tomato chromosomes can be distinguished morphologically at pachytene. The work of Young and MacArthur (1947) and Butler (1952) reported several new genes and established several linkage groups. A new series of primary trisomics was isolated and the extra chromosome of each was identified by Rick and Barton (1954) and Rick *et al.* (1964). The time was now ripe for using these newly established trisomics for testing the independence of newly described linkage groups and associating them with their respective chromosomes. These workers, using the data on segregating progenies, assigned ten linkage groups to their chromosomes (Table 5.VI) . All the segregating progenies were separated into disomic and trisomic fractions. The proportion of recessive individuals within the disomic fraction of the progenies varies between 8.0 and 12.0%. These frequencies differ significantly from the 25% expected on the basis of disomic inheritance, but agree rather closely with 11.1% (8:1 ratio) expected on the basis of trisomic inheritance. Only one family, that for triplo-8 and *l*, has an excess of recessive plants in the disomic fraction, but the absence of any recessive trisomics among 38 clearly shows that it could not be a disomic inheritance. A look at the trisomic fractions of other progenies also fortifies the conclusions regarding trisomic inheritance. In ten of the twelve families all the trisomics had normal phenotype, and in the other two, one trisomic with the recessive phenotype appeared. Finally, the percentage of recessives in the total of each progeny is given in the last column of Table 5.VI. The transmission frequency of tomato trisomics is quite low. Had the proportion of $2n + 1$ individuals been higher, the percentage of recessives in progenies would have been even lower. Two of the progenies were from backcrosses, and their observed ratios are very close to the expected ratio of 2:1.

Four linkage groups previously considered independent were assigned to two chromosomes, and the independence of the remaining linkage groups was established, as the tested markers gave trisomic inheritance with only one trisomic and disomic inheritance with the others.

Through the use of primary trisomics, Rick and Barton (1954) and Rick *et al.* (1964) were able to assign thirty genes to ten different chromosomes. The original collection of primaries did not include the triplo-11; the linkage

TABLE 5.VI

Summary of Data from Segregating Progenies of Different Trisomics of *AAa* Constitution which Led to the Establishment of Association between Chromosomes and Linkage Groups of Tomato[a]

Chromosome	Gene	F_2 or BC[b]	Total No. of plants	2n Normal	2n Recessive	2n (%) Recessive	2n + 1 Normal	2n + 1 Recessive	2n + 1 (%) Recessive	Total Normal	Total Recessive	Total (%) Recessive
1	*y*	BC	101	72	29	28.7	0	0	0.0	72	29	28.7
2	*d*	F_2	158	130	14	9.7	13	1	7.1	143	15	9.4
3	*wf*	F_2	134	103	11	9.6	20	0	0.0	123	11	8.2
4	*e*	F_2	121	90	11	10.8	20	0	0.0	110	11	9.0
5[c]	*mc*	F_2	247	180	20	10.0	47	0	0.0	227	20	8.0
	wt											
6	*yv*	F_2	202	160	22	12.0	19	1	5.0	179	23	11.3
7	*gs*	F_2	230	180	18	9.0	32	0	0.0	212	18	7.8
8[c]	*l*	F_2	102	54	10	15.6	38	0	0.0	92	10	9.8
	al	F_2	274	189	24	11.2	61	0	0.0	250	24	8.7
9	*wd*	F_2	119	92	8	8.0	19	0	0.0	111	8	6.7
10	*h*	BC	175	95	52	35.3	28	0	0.0	123	52	29.7
11[d]												
12[d]												

[a]From Rick and Barton, 1954; Rick *et al.*, 1964.

[b]BC, backcross.

[c]*mc* and *wt* as well as *l* and *al* were considered in independent linkage groups by Young and MacArthur (1947) and Butler (1952) in the latest preceding linkage summary.

[d]The linkage groups of chromosomes 11 and 12 were assigned by means of radiation-induced deficiencies (Rick and Khush, 1961; Khush and Rick, 1966b).

group of this chromosome was determined with the use of X ray-induced deficiencies (Rick and Khush, 1961). Of the forty-two genes tested against triplo-12, none segregated in a trisomic fashion; however, two seedling markers were found on this chromosome by means of neutron-induced deficiencies (Khush and Rick, 1968a). The twelve linkage groups of tomato and their respective chromosomes are now known (Fig. 5.2).

An extensive use of multiple gene stocks for studies of trisomic inheritance was made in tomato. Their several important advantages are discussed by Rick and Barton (1954) . In addition to reducing the number of progenies required, the multiple gene stocks aid in detecting contamination by foreign pollen or by undesired seed mixtures. In either type of inheritance, the contamination can be detected by simultaneous deflection in the same direction of the ratios of all genes. Indications of false trisomic segregation that might result from contamination by the seed or pollen with dominant alleles can often be discovered.

In the trisomic inheritance studies in tomato, F_2 segregations were preferred over backcrosses (Table 5.VI) . Rick and Barton (1954) point out several advantages in growing second generation progenies. A somewhat smaller population is needed to discriminate between disomic and trisomic segregation. If no $2n + 1$ plants appear in the progenies, the number of plants needed to discriminate the fiducial limits of the two types of segregation at the 5% level in F_2 is 122; in the backcross it is 137. Since $2n + 1$ plants, in which trisomic segregation is diagnostic, appear in nearly all progenies of tomato trisomics, the smaller population of 100 was considered sufficient for the F_2's. Another advantage of the F_2 in tomatoes is that it is easier to produce and allows less opportunity for contamination than the backcross. In only a few instances, in which pollen sterility of the trisomic hybrid proved too high to permit selfing, was it necessary to resort to backcrossing.

Even after the assignment of the linkage groups to their chromosomes, the primary trisomics of tomato continued to be used for assigning the unlocated genes to the linkage groups. This permitted the rapid accumulation of useful markers for most of the chromosomes, and these markers could be used for synthesizing tester stocks for linkage studies. However, when the appropriate linkage testers for all the chromosomes became available, the use of primary trisomics for locating genes was discontinued.

Another species in which the primary trisomics have been employed extensively for gene location is *Datura stramonium*. The first report of trisomic inheritance in any plant species was, in fact, made for *Datura*. Blakeslee *et al.* (1920) and Blakeslee and Farnham (1923) presented theoretical ratios expected from primary trisomics of different genotypes on the basis of chromosome segregation. They also reported detailed data on the inheritance of the gene *white* and the Poinsettia trisomic ($2n + 17 \cdot 18$). A good fit between observed ratios and the expected on the basis of trisomic inheritance was obtained, and it was concluded that the gene in question was located on the extra chromosome in the Poinsettia trisomic. It is interesting that the authors explained the occurrence of a single recessive trisomic on the basis of a $2n$ pollen grain carrying two recessive alleles fertilizing an $n + 1$ gamete with a recessive allele. The occurrence of such trisomics can now be readily explained, as previously discussed, on the basis of chromatid crossing over. Later the gene, *curled,* was also associated with the Poinsettia chromosome, *tricarpel,* with the Reduced chromosome (Blakeslee *et al.,* 1927; Buchholz and Blakeslee, 1927), *swollen,* with the Ilex trisomic (Gager and Blakeslee, 1927), and *pale-1, baltimore bad pollen,* and *slender capsule* with Globe trisomic (Blakeslee and Avery, 1934) by means of trisomic tests. According to Avery *et al.* (1959), thirty-eight of the eighty-one genes definitely located in specific chromosomes were located by means of primary trisomic tests. Apparently all the associations between the chromosomes and the linkage maps were established by the use of primary trisomics.

Instead of making twelve crosses to render all the primary trisomics heterozygous for the gene to be located, Blakeslee and his associates used the pollen from the homozygous recessive to pollinate a $3n$ clone, which they apparently maintained by vegetative propagation. Among the offspring of such a cross, appeared all the primary trisomics which were heterozygous for the gene to be located. These trisomics could then be used for producing F_2 or backcross populations (Blakeslee and Avery, 1934; Avery *et al.,* 1959).

Although this method of producing heterozygous trisomics saves tremendous labor, it can be employed only for those species in which the triploids are sufficiently female fertile so that large progenies can be obtained from them.

In barley, the primary trisomics have been used successfully in establishing the relationships between the linkage groups and the chromosomes. Of seven linkage groups known, some of the associations between them and the chromosomes were determined by the use of reciprocal translocations (Hanson, 1952; Hanson and Kramer, 1949, 1950; Burnham, 1956). These determinations were far from clear, and some doubt regarding the independence of some of the linkage groups remained (Kramer *et al.,* 1954).

However, Tsuchiya (1954, 1958) was able to establish all the seven primary trisomics of barley. The extra chromosome of each was identified and the representative markers of the known linkage groups were tested with the seven primaries for disomic or trisomic inheritance (Tsuchiya, 1956, 1958, 1959, 1963; Tsuchiya *et al.,* 1960). These studies established the relationships between the linkage groups and the chromosomes as numbered by Burnham and Hagberg (1956) . Two of the linkage groups were found to belong to one chromosome, and one of the trisomics gave only disomic segregations with the markers of known linkage groups. A gene which was located on the unmarked chromosome was discovered by Ramage and Suneson (1958a).

Five primary trisomics of *Arabidopsis thaliana* ($n = 5$) have been produced by Steinitz-Sears (1963) and Robbelen and Kribben (1966). Lee-Chen and

Steinitz-Sears (1967) and Sears and Lee-Chen (1970) were able to associate five linkage groups with their respective chromosomes by means of trisomic tests.

Of the eight primary trisomics of *Antirrhinum majus* L., Stubbe (1934) and Propach (1935) studied five that originated spontaneously. Rudorf-Lauritzen (1958) and Sampson *et al.* (1961) obtained all eight of primary trisomics from the progenies of $3n \times 2n$ crosses. The extra chromosomes have not been identified cytologically, but Rudorf-Lauritzen (1958) reported using these trisomics for testing the independence of six linkage groups.

Four primary trisomics of spinach have been obtained from $3n \times 2n$ crosses by Tabushi (1958), and Janick *et al.* (1959) isolated the six possible trisomics. The sex-determining factors XY were placed on the longest chromosome (number 1) by trisomic analysis (Ellis and Janick, 1960; Suto and Sugiyama, 1961a, b) .

The seven primary trisomics of *Oenothera blandina* were produced by Catcheside (1954), and he was able to locate the gene *br (brevistylis)* on chromosome 11·12.

In *Matthiola incana,* primary trisomics appear spontaneously by nondisjunction at a high frequency (Frost and Mann, 1924; Frost, 1927) . Frost (1931) obtained trisomic inheritance for the recessive gene *b* (red color compared to purple for the normals) with the trisomic called Narrow. The efforts of Philp and Huskins (1931) to associate *double flower* with a particular trisomic were not successful.

Thirteen morphologically indistinguishable trisomic lines of *Collinsia heterophylla* ($n = 7$) were established by Dhillon and Garber (1960). Rai and Garber (1961) studied the genetic ratios in the progenies of these trisomic lines segregating for eleven genes, but the results were inconclusive.

All the ten primary trisomics of *Sorghum vulgare* have been produced (Schertz, 1966; Lin and Ross, 1969a). Hanna and Schertz (1970) located five genes on four chromosomes by trisomic tests.

Genetic Segregation in Tertiary Trisomics

After a genetic marker has been assigned to a particular chromosome, its arm location and approximate distance from the centromere can be determined through the tertiary trisomic or the telotrisomic test. This test also permits the localization of the centromere on the linkage map and reveals the orientation of the linkage map. These analyses are possible because the genetic ratios in the segregating progenies of tertiary trisomics are modified, but not in the same manner as in primary trisomics. Since a tertiary trisomic has only one extra arm, or part of an arm, of a particular chromosome, the genetic ratios are modified only for the genes located on that arm. In the diploid species, the tertiary chromosome cannot replace a normal chromosome in the gametes, so any gamete receiving a tertiary chromosome in place of a normal chromosome aborts because of the deficiency of part of the chromosome. Consequently, the n gametes and $2n$ zygotes produced by tertiaries can consist only of normal homologs. If we ignore, for the time being, the effect of crossing over, there would be two alleles at the locus under study which assort to the n gametes compared to three in the primary trisomics.

In making a tertiary trisomic test, the trisomic with normal alleles is used as the female parent and a $2n$ homozygous mutant stock as the male. Therefore, one of the two normal homologs of the heterozygous trisomic carries the recessive allele, and the other homolog, as well as the tertiary chromosome, carry the normal alleles. In the F_2 or backcross progeny of such a trisomic, the disomic portion would segregate in a normal 3:1 or 1:1 fashion, and all the trisomics would be of normal phenotype. This 3:1::1:0 ratio indicates that the locus of the gene under study is in the arm, or part, which is duplicated in the trisomic. However, if the gene is not situated in the duplicated arm the disomic portion as well as the trisomic portion of the progeny would give normal disomic ratios. In this way, the observed ratios, whether 3:1::1:0 or 3:1::3:1, can identify the appropriate arm for the marker. Such ratio sets are obtained only when the trisomics can be distinguished from the disomics on the basis of morphology or chromosome counts. If the tertiary trisomic is not phenotypically distinct and chromosome counting is not feasible, ratios for the total progeny can be evaluated, but with poorer discrimination. If one-half the female gametes transmit the tertiary chromosome, the trisomic ratio among the total progeny is 7:1; if one-third transmit, the ratio is 5:1. Clearly, the test is more efficient and much smaller populations are needed when disomics and trisomics can be identified.

In calculating the above ratios, it was assumed that crossing over between the centromere and the locus of the marker does not take place and that only two alleles, one normal and one recessive, segregate to the n gametes in a 1:1 ratio. It was also assumed that the tertiary chromosome always carries the normal allele, so that all the trisomics are of normal phenotype. However, crossing over between the tertiary chromosome with the normal allele and the normal chromosome with the recessive allele can modify the above ratios. When crossing over occurs, the recessive allele is transferred to the tertiary chromosome, and trisomic individuals with recessive phenotype are obtained. Another consequence of this event is the transfer of the normal allele from the tertiary chromosome to the normal chromosome, resulting in increase in the proportion of gametes with normal alleles over those with recessive alleles. This, of course, lowers the yield of mutants in the $2n$ fraction. However, modifications of the 3:1::1:0 ratio due to crossing over are generally small and the distinction between a trisomic and disomic segregation can still be made. The degree of reduction

TABLE 5.VII
Frequency of Spores of Different Kinds Expected from Three Types of Pairing and Three Types of Disjunctions in a Tertiary Trisomic of *AAa* Genotype

Type of pairing	Type of disjunction	Genotype of *n* spores		Genotype of $n + 1$ (tertiary) spores			Genotype of $n + 1$ (primary) spores		
		A	*a*	*AA*	*Aa*	*aa*	*AA*	*Aa*	*aa*
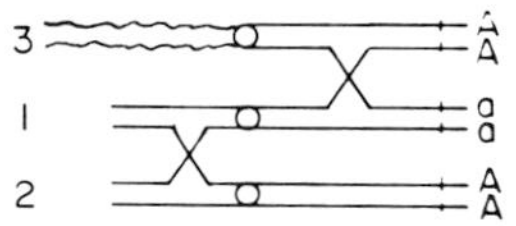	1, 3 versus 2	4		1	2	1			
	1 versus 2, 3	2	2	2	2				
	1, 2 versus 3						2	2	
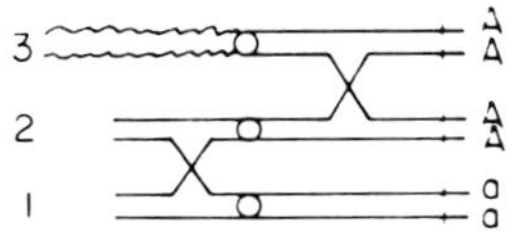	1, 3 versus 2	4		1	2	1			
	1 versus 2, 3		4	4					
	1, 2 versus 3						1	2	1
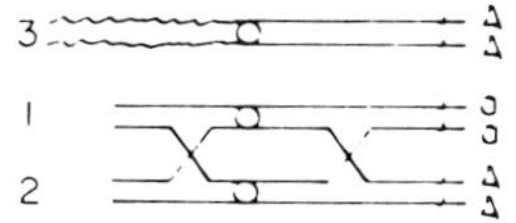	1, 3 versus 2	2	2	2	2				
	1 versus 2, 3	2	2	2	2				
	1, 2 versus 3						1	2	1
		—	—	—	—	—	—	—	—
		14	10	12	10	2	4	6	2
	Ratio	7	5	6	5	1	2	3	1

of the mutant category in the disomic fraction and the frequency of trisomics with recessive phenotype depend on (a) the distance of the locus from the centromere, (b) the proportion of cells in which the locus-carrying arm of the tertiary chromosome recombines with the homologous arms of the normal pair, (c) the chromosomes which participate in the crossover, and (d) the type of disjunction at anaphase I (Reeves *et al.*, 1968). If we assume that (a) the distance of the marker from the centromere is such that one crossover always occurs between the locus and the centromere (50 crossover units), (b) each of the three chromosomes recombines with the other two at random, (c) disjunction of these three chromosomes from the trivalent or the pentavalent or other associations is at random, and (d) the three chromosomes form trivalents in all cells which permit the aforementioned random events to occur in all cells, then the expected gametic frequencies can be calculated as shown in Table 5.VII. Based on these assumptions, 41.8% (5/12 mutant *n* spores are produced rather than the 50% expected on the basis of chromosome segregation or the 33% expected on the basis of random chromatid segregation. On selfing or test crossing, 3:1 and 1:1 ratios of the disomic fraction are modified to 4.7:1 and 5.8:4.2. The proportion of the tertiary trisomics with normal and recessive phenotypes in the absence of male transmission of the extra chromosome is 11:1 in the backcross and 27.3:1 in the F_2, instead of 1:0.

The third type of disjunction (Table 5.VII), which produces abortive spores and $n + 1$ spores with the extra primary chromosome instead of the tertiary chromosome, occurs less frequently than the other two types. This does not influence the genotypic frequency of the spores with n and $n + 1$ (tertiary). The spores with $n + 1$ (primary) occur in the proportion of 2*AA*:3*Aa*:1*aa* (Table 5.VII), and the phenotypic ratios among the primary trisomic fractions are 13.3:1 and 5:1 in the F_2 and backcross, respectively.

If the locus of the marker is more than 50 crossover units from the centromere or the point of rearrangement of the tertiary chromosome, even stronger modifications of these ratios can be expected. Under the same set of assumptions, two randomly situated crossovers for a locus 100 units from the centromere would result in 37.5% mutant *n* spores, so the proportion of the recessives among the disomic portion of the progeny may be as low as 14% if the gene is 100 crossover units from the centromere.

A tertiary trisomic in which the extra chromosome consists of one complete arm of one chromosome and one complete arm of another can be used for determining arm location of markers, position of centromeres, and orientation of the linkage groups of two chromosomes. For a species with twelve pairs of chromosomes, a set of six tertiary trisomics is theoretically sufficient for these purposes. Suppose we have a linkage group of three markers, *a–b–c*, and if we can determine which markers are in the short arm of the chromosome and which in the long arm, we would know whether the centromere is between *a* and *b* or between *b* and *c*, and that the orientation of the map is *a–b–c* or *c–b–a*. If a tertiary trisomic having one complete extra arm of the chromosome which carries the linkage group in question is made heterozygous for these three genes, the segregating progenies can be studied. For example, if a tertiary trisomic in which the short arm of the chromosome is duplicated gives tertiary trisomic ratios for *c* but not for *a* and *b*, it can be concluded that gene *c* is located on the short arm, *a* and *b* are on the long arm, the centromere is between *c* and *b*, and the orientation of the map is *c–b–a*.

The position of the centromere can be located if the tertiary chromosome consists of two complete arms of nonhomologous chromosomes joined together at the centromere region. The arm location of the marker can be determined only if the points of rearrangement which produced the tertiary chromosome are accurately determined cytologically. Only a few species lend themselves to this type of study. It is no wonder that tertiary trisomics have rarely been used for this type of genetic analysis.

A thorough study of the breeding behavior of tertiary trisomics was made by Blakeslee and co-workers. According to Avery *et al*. (1959), thirty tertiary trisomics were obtained in *Datura*. Some of these apparently were used for determining the arm location of the markers, but no detailed data have been published except for the following example. The marker *pale-7* was located on chromosome 1·2 by a primary trisomic test. Tertiary trisomic $2n + 1 \cdot 18$ was rendered heterozygous and backcrossed as female with *pale-7*. Among the 282 disomic progeny there were 147 normal green to 135 *pale-7* plants, and all the 51 tertiary trisomics of the progeny were normal green. This was interpreted as a 1:1::1:0 ratio and accepted as evidence that the gene for *pale-7* is in the ·1 arm of the 1·2 chromosome. This is the only published example of the use of tertiary trisomics of *Datura* in determining the arm location of the marker. In this species it has not yet been possible to determine the points of rearrangement along the chromosome length; therefore, it is impossible to be certain of the amount of duplication caused by the tertiary trisomic. As pointed out by Avery *et al*. (1959) there is no way of knowing whether the extra chromosome carries one complete arm or less of the particular chromosome. Under the circumstances, if the tertiary trisomic ratios are obtained, one can assume that the gene is on the extra arm. However, if disomic ratios are observed, one could not say that the gene is not on the extra arm, as the duplicated arm may not be represented in its entirety.

Another species in which a tertiary trisomic has been used to determine the arm location of the marker is *Oenothera blandina*. The gene *br* gave a primary trisomic ratio with the $2n + 11 \cdot 12$ trisomic and a tertiary trisomic ratio with the $2n + 4 \cdot 12$ trisomic. Catcheside (1954) was thus able to place *br* on arm ·12 of the 11·12 chromosome.

In barley, Ramage (1960) obtained several trisomics from translocation heterozygotes. Whether they were tertiary or primary was not determined. He reportedly obtained trisomic ratios for the genes *K*, *lg*, and *V* but presented no data, so it is not known whether he obtained primary or tertiary trisomic ratios. Also, in barley, it is difficult to be sure about the exact amount of duplication caused by the extra chromosome, hence, the tertiary trisomics of barley may not prove very useful for cytogenetic analysis.

The two genetically well-known species whose tertiary trisomics can be used profitably in linkage analysis are maize and tomato. The pachytene analysis of the translocation heterozygotes from which the tertiary trisomics originated, or of the tertiary trisomics themselves, permits the identification of points of rearrangement and the exact amount of duplication caused by the extra chromosome. Two tertiary trisomics of maize were obtained by Burnham (1930) but were not employed in genetic studies. In fact, in maize the orientation of linkage groups and the arm locations of the genes were determined by the use of radiation-induced deficiencies (Rhoades, 1955).

In tomato, the radiation-induced deficiencies have proved equally useful for these purposes (Rick and Khush, 1961, 1966, 1969; Khush *et al.,* 1964; Khush and Rick, 1968a). This method is unsatisfactory, however, for locating genes which affect mature plant characters and cannot be used for completely dominant genes which are lethal when homozygous or for ones that are inviable in a hemizygous condition. For determining the arm location of such markers, a set of seven tertiary trisomics was synthesized. Arms of nine different chromosomes out of a total of twelve are involved in the extra chromosomes of these tertiaries (Khush and Rick, 1966a, 1967b). In synthesizing these tertiary trisomics, care was taken to make sure that the extra chromosome consisted of two complete arms of the nonhomologous chromosomes (Chapter 2). Theoretically, these tertiary trisomics could have been used to determine the orientation of nine linkage groups. Because some of the linkage groups had already been thoroughly investigated by means of radiation-induced deficiencies, these trisomics were employed in studying the markers of six chromosomes (Table 5.VIII). A glance at the last three columns of the table will readily identify the genes which gave tertiary trisomic ratios and those which gave disomic ratios. For instance, genes *br* on chromosome 1, *clau* and *w-4* on 4, *wt* on 5, and *ah* and *marm* on 9 gave disomic ratios among the trisomic fractions of their progenies. The proportions of *clau* and *wt* (31.8 and 35.7%) were somewhat higher than the 25% expected in a disomic ratio, but the departures from 3:1 were not significant. All the other genes gave trisomic ratios and either none or very few tertiary trisomics with recessive phenotypes appeared in these progenies. The departures from the 3:1 ratio were highly significant. The occurrence of few trisomics with recessive phenotypes is in agreement with known map distances of the markers in question from their respective centromeres, which happen to be points of rearrangement as well. It may be noted that none of these markers is very close to the centromere (Fig. 5.2). Gene *di* on 4L is at least 61 crossover units from the centromere, an interval

TABLE 5.VIII
Genetic Ratios in the F_2 Progenies of Tertiary Trisomics of Tomato of *AAa* Genotype[a,b]

				Progeny							
				2*n*				2*n* + 1			
Trisomic	Gene	Chromo-some	Total	Normal	Recessive	Recessive (%)	X^2 3:1	Normal	Recessive	Recessive (%)	X^2 3:1
2*n* + 1L·11L	*scf*	1	500	334	84	20.0	5.36	82	0	0.0	24.72***[c]
	inv	1	500	340	78	18.6	8.92	81	1	1.2	22.25***
	br(BC)	1	230	95	92	48.1	0.04	23	18	46.3	0.61
2*n* + 4L·10L	*clau*	4	496	263	95	26.2	0.44	94	44	31.8	3.58
	ra	4	496	272	86	24.0	0.17	138	0	0.0	46.00***
	di	4	496	300	58	16.2	14.70***	136	2	11.4	40.81***
	ful	4	239	107	33	23.5	0.14	73	26	26.2	1.17
	w-4	4	239	107	33	23.5	0.14	99	0	0.0	33.00***
	ag	10	202	91	25	21.5	0.73	83	3	3.4	21.80***
	t	10	202	89	27	23.2	0.17	83	3	3.4	21.80***
2*n* + 5S·7L	*mc*	5	150	82	26	24.0	0.04	42	0	0.0	14.00***
	tf	5	150	87	21	19.4	2.15	41	1	2.3	11.45***
	wt	5	150	72	36	33.3	4.00	27	15	35.7	2.57
2*n* + 7S·11L	*gs*	7	152	67	22	24.2	0.00	61	1	1.6	18.08***
2*n* + 9S·12S	*ah*	9	192	88	34	27.8	0.53	51	19	27.1	1.72
	marm	9	192	91	31	25.4	0.00	54	16	22.8	1.72

[a] *a* on one of the normal chromosomes.
[b] From Khush and Rick, 1967b.
[c] Significant at 0.1% level.

which should permit at least one crossover. According to the assumptions enumerated earlier, at least one tertiary trisomic out of twenty-eight should be of recessive phenotype. The actual proportion is much lower, and it is evident that some of the assumptions are not valid for this trisomic. The observed frequencies of the tertiary trisomics recessive for *ag* and *t*, however, are in remarkable agreement with the expected for a gene 50 crossover units from the centromere (27.7:1 versus 27.3:1).

The ratios in the disomic fraction of the progenies in Table 5.VIII can be considered in the light of their segregation from disomic heterozygotes. The somewhat low yield of *inv*, which is not far from the centromere, can be explained by its normal, consistent tendency to be deficient in disomic segregations. The low yield of *di* undoubtedly is due to the combined effects of double reduction and its tendency to be somewhat deficient in normal segregations. On the whole, the departures from the 3:1 ratio for all the genes are not large, and it can be safely assumed that maximum equational segregation did not occur for any of the genes.

The genetic analysis of tomato carried out by Khush and Rick (1967b) with the five tertiary trisomics and sixteen markers belonging to six linkage groups was very rewarding (Table 5.VIII). This study permitted the localization of centromeres of six chromosomes on their linkage maps. The orientations of five linkage maps were determined. The arm locations of five genes were verified, and eleven new genes were assigned to respective chromosome arms. The following two examples from this study show how the data from tertiary trisomic segregations can be helpful in linkage mapping.

The linkage map of chromosome 4 was known to be *clau–afl–w–ful–w-4–ra–ven–e–di–sl*.

The markers *clau* and *ful* were located on the short arm of chromosome 4 by means of radiation-induced deficiencies; *e* was similarly located on 4L (Khush and Rick, 1967c). The markers *ra* and *di* were located on 4L by means of the tertiary trisomic test. However, it was not known whether the centromere was located between *ful* and *w-4* or between *w-4* and *ra*. The tertiary trisomic $2n$ + 4L·10L was made heterozygous for *ful* and *w-4* and F_2 ratios for this combination were studied. Marker *ful* segregated in a normal disomic fashion, but *w-4* gave trisomic ratios (Table 5.VIII). It was concluded that *w-4* is located on 4L and the centromere is between *ful* and *w-4*. Had *w-4* segregated in disomic fashion, the centromere would have been located between *w-4* and *ra*. Markers *ful* and *w-4* are only 4 crossover units apart, yet, as revealed by this analysis, they lie on different arms of the chromosome.

The linkage map of chromosome 5 was known to be *gq–mc–tf–wt*. The marker *tf* had been located on 5S, but the arm location of other markers, the position of the centromere, and the orientation of the linkage map was not known. The marker stock *mc–tf–wt* was crossed with $2n$ + 5S·7L and F_2 obtained.

The genes *mc* and *tf* gave trisomic ratios, *wt* gave disomic. These results showed that *mc* and *tf* are on 5S and *wt* on 5L, the centromere is between *tf* and *wt*, and the map order is *mc*–*tf*–centromere–*wt*–*gq*, not the reverse, as was shown in the last published map (Clayberg *et al.*, 1966).

The localization of genes to a particular segment of an arm can also be achieved through the use of two or more tertiary trisomics with rearrangements occurring at different locations of the same arm. For example, if a tertiary trisomic carrying the complete arm of a chromosome gives tertiary trisomic ratios for a particular gene but another tertiary trisomic carrying only the distal half of that arm does not give the tertiary trisomic ratio for the same gene, then the gene must be located in the proximal half of the arm. However, if the tertiary trisomic ratios are obtained with both the trisomics, then the gene must be in the distal half. This type of analysis can be extended to explore smaller segments of the chromosome arms.

Dominant mutants with dosage effects can be assigned to their respective arms simply by observing the phenotype of F_1 hybrid tertiary trisomics. For example, $2n$ + 4L·10L was made heterozygous for the dominant mutant *Xa* of linkage group 10. All vegetative parts of *Xa*/+ disomics are bright yellow at all stages of growth. Heterozygous trisomics were greener and clearly intermediate between +/+ and *Xa*/+ — evidence that trisomics had *Xa*/+ + genotype, hence, that the extra chromosome 4L·10L carried an additional + allele and, consequently, that *Xa* lies on 10L.

Genetic Segregation in Telotrisomics

The genetic ratios obtained in the segregating progenies of telotrisomics are very similar to those obtained in the segregating progenies of the tertiary trisomics. The ratios are modified only for the markers which are located on the telocentric arm, not for the markers of the other arm. In making a telotrisomic test, the trisomic parent carrying the normal alleles is used as the female parent and the homozygous recessive stock as the male. The heterozygous telotrisomic is selfed or backcrossed to the homozygous recessive stock. The observed trisomic ratios of 3:1::1:0 or 1:1::1:0 for the F_2 or the backcross would indicate the marker to be on the telocentric arm. However, if the disomic ratios of 3:1::3:1 for the F_2 and 1:1::1:1 for the backcross are obtained, the gene is on the other arm. After two or more genes of a particular chromosome are thus located, the orientation of the linkage map and the position of the centromere can be ascertained.

The trisomic ratios mentioned above, like the trisomic ratios obtained in the progenies of tertiary trisomics, can be modified by double reduction. The conditions favoring double reduction are similar to those described for the tertiary trisomics. Since the telocentric chromosome has pairing affinity with one homologous normal pair, compared to the tertiary chromosome which has a pairing affinity with two normal homologous pairs, the chances of trivalent formation in the telotrisomics are somewhat better. Therefore, other conditions being similar, the amount of double reduction and the modifications of the 3:1::1:0 ratio would be stronger in the telotrisomic progenies than in those of tertiary trisomics.

In the telotrisomics, as in tertiary trisomics, the distinction between the 3:1::1:0 ratio and the normal disomic ratio is possible only when $2n + 1$ types can be recognized morphologically. However, if the telotrisomic is indistinct from the disomic sibs or if scoring is done before the plants are large enough to be classified into $2n + 1$ and $2n$ types, the overall ratio can be studied. With 50% transmission of the $2n + 1$ type, the trisomic ratio would be 7:1, and with 33.3% transmission, the ratio would be 5:1.

The telotrisomics thus serve the same prupose as tertiary trisomics in linkage mapping. Theoretically six tertiary trisomics should be enough to map the centromeres of the twelve chromosomes, but twelve telotrisomics, one for each of the chromosomes, are required for the same purposes. However, there are certain advantages in the use of telotrisomics. They are generally more vigorous and fertile and easier to manipulate in crossing work compared to tertiary trisomics. Their transmission rates are generally much higher and, consequently, comparatively smaller progenies are required to differentiate between disomic and trisomic inheritance (Khush and Rick, 1968b).

Rhoades (1936) was the first to use a telotrisomic for determining the arm location of the markers. He was able to locate V_2, Ys_1, *pr*, v_3, and *bt* loci of linkage group 5 of maize on the long arm, and bm_1 and a_2 loci on the short arm by the use of the telotrisomic for 5S. Similarly, Moseman and Smith (1954) studied the inheritance of *y* and *js* with a telotrisomic of *Triticum monococcum* and obtained a normal 3:1 segregation for both the genes. They concluded that these genes were located on the arm not represented by the telocentric.

More recently, Khush and Rick (1968b) have isolated six telotrisomics of tomato and used three of them for determining the arm locations of six markers belonging to three chromosomes. The results (Table 5.IX) show clear-cut disomic or trisomic segregations. The genes *r* and *wf* of linkage group 3 gave disomic ratios with $2n + \cdot 3L$ and were assigned to 3S, but *rv* and *sf* gave trisomic ratios with the same telotrisomic and were placed on 3L. These results helped determine the orientation of the linkage map and the position of the centromere. One telotrisomic recessive for *rv* and one recessive for *sf* were obtained. The proportion of recessives in disomic portions of these two progenies was significantly lower than 3:1. As pointed out by Reeves *et al.* (1968), the deviation for *rv* is spurious, as this gene segregates in approximately this proportion in the disomic heterozygotes. However, the magnitude of deviation for *sf* (13.3

TABLE 5.IX
Genetic Ratios in the F_2 Progenies of Telotrisomics of Tomato of *AAa* Constitution[a,b]

			Progeny							
			2n				2n + 1			
Telotrisomic	Gene	Total	Normal	Recessive	Recessive (%)	X^2 3:1	Normal	Recessive	Recessive (%)	X^2 3:1
2n + ·3L	*r*	184	94	38	28.7	0.17	39	13	25.0	0.00
	wf	184	99	33	25.0	0.00	40	12	23.1	0.10
	rv	180	102	18	15.0	5.40	59	1	1.6	17.41
	sf	180	104	16	13.3	8.70	59	1	1.6	17.41
2n + ·7L	*var*	176[c]	77	31	28.7	0.78	52	15	22.3	0.22
2n + ·8L	*cpt*	131[d]	42	16	27.5	0.20	65	1	1.5	19.39

[a] *a* on one of the normal chromosomes.
[b] From Khush and Rick, 1968b.
[c] One plant in the family was triplo-7.
[d] Nine plants in the family were 2n + 2 (2n + ·8L + ·8L).

instead of 25%) suggests that it is situated at a considerable crossover distance from the centromere. This agrees well with the terminal location of *sf* on rather long 3L, determined by means of radiation-induced deficiencies (Khush and Rick, 1968a). Therefore, the map distance of 37 units for *sf* from the centromere (Fig. 5.2) must be underestimated.

The gene *var* of linkage group 7, similarly tested against $2n + \cdot 7L$, segregated in a disomic fashion and must, therefore, be on 7S. However, *cpt*, of linkage group 8, giving a trisomic ratio with $2n + \cdot 8L$ was mapped on 8L (Table 5.IX).

A dominant gene with dosage effect can be allocated to its chromosome arm simply by comparing the phenotypes of heterozygous telotrisomics with those of disomics with known dosages. Thus *La (Lanceolate)*, a dominant mutant on chromosome 7, produces narrow, entire leaves with 0 to 2 small lateral segments when heterozygous in disomics (*La*/+). The leaves of triplo-7L with one dose of *La* were broader and usually had two pairs of lateral segments at the base of the terminal segment (Fig. 5.3). This departure from the typical *La*/+ phenotype is in the direction expected if the *La* locus is on 7L and the heterozygotes, therefore, are genetically *La*/+/+. The phenotype of these telotrisomics closely approximated that for *La*/+/+. determined earlier by Stettler (1964) from triploid heterozygotes.

The genetic studies with the telotrisomics of the three species described above point out the usefulness of these chromosomal variants for the cytogenetic analysis of diploid species. Concerted efforts are now being made to establish telotrisomic series in barley for use in centromere mapping (Tsuchiya, 1969; Fedak *et al.*, 1971).

Genetic Segregation in Secondary Trisomics

The genetic ratios in the progenies of secondary trisomics are modified in the same fashion as in the progenies of tertiary trisomics and telotrisomics (Khush and Rick, 1969). The nature of the modified ratios permits the determination of arm locations of markers. As discussed earlier, in a secondary trisomic of a diploid species, all the spores that receive an isochromosome in place of a normal chromosome abort because one arm is deficient; consequently, the *n* gametes and resultant $2n$ zygotes produced by such secondary trisomics can only consist of normal homologs. Therefore, if a secondary trisomic carries a recessive marker on one normal chromosome and the dominant allele on the normal homolog, in the absence of double reduction, the disomic fraction of its F_2 and backcross progeny would segregate 3:1 and 1:1, respectively. However, if the locus of the marker is far enough from the centromere to permit double reduction, the disomic ratios may be modified as in tertiary

Fig. 5.3. Representative mature leaves, upper surface of tomato. Left, normal (+/+), center, $2n + \cdot 7L$ with one dose of *La* (*La*/+/+); right, diploid heterozygous for *La* (*La*/+). (From Khush and Rick, 1968b.)

trisomics and telotrisomics. The magnitude of the deviation, of course, depends upon the distance of the locus from the centromere and the frequency of trivalent formation.

If the locus of the marker is on the isochromosome arm, all of the secondary trisomic progeny will be of normal phenotype, if it is not on the isochromosome arm, the secondary trisomic fraction will have normal disomic ratios. In a secondary trisomic progeny segregating the trisomic fashion, no secondary trisomic with recessive phenotype should appear, as only one arm of the isochromosome can receive the recessive allele by crossing over. The detection of a single secondary trisomic individual with recessive phenotype indicates conclusively that the progeny is segregating in disomic fashion. Since the progenies of most of the secondary trisomics include related primary trisomics in certain proportions, the genetic ratios among this fraction can be utilized for verifying any questionable chromosomal location of the marker. If the marker is situated near the centromere, all the primary trisomics will have the normal phenotype. However, if the marker is 50 crossover units away from the centromere, one out of fourteen primary

TABLE 5.X
Phenotypic Ratios Expected in the Progeny of a Secondary Trisomic of *AAAa* Constitution

Location of the marker	Ratios in F_2 and BC			
		2*n*	2*n* + 1 (secondary)	2*n* + 1 (primary)
On the isochromosome arm	F_2	3:1[a]	1:0	1:0[a]
	BC	1:1[a]	1:0	1:0[a]
Not on the isochromosome arm but on the other arm	F_2	3:1	3:1	1:0[a]
	BC	1:1	1:1	1:0[a]
Not on that chromosome	F_2	3:1	3:1	3:1
	BC	1:1	1:1	1:1

[a]These ratios may be modified somewhat by double reduction.

trisomics may have a recessive phenotype. In case the marker is not located on the chromosome under investigation, the normal disomic ratio would hold among the primary trisomic fraction also.

The ratios observed may fall into any of the categories shown in Table 5.X depending on the chromosomal or arm location of the marker. Of course, these ratio sets are possible only if the two types of trisomics can be distinguished from each other and from the disomics.

As an additional aspect of the genetic studies with secondary trisomics, unlocated markers can be assigned simultaneously to the respective chromosomes and chromosome arms if a set of secondary trisomics is available and they yield an appreciable number of primaries in their progenies (Khush and Rick, 1969).

According to Avery *et al.* (1959), the secondary trisomics of *Datura stramonium* were utilized for determining the arm location of several markers. Earlier, Blakeslee and Avery (1934) stated that the arm locations of three genes of chromosome 17·18 were determined by means of secondary trisomic ratios. As far as the author is aware, however, no published data are available on the secondary trisomic segregations obtained in *Datura stramonium* or any other species except for those in tomato elaborated below. Blakeslee (1924) presented data on the segregating progenies of two secondary and the two related primary trisomics. Both of the primary trisomics gave trisomic ratios with their respective genes, but, according to the author, only one secondary gave trisomic ratios. These data are unsatisfactory for the following reasons.

First, one of the two supposed secondaries, Wiry, was later found to be tertiary trisomic. Second, at that time the cytological nature of the secondary trisomic was not clear; Blakeslee's supposition that the ratios for the primary and secondary trisomics should be similar, we now know, for the reasons outlined above, to be erroneous. Third, for the sake of "convenience," he presented data only on the disomic fraction of the progenies. We now know that ratios in the disomic portions of the progenies of secondary and tertiary trisomics remain largely unmodified; therefore, Blakeslee's (1924) data on secondary trisomic segregations were improperly analyzed. He obtained an 8:1 ratio for *p* in the disomic portion of progeny of Poinsettia (a primary trisomic) and in the disomic portion of the progeny of Wiry (a tertiary trisomic), a ratio expected for the former trisomic, but not for the latter. He also obtained a ratio of 8:1 for *a* in the disomic portion of Cocklebur (a primary trisomic) and a ratio of 3:1 for the same gene in the disomic fraction of Wedge (a related secondary). The trisomic nature of the former segregation is unquestionable, but his interpretation of the disomic nature of the latter is invalid. All of the secondary trisomics of the progeny may have had normal phenotype with respect to *a,* and the ratio may have been trisomic.

The only published data on genetic segregation in secondary trisomics of any species are those of Khush and Rick (1968c, 1969) on five secondary trisomics of tomato. Nine secondary trisomics were isolated, and five of them were used to study segregation of eleven genes belonging to four chromosomes of the complement (Table 5.XI). As can be seen from the table, all the progenies except the one for *ag* gave disomic ratios among the $2n$ fractions, but for *ag* and *h*, the ratios among the secondary trisomic fraction were also disomic. The percentage of recessive secondary trisomics was lower than expected, because of the lowered viability of the secondary trisomics with recessive phenotypes. Finally, the ratios among the primary trisomic fractions, except for *lg*, were trisomic. These ratios permit the following conclusions.

The gene *re* is not on 8L, but on 8S. Since all the triplo-8 of this family were of normal phenotype, the chromosomal location of this marker was confirmed. The genes *pum, ah,* and *marm* must be on 9L. The number of primary trisomics in the progeny was too small to permit any conclusion about the chromosomal location of these markers, but the association of these markers with chromosome 9 was definitely known earlier from the primary trisomic tests (Rick *et al.*, 1964). The marker *wd* was placed on 9S, and its location on chromosome 9 was confirmed, as it gave a trisomic ratio among the primary trisomic fraction. The genes *h* and *ag* are the only two markers which gave trisomic segregation among the secondary trisomic fraction and, as expected, no secondary trisomic with recessive phenotype was present. These genes were assigned to 10L. Since *lg* gave a ratio of 1:1::1:1::1:1 with $2n$ + 10L·10L, it cannot be on chromosome 10. Finally, *u* was assigned to 10S, and *fd* and *alb* to 12S.

With the help of these analyses, the centromere of chromosome 8 was mapped precisely between *re* and *cpt*, and that of chromosome 9 between *wd* and *pum*, (Fig. 5.2). The analysis of linkage group 10 was very rewarding. This linkage

TABLE 5.XI

Genetic Ratios in the Backcrosses of Secondary Trisomics of Tomato of *AAAa* Constitution[a]

			Progeny									
			Disomic				Parental secondary trisomic			Related primary trisomic		
Trisomic	Gene	Total	Normal	Recessive	Recessive (%)	X^2 1:1	Normal	Recessive	Recessive (%)	Normal	Recessive	Recessive (%)
$2n$ + 8L·8L	*re*	268	114	102	47.2	0.66	9	6	40.0	34	0	0.0
$2n$ + 9S·9S	*pum*	298	119	115	49.1	0.06	44	17	27.8	3	0	0.0
	ah	298	119	115	49.1	0.06	44	17	27.8	3	0	0.0
	marm	298	119	115	49.1	0.06	38	23	37.7	3	0	0.0
$2n$ + 9L·9L	*wd*	397	147	153	51.0	0.12	37	30	44.7	29	0	0.0
$2n$ + 10L·10L	*h*	392	136	126	48.0	0.20	78	0	0.0	50	0	0.0
	ag	392	162	100	38.1	14.20***	78	0	0.0	47	3	6.0
	lg	289	98	87	47.3	0.62	26	26	50.0	23	29	55.7
	u	134	55	64	53.8	0.68	11	4	26.7	–	–	–
$2n$ + 12L·12L	*alb*	300	121	128	51.4	0.18	24	18	42.8	8	1	11.1
	fd	294	122	125	50.6	0.02	25	14	35.9	8	0	0.0

[a]From Khush and Rick, 1969.

group was known to be *lg–pe–u–oli–h–l-2–t–ag–Xa* (Clayberg *et al.*, 1966). The markers *t, ag,* and *Xa* had been located on 10L by means of a tertiary trisomic test (Khush and Rick 1967b), but these determinations did not reveal the orientation of the linkage map or the position of the centromere. To accomplish this, *lg, u, h,* and *ag* were tested against $2n$ + 10L·10L. As discussed earlier, *lg* gave disomic segregation in all the subfamilies and as such cannot be situated on chromosome 10. The latest map distance between *pe* and *u* was reported as 43 cM by Clayberg *et al.* (1966). Bona fide linkages of this magnitude are notoriously difficult to verify. It became evident that neither *pe* nor *lg* was located on chromosome 10 and that the reported linkage between *pe* and *u* must have been in error. In addition, *h* and *ag* were assigned to 10L, and *u* to 10S. These conclusions proved that the orientation of the linkage map shown above was correct with *ag* toward the distal end of the long arm, and the centromere was mapped between *u* and *h*.

Additional evidence of a distal locus for *ag* was provided by its segregation in the disomic fraction. The observed proportion of *ag*/*ag* (38.1%) is significantly lower than expected (X^2 = 14.2***) for a locus close to the centromere. It closely approximates 35%, expected for a distant freely recombining locus if account is taken of the fact that 60% of the sporocytes of this trisomic were observed to form trivalents (Khush and Rick, 1969).

The genetic studies described above show how the secondary trisomics can be used as efficient tools in linkage mapping. The segregating progenies of these trisomics can give simultaneously information about the chromosomal and arm location of the marker and the position of the centromere and some idea of the proximity of the marker to the centromere.

Genetic Segregation in Compensating Trisomics

The genetic ratios obtained in the segregating progenies of the compensating trisomics differ from those obtained in any other trisomic. These trisomics have only one normal chromosome for a particular pair of the disomic complement. As a result, all the n gametes on the male and female sides produced by a trisomic of this type in diploid species possess this chromosome. When the trisomic is made heterozygous and the marker under test is located on this chromosome, all the n gametes receive the recessive allele except those which receive the normal allele from the compensating chromosome by double reduction. All the $2n$ offspring in the F_2 or backcross are homozygous recessive and all the compensating trisomics have normal phenotype. The observed ratios, 3:1::3:1 (disomic ratio) or 0:1::1:0 (a compensating trisomic ratio), will indicate whether or not the marker is located on the compensated chromosome.

If twelve compensating trisomics are available in a species with twelve pairs of chromosomes, they can be employed profitably in gene location work. They can all be made heterozygous for the gene to be located and the genetic ratios in the F_2 or backcross tested for a fit to the disomic or trisomic ratios calculated above. Eleven families would segregate in a disomic fashion and one would give trisomic ratios. Thus, the gene can be located on the chromosome being compensated in that trisomic. In gene location work, the most important advantage of using the compensating trisomics compared to primary trisomics is that much smaller populations are needed to differentiate between the disomic and compensating trisomic ratios.

If the trisomics are indistinguishable from the disomic sibs, or if the progenies are scored before the $2n + 1$ types can be recognized morphologically, the expected and observed ratios for the entire progeny can be compared. For the trisomic inheritance with 50% transmission of the extra chromosome, the F_2 ratio would be 1:1, and with 33.3% transmission, 1:2.

Smith (1947) studied the genetic ratios for eleven genes belonging to six of the seven linkage groups of diploid wheat, *Triticum monococcum*, in the F_2 generation of a compensating trisomic. The ratios were disomic for eight genes of five linkage groups, but the remaining three genes of the sixth linkage group segregated in a trisomic fashion. These three genes, *c-2*, *cx-1*, and *y*, of linkage group D gave 3:1 ratios among $2n$ controls, but segregated in a 1:2 ratio in the F_2 of the compensating trisomic (Table 5.XII). The observed trisomic ratios agree well with those expected on the basis of 33% transmission of the compensating trisomic reported by Smith (1947). These segregations established the relationship between linkage group D and its chromosome and proved the independence of this group from the other five groups.

The cytogenetic potential of the compensating types was realized by Blakeslee and his associates in the late 1920's (Blakeslee *et al.*, 1930). They were able to experimentally produce at least two compensating trisomics for each of the twelve chromosomes of *Datura stramonium* (Avery *et al.*, 1959). These types were used extensively for assigning the unlocated genes to their respective chromosomes. In an example cited by Avery *et al.* (1959), a compensating trisomic for chromosome $1\cdot2$ ($2n - 1\cdot2 + 1\cdot9 + 2\cdot5$) was made heterozygous for *albino*-11, and the heterozygous trisomic used as a male parent in a backcross to *albino*-11, producing a family with 36 normal plants and 199 albinos. This was interpreted as a compensating trisomic ratio. The appearance of 36 normal individuals in the family signifies that either *albino*-11 is located away from the centromere, so that n gametes with normal allele were produced by double reduction, or the compensating chromosomes were transmitted through the pollen at a low frequency. In the absence of both of these events, all the progeny should have been mutant.

An important cytogenetic feature of the compensating trisomics not appreciated by the earlier workers was pointed out by Khush and Rick (1967a). The compensating trisomics produce related tertiary, telo, or secondary trisomics in their

TABLE 5.XII
Segregation of Eleven Genes Belonging to Six Linkage Groups in the F_2's of a Compensating Trisomic of *Triticum monococum* and in Controls ($2n$ Heterozygotes)[a]

Linkage group	Gene	Het. parent $2n + 1$ versus $2n$	Progeny: Total no. of plants	Progeny: Recessive (%)
A	gl_2	$2n$	3390	22
A	gl_2	$2n + 1$	398	19
A	*c*-1	$2n$	2920	26
A	*c*-1	$2n + 1$	143	27
B[b]				
C	*e*-2	$2n$	3683	21
C	*e*-2	$2n + 1$	48	27
C	*yx*-2	$2n$	1413	22
C	*yx*-2	$2n + 1$	257	24
D	*c*-2	$2n$	1160	26
D	*c*-2	$2n + 1$	35	*77*
D	*cx*-1	$2n$	804	20
D	*cx*-1	$2n + 1$	404	*64*
D	*y*	$2n$	3090	25
D	*y*	$2n + 1$	1108	*66*
E	*Gp*	$2n$	2832	28
E	*Gp*	$2n + 1$	42	21
E	*G*	$2n$	1434	26
F	*G*	$2n + 1$	106	25
G	*e*-1	$2n$	5227	15
G	*e*-1	$2n + 1$	42	19
G	*ar*-2	$2n$	1157	17
G	*ar*-2	$2n + 1$	109	19

[a]From Smith, 1947.
[b]No marker of this linkage group was tested.

progenies at a low but consistent frequency (Chapter 4). Examination of the genetic ratios within these types yields information on the arm location of the marker under study. For example, among the progeny of the $2n - 1 \cdot 2 + 1 \cdot 9 + 2 \cdot 5$ compensating trisomic of *Datura* there appear, in addition to the parental trisomic, two tertiaries, $2n + 1 \cdot 9$ and $2n + 2 \cdot 5$. If the gene under study is located on $\cdot 1$ arm, all the $2n + 1 \cdot 9$ tertiaries would have the normal phenotype (1:0 ratio), while one-fourth of the $2n + 2 \cdot 5$ tertiaries would have the recessive phenotype (3:1 ratio). Thus, four ratios among a single progeny of a compensating trisomic can be studied—0:1 for the $2n$ fraction, 1:0 for the compensating trisomic fraction, 1:0 for one tertiary trisomic fraction, and 3:1 for the other tertiary trisomic fraction. The advantages of using twelve compensating trisomics over using primary trisomics are overwhelming. The segregation ratios of compensating trisomic progenies can yield information simultaneously about the chromosome and arm location of the marker.

However, the limitations of this type of analysis should be pointed out. First, it can be carried out only if the three trisomics are morphologically distinct from each other as well as from the disomics.This requirement is met in *Datura* and tomato. Second, because of the low frequency, generally between 2 and 3%, of the related tertiaries, secondaries, and telotrisomics in the progenies of compensating $2n + 1$ types, rather large populations are needed. If a minimum of twelve plants of each related trisomic is considered sufficient to test for a 1:0 or 3:1 ratio, at least 400 to 600 plants would be needed in the progeny. One need not grow this many plants for all the twelve progenies; small families would establish the relationship between the gene and the chromosome, and a large family of the critical trisomic can be grown later to determine the arm location of the marker.

This type of analysis was attempted by Khush and Rick (1967a) with a backcross progeny of the *ru* marker and the $2n - 3S\cdot3L + \cdot3S + 3L\cdot3L$ trisomic of tomato. The 103 disomic plants were all of the recessive *ru* phenotype and all of the 74 compensating trisomics were of ru^+ phenotype, a perfect fit to the 0:1::1:0 ratio. In addition, five $2n + \cdot3S$ telotrisomic plants were obtained, and all were ru^+. Although the number of $2n + \cdot3S$ plants was too small to test for a fit to the 1:0 or 1:1 ratio in the $2n + \cdot3S$ fraction, it gave a strong indication that *ru* is located on 3S.

Because of the somatic instability of the $\cdot3S$ telocentric of $2n - 3S\cdot3L + \cdot3S + 3L\cdot3L$, this compensating trisomic could be employed in monosomic analysis for the *ru* gene. The $\cdot3S$ is lost sporadically in the somatic tissues. When the compensating trisomics heterozygous for *ru* were grown, their leaves were mosaics of normal and *ru* tissues (Fig. 5.4). Cytological examination of certain purely *ru* tissues revealed the loss of the telocentric $\cdot3S$, thereby permitting *ru* to be expressed in a hemizygous condition. From this evidence *ru* was assigned to 3S. In effect, this is monosomic analysis, i.e., obtaining the critical information from the examination of F_1 progeny. A model for the monosomic analysis of diploid species was thereby provided (Khush and Rick, 1967a).

The segregating progenies of certain compensating trisomics yield information about the map distance of the marker from the centromere simultaneously with its chromosomal and arm location. For example, in $2n - 3S\cdot3L + \cdot3S + 3L\cdot3L$, the presence of only two 3S arms gives this trisomic a special advantage over the others, for any crossover between the two arms will affect one-half the normal chromatids of that arm. Consequently, since the $\cdot3S$ telocentric with the normal allele must always pair with $3S\cdot3L$, carrying the recessive allele, the percentage of dominants in the disomic progeny gives a measure of the distance of

Fig. 5.4. Tomato compensating trisomic for chromosome 3 ($2n$ − 3S·3L + ·3S + 3L·3L) heterozygous for recessive marker *ru* of 3S. Right, plant in which the first three leaves have ru^+ phenotype, all subsequent growth is *ru*, presumably as a result of complete loss of ·3S. Left, leaf with mosaic pattern resulting from elimination of ·3S. (From Khush and Rick, 1967a.)

the locus from the centromere. In the heterozygote studied by Khush and Rick (1967a), ·3S carried ru^+ and 3S·3L carried *ru*. The fact that all of the 103 disomic progeny were *ru*/*ru* argues for a locus very close to the centromere, a site supported by deficiency studies (Khush and Rick, 1968a; Fig. 5.2).

From the foregoing paragraphs it is evident that genetic segregations obtained in different trisomics can delimit the markers to the chromosomes, or to a chromosome and a single chromosome arm, or to two chromosomes and single arms of these two chromosomes, or to a chromosome and both of its arms. These comparisons are summarized in Table 5.XIII.

Other Special Uses of Trisomics

In addition to their usefulness in genetic studies, the various kinds of trisomics can be employed in studying the effect of duplication of a whole chromosome, a chromosome arm or parts of chromosome arms on the morphology, anatomy, and physiology of the organism. These investigations can throw light on the basic nature of the genome of the species. For example, Gottschalk (1954) suspected tomato to be a secondary polyploid and pointed out certain similarities

TABLE 5.XIII
Summary Comparison of Locus Delimitations Permitted by Segregation in the Different Trisomic Types[a]

	Disomic progeny		Delimitations in trisomic progeny		
Trisomic type	Segregation[b]	Delimitation	Parental	Related primary	Related secondary, telo, or tertiary
Primary	8:1	Chromosome	Chromosome	–	–
Secondary	3:1	None	Single arm	Chromosome	–
Tertiary	3:1	None	Single arms of two chromosomes	Chromosomes (2)	–
Telo	3:1	None	Single arm	Chromosome	–
Compensating	0:1	Chromosome	Chromosome	–	Both arms

[a] Assuming simplex genotype with recessive gene on normal chromosome.
[b] Assuming no double reduction.

in gross morphology of a few chromosome pairs. When the morphology of the primary trisomic plants of these "similar" pairs was compared by Rick and Khush (1966), no similarities were found. Gottschalk's hypothesis is therefore improbable. Generally, the trisomics of the diploid species are morphologically well distinguished from the disomic sibs and from each other. If the trisomics are not well defined morphologically, the species is suspected to be polyploid. This subject is treated in greater detail in Chapter 6.

Primary trisomics are useful in identifying the chromosomes involved in translocations. All the primary trisomics are crossed as female parents with translocation stock, and one trisomic plant from each cross is examined cytologically. A ring of four plus a separate trivalent indicate that the extra chromosome of the trisomic is not the one involved in the translocation, while a chain of five indicates that it is. The chromosomes involved in "semisterile-3" in maize (Burnham, 1948) were identified in this way. Conversely, the extra chromosomes of the primary trisomics may be identified by crossing with known translocation testers as was done in barley by Tsuchiya (1961). For example, F_1 trisomics of Bush crossed with *a–b* and *b–d* translocations showed the chromosome configuration of 1 V + 5 II, indicating that the chromosome which is present in triplicate in Bush is involved in both the translocations. Since *b* chromosome is involved in both the translocations, the extra chromosome in Bush must be *b*. Similar identifications were made by Iwata *et al.* (1970) in rice and by Hanna and Schertz (1971) in *Sorghum vulgare*.

The primary trisomics and tetrasomics of wheat have been used for determining the homeologous series in wheat. This subject is treated in Chapter 9.

In species in which vegetative propagation is the normal means of multiplication, some of the $2n + 1$ types are introduced as cultivars. Many hyacinth varieties are aneuploids (Darlington and Mather, 1944).

The trisomics of each type are good sources for the related trisomics of a different category (Chapter 2).

The telotrisomics provide an opportunity for studying the stability of the terminal centromeres (Khush and Rick, 1968b).

The tertiary trisomics are useful for maintaining genetic stocks unfruitful in homozygous condition such as male steriles and other mutants which produce flower abnormalities. A tertiary trisomic with mutant alleles on the two normal chromosomes and the wild-type allele on the tertiary chromosome will have normal phenotype and be fruitful. In the absence of double reduction, all the $2n$ progeny of this trisomic will be homozygous for the mutant gene, and the trisomic progeny of the parental constitution. Thus the stock can be maintained indefinitely by selfing the tertiary trisomics in each generation. In the absence of male transmission of the tertiary chromosome, and if the marker is located close to the interchange point, all the pollen produced by the heterozygous trisomic is n and carries the mutant allele. Crosses of these trisomics as male parents produce F_1 progeny heterozygous for the mutant gene. Ramage and Tuleen (1964) established two tertiary trisomics of barley which were heterozygous for the homozygous lethal albino gene. The tertiary chromosome in each trisomic carried the normal allele of albino while the normal chromosomes carried the albino alleles. Pollen from these trisomics could be used for producing the heterozygous disomics and the albino stock was maintained through self-pollination of the heterozygous trisomics.

A scheme in which the tertiary trisomic of tomato can be employed as a fertility restorer in a program of hybridization utilizing male sterility was suggested by Khush and Rick (1967b). A tertiary chromosome carrying the normal allele of a recessive male sterility gene, which is present on both members of a pair of related homologs, forms a trisomic which can be used as the pollen parent in crosses with homozygous male-sterile stock. Since all the progeny from such a cross will be male sterile, no roguing is necessary. The prerequisites for such a program—good pollen fertility of the trisomic and no transmission through the male—are satisfied by many tomato tertiaries. From a collection of nearly fifty male sterility genes available in tomato, appropriate ones can be selected that lie on one arm or the other of certain tertiary chromosomes. Loci near the centromere serve these purposes best, as they suffer the least chance of yielding haploid gametes with the normal allele as a result of double reduction. A more complex scheme using balanced tertiary trisomics of barley in hybrid seed production has been proposed by Ramage (1965).

Certain trisomics of *Datura stramonium* promote the production of hybrid embryos in interspecific crosses. When *Datura stramonium* is crossed with *D. ceratocaula,* pollen germination and pollen tube growth are normal and fertilization takes place normally, but there is a high mortality of embryos in the early stages of development. When the $2n + 17 \cdot 17$ trisomic of *D. stramonium* was used as a female in the cross, normal seeds were obtained in 100% of the capsules. These seeds produced hybrids which could be grown to maturity. The primaries $2n + 19 \cdot 20$ and $2n + 23 \cdot 24$ and the secondary $2n + 11 \cdot 11$ also considerably increased the number of seeds containing embryos in this interspecific cross (Cole, 1956).

From the foregoing discussion it is evident that the various kinds of trisomics are useful in a wide variety of cytogenetic investigations. To date the most important of these uses are determining gene and chromosome relationships, assigning genes to chromosome arms, centromere mapping, determining the independence of linkage groups, and finding the orientation of linkage maps. As discussed earlier in this chapter, trisomics have been investigated most extensively in *Datura* and tomato. The overall summary of *Datura* work has been given by Blakeslee *et al.* (1940) and Avery *et al.* (1959). The summary of gene locations carried out with the aid of trisomics of tomato is given in Table 5.XIV. In tomato, induced deficiencies were employed rather extensively in

TABLE 5.XIV
Summary of Tomato Genes Located by Means of Different Trisomics and Induced Deficiencies

Chromosome	Primary trisomic	Genes located by: Tertiary trisomics		Secondary trisomics		Telotrisomics		Compensating trisomics		Induced deficiencies	
		Short arm	Long arm	Short arm	Lcng arm	Short arm	Long arm	Short arm	Long arm	Short arm	Long arm
1	*y*	*br*	*inv,scf*							*au*	
2	*d,lx*										*aw,d*
3	*r,wf*					*r,wf*	*rv,sf*	*ru,sy*		*ru,sy*	*bls,sf*
4	*e,ls,vg*	*ful,clau*	*ra,di* *w-4*							*clau,ful*	*ra,e*
5	*wt,mc,sd*	*mc tf*	*wt*							*tf*	
6	*c,cl2,og,sp,yv*									*tl*	*yv,c*
7	*gs*	*gs*				*var*				*var*	*not,deb*
8	*l,dl,al,ch,cpt*			*re*			*cpt*			*l,spa*	*bu,dl,al*
9	*wd,ah*		*ah,marm*	*wd*	*pum,ah* *marm*						*nv,ah,* *lut*
10	*h*		*ag,tv*	*u*	*h,ag*						*ag,tv*
11										*a*	*hl*
12				*alb,fd*							*alb,fd*

determining the cytological loci of the markers. The genes located by induced deficiencies are enumerated in the last two columns of the table. The latest linkage map of tomato, based on the information obtained with the aid of trisomics and induced deficiencies, is shown in Fig. 5.2. The linkage groups of tomato are the best known of any of the higher plant species, thanks to the applicability of the trisomic technique in exploring the genome of this species.

6

Morphology, Anatomy, Physiology, and Biochemistry of Trisomics

When present in the trisomic condition, the individual chromosomes of the complement have a distinct effect on cellular processes and developmental rhythm of the individual. These effects are detected in the altered morphology, anatomy, and physiology. Compared to the disomic sibs, the trisomics also have reduced viability and vigor and are less fruitful. Thus the trisomics of a species can be distinguished from each other and from the disomics by the morphological, anatomical, and physiological deviations and their reduced vigor and fertility.

Morphology of Trisomics

The morphological differences between the trisomics and the disomics are of the same nature as the differences between related species and subspecies. In fact, many "mutants" of *Oenothera lamarckiana* to which de Vries (1901) gave specific names and from whose study he formulated his famous mutation theory were trisomics. The "mutants" *Albida, Oblonga, Lata,* and *Scintillans* described by him as separate species were later found to have an extra chromosome each (Emerson, 1935).

Similarly, the trisomics of *Datura* differ from each other and from the disomics in many traits by which species of this genus are separated. These include differences in capsule size and shape; size and length of the spines; size of plant and growth habit; and size, shape, and form of leaves, flowers, and stigmas (Blakeslee, 1922, 1930, 1934; Blakeslee and Belling, 1924a; Avery and Blakeslee, 1948). The trisomics were given laboratory names by the *Datura* workers generally signifying some morphological aspect such as fruit shape or growth habit. Globe trisomic designates the globose shape of the fruit; Slender,

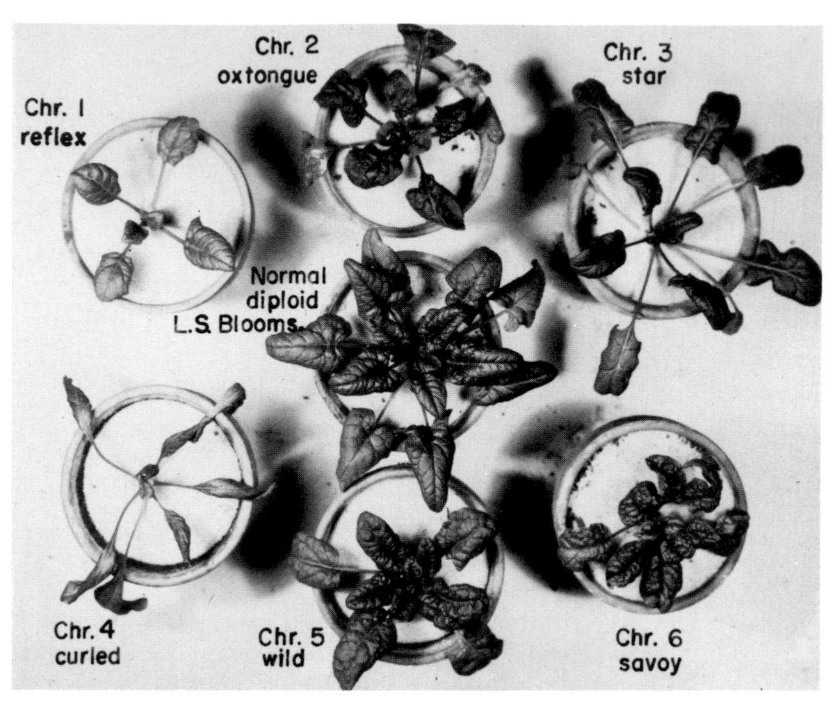

Fig. 6.1. The primary trisomics and a diploid of *Spinacia oleracia*. (From Ellis and Janick, 1960.)

the slender and elongate shape of the capsules. Laboratory names signifying morphological features have also been given to the trisomics of *Matthiola incana* (Frost, 1927; Prakken, 1942), *Antirrhinum majus* (Stubbe, 1934; Propach, 1935), *Nicotiana sylvestris* (Goodspeed and Avery, 1939, 1941), *Beta vulgaris* (Levan, 1942), *Oenothera blandina* (Catcheside, 1954), *Hordeum spontaneum* (Tsuchiya, 1960), *Secale cereale* (Kamanoi and Jenkins, 1962), *Sorghum vulgare* (Schertz, 1966), rice (Hu, 1968), and *Pennisetum typhoides* (Gill *et al.*, 1970a).

As pointed out by Goodspeed and Avery (1939) and Rick and Barton (1954), a particular extra chromosome exerts similar modifications in the different parts of the individual. The trisomics for chromosomes 3 and 4 of tomato cause elongation in the leaf segments, the internodes of the stem, and the inflorescences, flowers, and fruits, and even seed size. In contrast, alterations of shape and size in the opposite direction are produced by triplo-7 and triplo-10—the leaves are broader and thicker, the internodes, inflorescences and flowers are shorter, and the fruits flatter. In triplo-1, all the plant parts, including the fruits, are extremely reduced in size. The trisomic Rolled of *Datura* produces narrow leaves, and small capsules with short spines; the trisomic Thistle has long, thin leaves, long and slender flowers, and elongated and slender capsules with

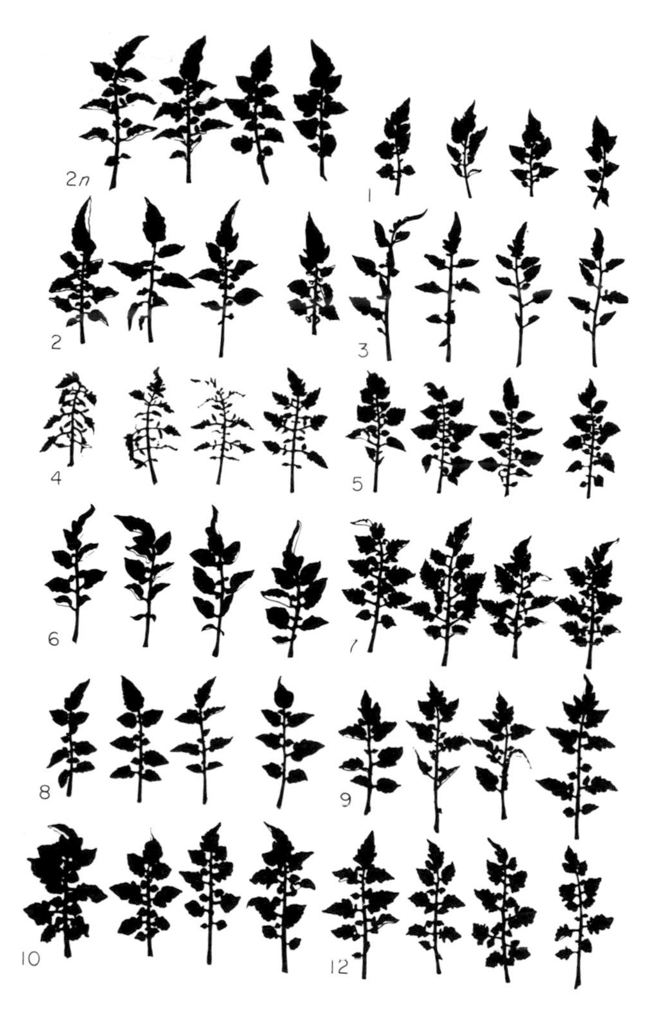

Fig. 6.2. Outline of representative leaves of diploid and eleven primary trisomics of tomato. Numbers correspond to those of extra chromosomes. (From Rick and Barton, 1954.)

many long spines (Avery *et al.*, 1959). From the descriptions of trisomics of other species, similar correlations are evident.

Generally the trisomics grow more slowly than the normals and are liable to be crowded out by the latter. They are usually less fruitful. These characteristics and others such as leaf shape, leaf color, texture of leaf surface, waviness of leaf margins (Fig. 6.1), internode length, degree of branching, hair density, spread and growth structure of the plant, modification of flower parts, and fruit size and shape are all helpful in distinguishing the $2n + 1$ individuals from the disomic sibs. Close examination reveals that most of the plant organs are modified. Sometimes the trisomics can be identified as soon as the seedlings emerge from the soil. In tomato, six trisomics can be recognized as soon as the cotyledons expand. In triplo-1, for example, the cotyledons are less than half the size of those of the disomics and are darker in color. In triplo-3, they are narrow and elongated; those of triplo-7 are reduced in size and the tip is obtuse; in triplo-8, they are thick, leathery, glossy, and wedge-shaped; and triplo-10 has cotyledons reduced in size and somewhat lighter green in color than the disomics. In all the primaries of tomato, the leaves are modified in a similar manner. The differences in size and shape of mature leaves of the eleven primaries and a disomic are shown in outline in Fig. 6.2. Differences in venation pattern and color and texture are not shown but are easily detected during most of the growth stages. The stem modifications of tomato trisomics are expressed in internode length, in thickness, and in spreading habit. For example, the internode length of triplo-2, triplo-3, triplo-4, and triplo-6 is increased, while that of triplo-1, triplo-7, and triplo-10, is decreased. The stems of triplo-2, triplo-3, and triplo-4 are thinner than the normal, those of triplo-7, and triplo-10 are thicker. The growth pattern of triplo-2, triplo-3, and triplo-6 is straggly, and that of triplo-1, triplo-7, triplo-9, and triplo-10 is upright and compact. The primaries and the disomics also differ in flower and fruit characteristics.

In a similar manner, the trisomics of *Datura* can be identified at all stages of growth. However, the $2n + 1$ types can be most easily distinguished by differences in flower parts and mature capsules (Figs. 6.3 and 6.4). The modifications caused by the extra chromosomes are evident from the figures. Note, for example, the modification in size and shape of the capsules and in the number and length of the spines. Most of the laboratory names of the trisomics of *Datura* are based on the deviant morphology of their capsules.

In addition to gross morphological changes of flower parts, male and female reproductive parts of a trisomic may also be affected. Thus, in trisomic type 8 of *Sorghum vulgare,* pollen grains have papilliform appendages (Lin and Ross, 1969a) and ovaries of trisomic type 9 have tumorous growths (Lin and Ross, 1969b).

The contribution of different arms of a chromosome to the deviant morphology of the primary trisomic can be determined by comparing the morphology of primaries with that of related telotrisomics, tertiary trisomics, or secondary trisomics. Tomato telotrisomics for the long arms are strikingly similar in gross morphology to the corresponding primary trisomics, while those for short arms are nondescript, being distinguishable from disomics only under optimum conditions. Thus $2n + \cdot 3L$, $2n + \cdot 4L$, $2n + \cdot 7L$ and $2n + \cdot 8L$ resemble the corresponding primary trisomic closely but $2n + \cdot 3S$ and $2n + \cdot 10S$ are very difficult to distinguish and chromosome counts are often necessary to separate them from the disomics (Khush and Rick, 1968b). This differential influence of the two arms is also evident from the examination of tomato tertiary trisomics

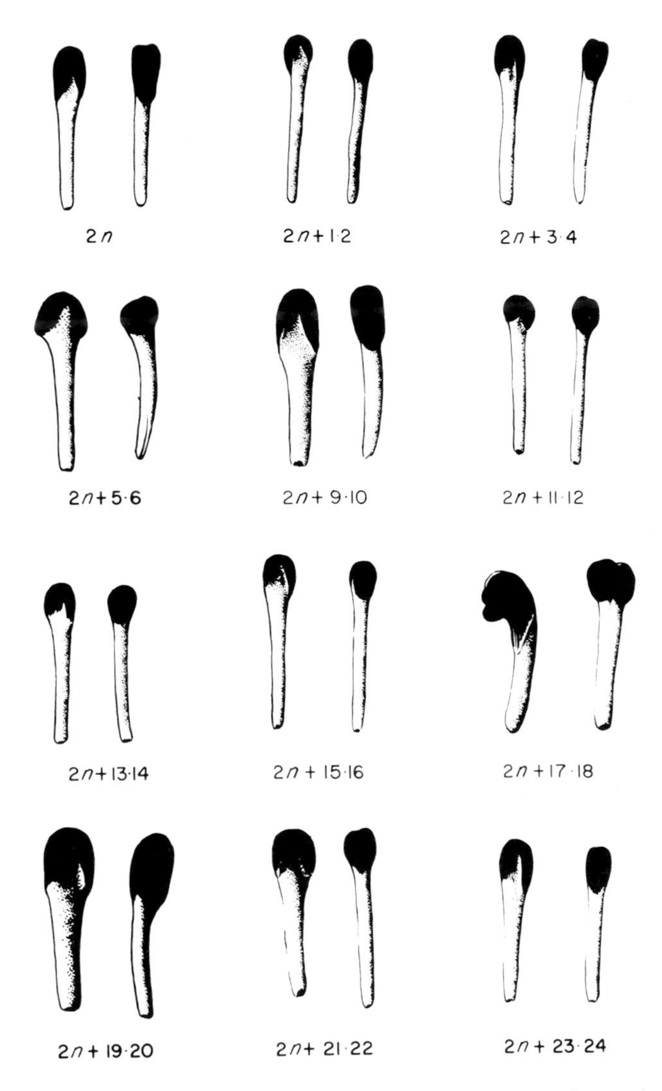

Fig. 6.3 Stigmas of *Datura stramonium*. Normal diploid and eleven primary trisomics. [From Avery, A. G., Satina, S., and Rietsema, J. (1959). "Blakeslee: The Genus *Datura*," 289 pp. © 1959, The Ronald Press Company, New York.]

(Khush and Rick, 1967b). For example, 2*n* + 5S·7L greatly resembles triplo-7, the contribution of 5S being relatively minor. Likewise, 2*n* + 7S·11L closely resembles triplo-11, the influence of 7S being barely discernible. If the tertiary chromosome consists of two long arms, as in 2*n* + 1L·11L, 2*n* + 2L·10L, 2*n* + 4L·10L, and 2*n* + 9L·12L, the longer of the two arms exerts the greater phenotypic effect. For example 1L, more than three times the length of 11L,

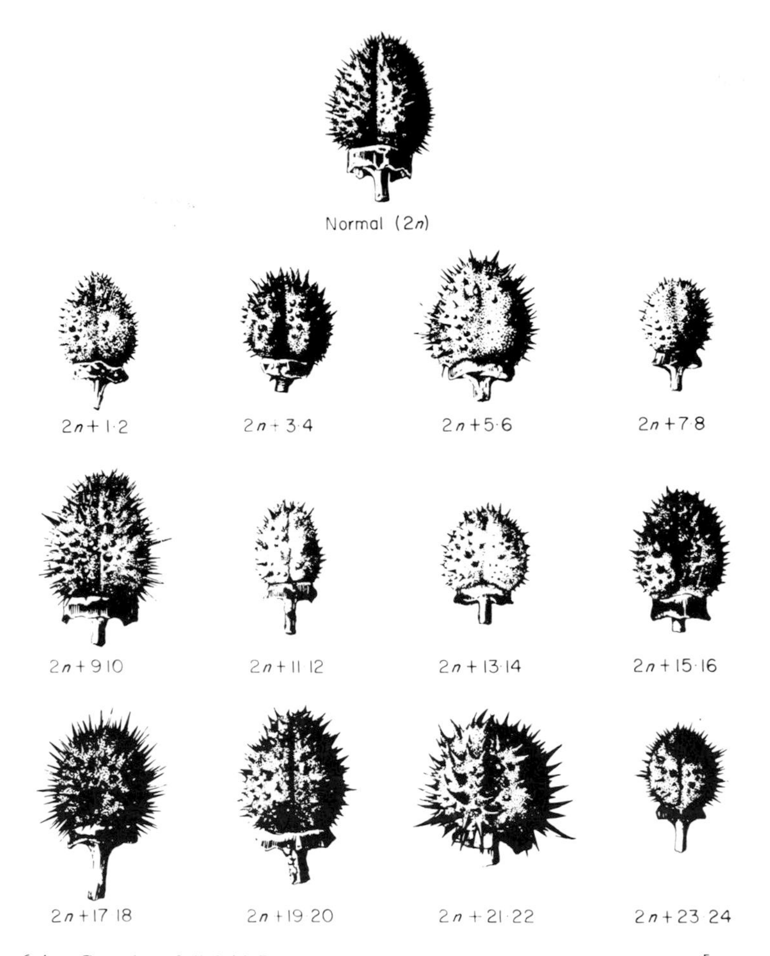

Fig. 6.4. Capsules of diploid *Datura stramonium* and twelve primary trisomics. [From Avery, A. G., Satina, S., and Rietsema, J. (1959). "Blakeslee: The Genus *Datura*," 289 pp. © 1959, The Ronald Press Company, New York.]

exerts a dominant influence on the morphology of $2n$ + 1L·11L (Fig. 6.5) which resembles triplo-1 in extremely slow growth and diminutive size, acuminate cotyledons, darker color at seedling stage, reduced number of epidermal hairs, chlorotic leaves at adult stage, fewer leaf segments with convex margins and obtuse terminal segments, long inflorescences, and complete lack of fruitfulness under field conditions. The only resemblance to triplo-11 is in the deeply cut and recurved leaves, the persistent styles, and the thickened pedicel joint.

Similarly, 4L is about twice as long as 10L and the prevailing influence of 4L on the morphology of $2n$ + 4L·10L is evident (Fig. 6.6). Thus

Fig. 6.5. Seedling tomato plants 5 weeks old of a tertiary trisomic, the two related primaries, and diploid. (From Khush and Rick, 1967b.)

$2n$ + 4L · 10L resembles triplo-4 in having longer internodes, slender, recurved, and plicate leaves with narrow segments, elongate inflorescence and flower parts, especially the calyx, turbinate flowers, depressed stigmas, and unconstricted, pyriform fruits with reduced locule number. It resembles triplo-10 only in its thicker stems and bright yellow-green leaves.

The tomato tertiary trisomic $2n$ + 9S · 12S bears only slight resemblance to either triplo-9 or triplo-12 and is difficult to distinguish from the disomics at the seedling stage. The short arms of the tomato complement appear to contribute very little to the phenotype of tertiary trisomics; in fact, their contribution is disproportionately small considering their relative total chromosomal or euchromatic lengths.

Also noteworthy is the fact that some characters in the trisomic phenotype must be controlled by interaction between the two arms of the extra chromosome. For example, in triplo-9 of tomato, a high proportion of the progeny lose their plumular meristems and cease growth in the early seedling stage, but this tendency is completely absent in seedlings of $2n$ + 9S · 12S as well as in those of $2n$ + 9L · 12L. If this character depends solely on the presence of a specific arm, it should have appeared in one of these two tertiaries (Khush and Rick, 1967b).

The tertiary trisomics of *Datura* display the morphological features of the two related primaries and secondaries in such a way that the arm constitution of the extra tertiary chromosome can be predicted. Generally the two tertiaries are characterized by some of the outstanding features of the two primaries. Unlike tomato, any disproportionate influence of the two arms of *Datura* chromo-

Fig. 6.6. Seedling tomato plants 5 weeks old of a tertiary trisomic, the two related primaries, and diploid.

somes has not been indicated. *Datura* chromosomes may not be strongly heterobrachial. The karyogram of the somatic chromosomes (Avery *et al.*, 1959) supports this conclusion. However, as discussed in Chapter 5, the somatic karyogram can be misleading.

Since, in a secondary trisomic, there are four doses of a chromosome arm, some of the characters of the primary trisomic for that chromosome are exaggerated. If the isochromosome arm happens to be the long one, the secondary trisomic may be weak and slow growing due to excessive duplication. Thus the tomato secondaries for the long arms 6L, 7L, 8L, 9L, and 10L have much slower growth rates at all stages of development (Khush and Rick, 1969). Secondaries for these arms are sometimes difficult to distinguish from the corresponding primaries. Comparison between the morphologies of primaries and secondaries indicates that there is hardly any trait of the primary which is not present in the secondary of the long arm. On the other hand, the seedling vigor of the secondaries for the short arms such as 3S, 7S, and 9S is scarcely reduced and the morphology of their seedlings is not appreciably modified. Seedlings of $2n$ + 3S·3S and $2n$ + 9S·9S cannot be separated from the disomics, and those of $2n$ + 7S·7S, only with difficulty. Under field conditions, $2n$ + 9S·9S is rather weak and distinct in growth habit, but $2n$ + 3S·3S and $2n$ + 7S·7S are almost as vigorous as disomics. The secondary for the metacentric chromosome 12 ($2n$ + 12L·12L) shows only about one-half of the characters of its primary. The conclusion from the study of tertiary trisomics and telotrisomics of tomato (Khush and Rick, 1967b, 1968b) that long arms have a disproportionately greater influence than short arms on the morphology of trisomic plants is thus confirmed from the studies of secondary trisomics.

Upon closer examination, the secondaries for two arms of the same chromosome are found to have a complementary relationship to their primary (Khush and Rick, 1969). For example, in tomato, the $2n + 7S \cdot 7L$ primary in contrast to the disomic has thick stems, broad leaves, short internodes, compact inflorescences, and a bushy growth habit. Some branches of the trisomic cease terminal growth and become blind. All these characters are magnified in the $2n + 7L \cdot 7L$ secondary which has thicker stems, broader leaves, shorter internodes, more compact inflorescences, and a bushier growth habit. The tendency for blindness is so enhanced that all the branches cease growth after producing two or three leaves. In contrast, the morphology of the $2n + 7S \cdot 7S$ secondary is modified in the opposite direction. This trisomic has slender stems and leaves, longer internodes and inflorescences, an open growth habit, and no tendency for blindness. All the plant parts of $2n + 9S \cdot 9L$, including internodes, leaves, flowers, and fruits, are somewhat smaller than the disomic, and in $2n + 9L \cdot 9L$ they are even smaller. However, in $2n + 9S \cdot 9S$, all these organs are larger than the disomic. The $2n + 9S \cdot 9L$ has compound inflorescences, and this character is exaggerated in $2n + 9L \cdot 9L$ so that the number of flowers per cluster is double that in the primary. However in $2n + 9S \cdot 9S$ the number of flowers is much reduced, and often there are only two or three flowers per cluster. It is therefore evident that, in general, the primary is intermediate between its two secondaries in most characters, and this intermediate condition occurs because of interaction of opposing factors.

The relationship of the two secondaries of *Datura* to their primary is similar (Avery *et al.*, 1959). For example, the $2n + 1 \cdot 1$ secondary has small capsules and narrow leaves, $2n + 2 \cdot 2$ has large capsules and relatively broad leaves, and $2n + 1 \cdot 2$ is intermediate in these two characters (Blakeslee, 1930, 1934). A clearer example is provided by the trisomic group $2n + 9 \cdot 9$, $2n + 9 \cdot 10$, and $2n + 10 \cdot 10$. The $2n + 9 \cdot 9$ plants have relatively broad, thick, and dark leaves, the flowers are short and dark purple, the capsules are globose with a few short spines, and the plants are short and compact in habit. The $2n + 10 \cdot 10$ plants have long, thin leaves, the flowers are very long and slender, the capsules are slender with many long spines, and the plants are tall and erect. In the primary $2n + 9 \cdot 10$, the expression of these characteristics is intermediate between those shown by the two secondaries. Figure 6.7 shows 3-week-old seedlings of the disomic, three primaries ($2n + 1 \cdot 2$, $2n + 5 \cdot 6$, and $2n + 9 \cdot 10$), and their respective secondaries.

The extra chromosomal material of the compensating trisomic is similar to that in primary, tertiary, secondary or telotrisomic. Therefore a compensating trisomic must be identical in morphology to one of the other kinds. In tomato, the $2n - 7S \cdot 7L + 7S \cdot 7S + 7L \cdot 7L$ has an extra 7S and 7L, and it is indistinguishable from the primary $2n + 7S \cdot 7L$ (Khush and Rick, 1967a). Similarly, $2n - 3S \cdot 3L + \cdot 3S + 3L \cdot 3L$ of tomato is identical to the $2n + \cdot 3L$ telo-

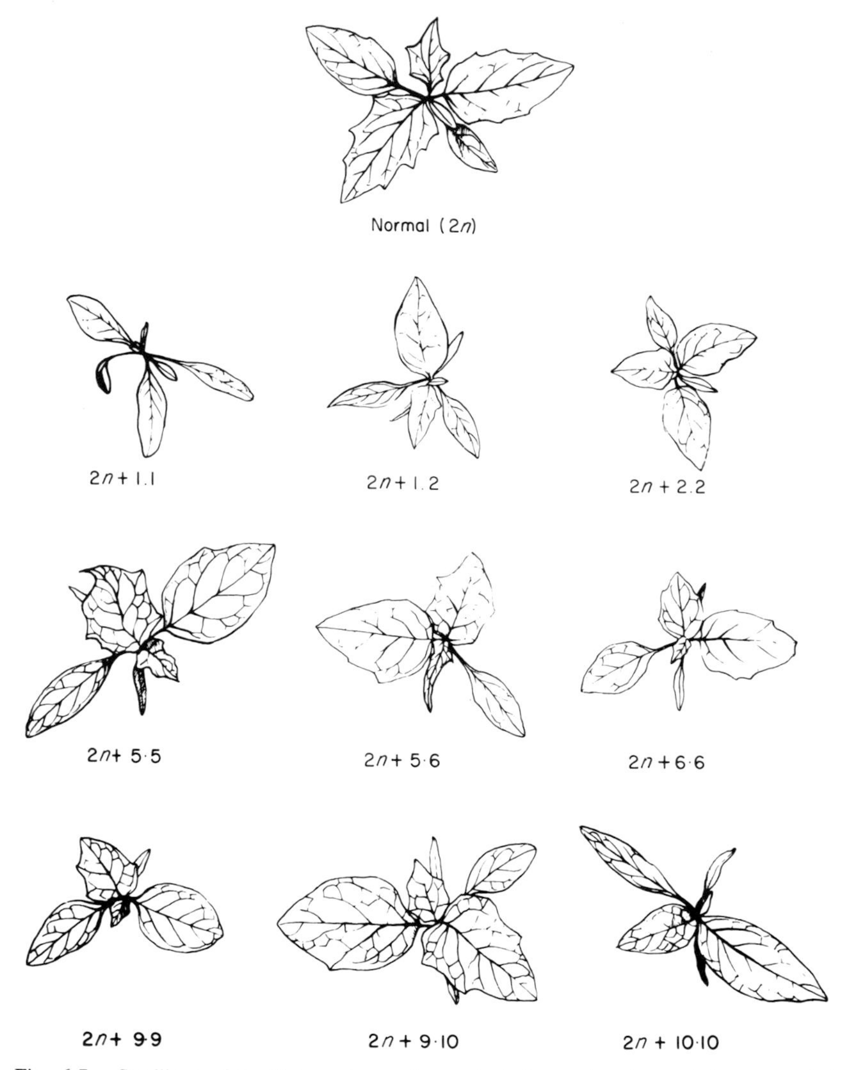

Fig. 6.7. Seedlings of 2*n* *Datura stramonium* and of three primaries and their secondaries. [From Avery, A. G., Satina, S., and Rietsema, J. (1959). "Blakeslee: The Genus *Datura*," 289 pp. © 1959, The Ronald Press Company, New York.]

trisomic. The compensating trisomic, $2n - 1\cdot 2 + 1\cdot 1 + 2\cdot 17$ of *Datura* (Blakeslee, 1934) resembles the tertiary $2n + 1\cdot 17$. This plant gets its erect habit of growth and narrow leaves from the extra $\cdot 1$ arm, but the droopy leaves from the extra $\cdot 17$ arm.

Anatomy of Trisomics

Although complete trisomic series have been established in more than twenty species, the comparative anatomy of trisomics has been studied only in *Datura stramonium*. Sinnott and Blakeslee (1922) and Sinnott *et al.* (1934) investigated the comparative anatomy of flower stalks and, to a lesser extent, of stems, leaves, roots, and styles of the disomic and in all the primary and all the known secondary trisomics of *Datura*. A large number of sections of the pedicels or flower stalks at a similar stage of development (day of flowering) and from different seasons were examined and many differences in various anatomical characters were observed. In material grown in different seasons the same differences were noted, indicating that deviations were not due to environment. The following measurements were taken from plants grown under uniform conditions: diameter of flower stalk; outer diameter of entire vascular cylinder; diameter of pith; depth and width of vascular bundles; depth, width, and length of epidermal cells; diameter and length of hypodermal cells; diameter and length of inner cortical cells; diameter and length of pith cells; diameter of pericycle fibers; diameter of vessels; and diameter of starch grains in the endodermis.

From these measurements, values for the following were determined: area of pedicel (in transverse section); area of cortex; area of pith; total area of all vascular bundles; ratio of length to width of vascular bundles; volume of epidermal, hypodermal, inner cortical, and pith cells; ratio of volume of cortical cells to volume of pith cells, transverse area of pericycle fibers and vessels; number of epidermal cells; number of pith cells; and the percentage of circumference of the vascular ring which consisted of interfascicular spaces between the xylem portions of the bundles.

The primary and secondary trisomics of *Datura* differ from the disomic as well as from each other in many of these anatomical features. The size and number of cells in the various tissues of the pedicel seem in many cases, to be determined independently of each other. In the primary trisomic $2n + 19 \cdot 20$, the cells of both pith and cortex are much larger than those of the disomic, though their number remains approximately same. In $2n + 23 \cdot 24$, the pith cells are smaller than those of any other trisomic, but their number is approximately the same as in the disomic. The pith of $2n + 19 \cdot 19$ has many more cells than the disomic, but they are essentially of the same size. In $2n + 15 \cdot 15$ and $2n + 15 \cdot 16$, cell number in the pith is greatly reduced, but the size of the cells is not significantly different from the disomic. In numerous instances both cell size and cell number differ from the disomic. Sometimes these increase or decrease together, but frequently they vary in opposite directions. In the pith of $2n + 17 \cdot 18$, the cells increase in both size and number compared with the disomic; in $2n + 5 \cdot 6$ and $2n + 3 \cdot 3$ there is a decrease in both size and number. In $2n + 1 \cdot 2$, $2n + 2 \cdot 2$, $2n + 7 \cdot 7$, and $2n + 13 \cdot 14$, on the other hand, cell size is increased and cell number decreased. The opposite condition is shown by the secondaries $2n + 6 \cdot 6$ and $2n + 9 \cdot 9$ in which the size of the pith cells is much smaller than in disomic, but the number is greater.

This analysis indicates that in the series of trisomics as a whole cell size and number are not necessarily related and may be controlled quite independently.

The differences in general anatomical plan and pattern between the trisomics are quite evident. They can be seen most simply in the relative development of vascular and fundamental tissues. In some trisomics, e.g., $2n + 9 \cdot 9$, the pedicel is much larger than in the disomic, and the increase in size is shared about equally by the two classes of tissues. Others, as $2n + 13 \cdot 14$, have a pedicel smaller than in the disomic, and the reduction is also shared about equally. In several cases, however, one tissue is relatively much larger than the other. In $2n + 6 \cdot 6$, the vascular bundle is larger than in any other trisomic, but the fundamental tissue is about the size as that in the disomic. The same tendency is evident in $2n + 10 \cdot 10$, but to a lesser degree. A directly opposite tendency is shown by $2n + 1 \cdot 1$, in which the area of pith and cortex (as seen in cross section) is not very different from that of the disomic, but the bundles are the smallest of any trisomic.

Within the bundles there are pronounced and characteristic differences in the distribution of cells so that bundle shape, as seen in transverse section, is modified. In $2n + 3 \cdot 3$ and $2n + 11 \cdot 11$, the bundles form an almost continuous but narrow ring of vascular tissue. In $2n + 1 \cdot 1$ and $2n + 17 \cdot 18$, on the other hand, the bundles form a thick ring. Even between different parts of the same tissue, specific differences in development are evident among the trisomics. Thus the relations of cortex to pith are by no means constant. In $2n + 3 \cdot 3$, $2n + 10 \cdot 10$, and $2n + 23 \cdot 24$, the amount of pith is relatively small and in $2n + 17 \cdot 18$ relatively large compared with the disomic.

Increased cell number characterizes all tissues of the pedicel in some trisomics, notably $2n + 6 \cdot 6$, $2n + 9 \cdot 9$, and $2n + 19 \cdot 19$; in others, especially $2n + 3 \cdot 3$, $2n + 7 \cdot 7$, and $2n + 13 \cdot 14$, the cell number is uniformly decreased. All the tissues may be equally affected by modifications of cell size. The cells of $2n + 3 \cdot 3$ are small throughout, approaching the size of the haploid; in $2n + 21 \cdot 22$, they are somewhat larger.

In the vascular system a similar relationship is evident among the different elements. Both vessels and pericycle fibers are larger in $2n + 2 \cdot 2$ than in the disomic, and smaller in $2n + 17 \cdot 17$. In many cases, the differences are in opposite directions as in $2n + 9 \cdot 9$ and $2n + 17 \cdot 18$, where the vessels are large compared with the pericycle fibers, and in $2n + 1 \cdot 1$, $2n + 11 \cdot 11$ and $2n + 15 \cdot 15$, relatively small.

The epidermal cells are large in some of the trisomics and small in others. This trait seems to be closely related to differences in cell size within the cortex between the inner zone and the outer collenchymatous one. Some trisomics, e.g., $2n + 1 \cdot 1$ and $2n + 13 \cdot 14$, which have a relatively large celled collenchymatous layer, also have large epidermal cells while $2n + 5 \cdot 6$, $2n + 14 \cdot 14$, and $2n + 19 \cdot 19$ have small celled collenchyma and small epidermal cells.

This comparative study of pedicel anatomy clearly showed that anatomical characters are influenced by the extra chromosomes in much the same way as the external traits. The data permitted certain conclusions of morphogenetic interest. These rather complex differences could be described in terms of three relatively simple categories—cell size, cell number, and anatomical "pattern" and their localization within the organ. By studying the effect of each of the extra chromosomes or the extra isochromosome arms on these three features, it was possible to conclude that the various elements in the anatomy of the *Datura* pedicel show considerable independence in manner of response to the influence of specific chromosomes.

It is regrettable that these pioneering investigations of Sinnott *et al.* (1934) have not been followed in any other species. Primary trisomic series of several species and tertiary and secondary trisomic series of tomato are now available in uniform genetic backgrounds. Many of the trisomics of these species show well defined morphological differences from each other and from the disomic. An investigation of the anatomical and biochemical basis of the morphological deviations caused by the extra chromosomes should be a fruitful field of inquiry for the morphogeneticist.

Physiology of Trisomics

As far as I am aware, not a single study dealing with the physiology of trisomics of any plant species has been reported in the literature, although there appears to be great potential in employing the extra chromosomal types in physiological investigations. They may prove even more useful in developmental studies.

The extent of physiological imbalance caused by an extra chromosome can be determined by the vigor of the trisomic plant, as the trisomics of most of the species are slower in growth than the normals. Triplo-1 of tomato, for example, is extremely weak and stunted and grows only about 5% as large as the disomic in a given period of time. In contrast, triplo-12 has nearly normal vigor and grows at about 80% of the rate of the normal. The vigor of the tomato trisomics is roughly correlated inversely with the length of the pachytene chromosome. Obviously the longer chromosomes produce greater imbalance due to the greater amount of duplication of genetic materials compared to the shorter ones.

This physiological imbalance also impairs the reproductive ability of the trisomics. In tomato, trisomics for chromosomes 2, 3, 6, and 12 give very poor fruit set even after careful hand pollinations, and when a few fruits are obtained the seed yields are extremely low, generally less than 15% of the normal. This clearly shows that a great proportion of female gametes, otherwise normal in respect to chromosomal constitution, are somehow physiologically too imbalanced to be functional. The same is true of the male gametes. These trisomics are poor pollen producers, and backcrossing instead of selfing is sometimes necessary to produce seeds for stock increases.

Differences in the physiological imbalance are also evident from the differential ability of the various trisomics to adapt to different ecological conditions. Triplo-1, -2, and -9 of tomato, for example, perform much better in greenhouse culture than under field conditions. When grown in the field, triplo-1 produces aborted flower buds, the few which develop anthers are greatly distorted, and no fruit set has ever been obtained. In the greenhouse, on the other hand, it does reasonably well. Poor performance of triplo-2 and -9 under field conditions is suspected to be due to increased susceptibility to wilt diseases which are more severe in the field than the greenhouse. Blossom end rot, a physiological disease associated with upset water relations, is much more severe in the fruits of triplo-3 and -4 than in any of the other types. The symptoms of this disease in these two trisomics are more severe under greenhouse than field conditions (Rick and Barton, 1954).

Biochemistry of Trisomics

Only one study dealing with biochemistry of trisomics has appeared to date. McDaniel and Ramage (1970) studied the seed proteins of disomic and seven primary trisomics of barley by disc electrophoresis. Embryonic and scutellar tissues of ungerminated seeds of the disomic and the trisomics were analyzed for comparing the proteins. Specific protein differences, attributable to specific extra chromosomes, in trisomics were detected by alteration of protein bands in the gels. At least three types of alterations caused by extra chromosomes were noted in the bands. A dosage effect was seen as illustrated by protein band 15 of trisomic 5 (Fig. 6.8). The presence of extra chromosome 5 resulted in production of 140 to 150% of normal quantity of specific protein. The second type of alteration involved production of a novel protein. This was in close proximity to, but not identical with a protein of normal disomic. The presence of a new protein was accompanied by loss of an adjacent protein band, for example, band 11 of trisomic 4 and band 8 of trisomics 1 and 6 (Fig. 6.8). The third type of alteration was the suppression of a protein band completely, such as band 9 in trisomic 6.

The first type of alteration was explained by the authors to be due to alleles with additive gene action, located on the extra chromosome. The second type of alteration was inferred to be due to differences between alleles involving

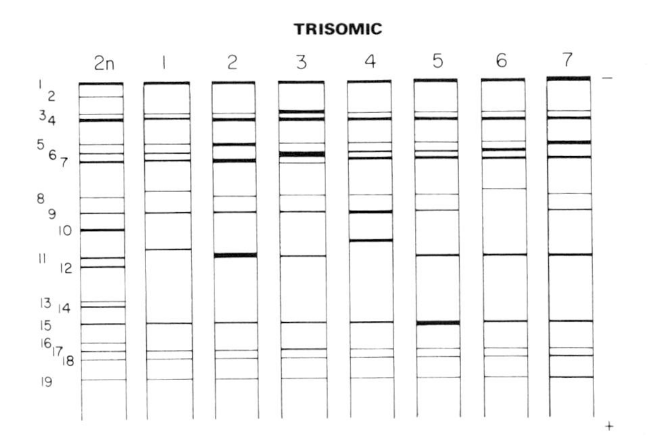

Fig. 6.8. Diagrammatic representation of the seven primary trisomics in the barley variety Betzes as identified by protein electrophoresis. Proteins presumably under the control of genes on each extra chromosome are seen to vary both quantitatively and qualitatively according to the specific trisomy when compared with proteins of the normal disomic. Protein bands which show no change from the disomic patterns in any trisomic are omitted but are shown in the disomic only. (From McDaniel and Ramage, 1970.)

only a single base pair. These differences act through trisomy to specify a protein with electrophoretically detectable differences. The third type of alteration manifested as a loss of specific protein could be explained due to suppression. The quantity of gene product specified by two alleles might be nearly ideal. This balance may be disturbed by the action of a third allele resulting in genic imbalance of an unspecified nature. This disturbance could impair or activate catalytic function of an enzyme associated with anabolism or catabolism of the protein resulting in repression.

The differences in electrophoretic patterns of disomic and the trisomics were so clear and consistent that the individual seeds obtained from a trisomic seed parent could be classified as $2n$ and $2n + 1$. The frequencies of trisomics determined on the basis of cytological and biochemical determinations were identical.

The Basis of the Trisomic Phenotype

Winkler's (1916) studies of polyploids led him to suggest that aneuploids are more likely to differ from the diploids than are euploids. His prediction has been amply verified by the study of aneuploids of several species. In an attempt to explain this phenomenon, Bridges (1922) advanced the theory of gene balance. This theory assumes that the genetic complement of a normal diploid permits the individual to develop and function as an integrated organism. Additions of whole genomes to the basic complement do not affect this gene

TABLE 6.I
List of Species in which Trisomics Have Been Studied

Species	Common name	n	References	Morphological distinctiveness of trisomics
1. *Triticum aestivum*	Wheat	21	Sears (1954); R. Riley (personal communication)	All the twenty-one trisomics, except for those in the homeologous group 2, are normal in appearance and indistinguishable from disomics. Those for the homologous group 2 are narrow leaved, small stemmed, and have longer awns.
2. *Avena sativa*	Oats	21	R. Riley (personal communication); R. McGinnis (personal communication)	Fifteen trisomic lines established by Riley and many trisomic individuals examined by McGinnis are morphologically indistinguishable from each other and from the disomics
3. *Nicotiana tabacum*	Tobacco	24	Clausen and Goodspeed (1924); D. R. Cameron (personal communication)	Only three trisomics, triplo-C, -F, and -N, differ slightly in flower size, but the rest are indistinct from each other and from the disomics.
4. *Gossypium hirsutum*	Cotton	26	Endrizzi, *et al.* (1963); Brown (1966); Kohel (1966); M. S. Brown (personal communication)	Some of the trisomics identified to date are morphologically quite distinct but others are less distinct.
5. *Zea mays*	Maize	10	McClintock (1929a); McClintock and Hill (1931); Rhoades and McClintock (1935)	Only two trisomics, triplo-5 and triplo-3, can be identified morphologically. The others although somewhat slower in growth cannot be distinguished from each other as well as from the disomics.
6. *Collinsia heterophylla*		7	Dhillon and Garber (1960); Garber (1964)	Trisomics induced by colchicine treatment are completely identical with disomics. Trisomics obtained from the progeny of the triploid are somewhat less vigorous, more slender, and shorter than disomics. The differences, however, are not large enough to enable one to identify different trisomics from each other and from the disomics.

7. *Clarkia unguiculata*		9	Vasek (1956, 1963)	Different trisomic lines are impossible to distinguish from each other and from disomics.
8. *Datura stramonium*	Datura	12	Blakeslee and Avery (1919); Blakeslee (1934); Avery *et al.* (1959)	All the trisomics are morphologically distinct from each other as well as from the disomics.
9. *Lycopersicon esculentum*	Tomato	12	J. Lesley (1928, 1932); Rick and Barton (1954)	All the trisomics are morphologically distinct from each other and from the disomics.
10. *Nicotiana sylvestris*	Tobacco	12	Goodspeed and Avery (1939, 1941)	All the trisomics are very distinct morphologically.
11. *Antirrhinum majus*	Snapdragon	8	Stubbe (1934); Propach (1935); Rudorf-Lauritzen (1958); Sampson *et al.* (1961)	All the eight primary trisomics are very distinct morphologically.
12. *Matthiola incana*	Stock	7	Frost and Mann (1924); Frost (1927); Prakken (1942)	At least five primary trisomics have been studied and all are very distinct.
13. *Oenothera lamarckiana*		7	Gates (1923); de Vries and Boedijn (1923; Emerson (1936)	All trisomics are very distinct phenotypically.
14. *Oenothera blandina*		7	Catcheside (1954)	All trisomics are very distinct morphologically.
15. *Crepis capillaris*		6	Babcock and Navashin (1930)	Five primary trisomics have been obtained and are morphologically distinct.
16. *Crepis tectorum*		4	Gerassimowa (1940)	All the four trisomics are very distinct.
17. *Lotus pedunculatus*		6	Chen and Grant (1968a)	Five primary trisomics have been obtained and all are very distinct.
18. *Spinacia oleracia*	Spinach	6	Tabushi (1958); Janick *et al.* (1959); Ellis and Janick (1960)	All the primary trisomics are very distinct from each other.
19. *Arabidopsis thaliana*		5	Steinitz-Sears (1963)	All the primary trisomics are morphologically distinct.
20. *Beta vulgaris*	Sugar beets	9	Levan (1942)	Five known trisomics are very distinct.
21. *Petunia*		7	Levan (1937)	All the trisomics are very distinct.

TABLE 6.I (Continued)

Species	Common name	n	References	Morphological distinctiveness of trisomics
22. *Capsicum annuum*	Chilie	12	Pal and Ramanujam (1940); Pochard (1968)	All the trisomics are very distinct.
23. *Hordeum spontaneum*	Barley	7	Tsuchiya (1958, 1960);	All the trisomics are very distinct.
24. *Secale cereale*	Rye	7	Kamanoi and Jenkins (1962)	All the primary trisomics are very distinct in appearance.
25. *Oryza sativa*	Rice	12	Sen (1965); Hu (1968)	All the primary trisomics are very distinct in appearance.
26. *Sorghum vulgare*	Sorghum	10	Schertz (1966); Poon and Wu (1967); Linn and Ross (1969a)	All the ten primary trisomics have been isolated and all are very distinct.
27. *Pennisetum typhoides*	Bajra	7	Gill *et al.* (1970a)	All the primary trisomics are morphologically distinct.
28. *Avena strigosa*	Oats	7	Dyck (1964)	Six of the seven primary trisomics are known and are very distinct.
29. *Verbena tenuisecta*	Moss	5	Arora and Khoshoo (1969)	Four of the five trisomics have been obtained and are morphologically well defined.

balance or very little if at all. However, when individual chromosomes are added to the normal complement of the species, the gene balance is greatly disturbed. Imbalance is reflected in physiological disturbance and morphological and developmental deviations. The addition of a single short chromosome in tomato produces less imbalance than the addition of a long chromosome or two or three chromosomes. Each chromosome affects the anatomy, physiology, and morphology of the plant in a distinctive way reflecting the differential gene contents of the different chromosomes.

Up till now we have been treating the subject of trisomic morphology as if the trisomics of all the species were morphologically well defined and distinct from each other. This, however, is not the case. The extra chromosomal types of some species are as vigorous and fertile as the normals and morphologically indistinct from them as well as from each other. A survey of literature dealing with the morphology of trisomics of twenty-nine species belonging to twelve families of plants is summarized in Table 6.I. On the basis of trisomic morphology these species can be grouped into three categories:

1. Species in the first group have trisomics that are morphologically distinct and can be distinguished from each other phenotypically. These species are *Datura stramonium, Lycopersicon esculentum, Nicotiana sylvestris, Petunia, Capsicum annuum, Antirrhinum majus, Matthiola incana, Oenothera lamarckiana, Oenothera blandina, Crepis capillaris, Crepis tectorum, Lotus pedunculatus, Spinacia oleracia, Arabidopsis thaliana, Beta vulgaris, Secale cereale, Hordeum vulgare, Oryza sativa, Sorghum vulgare, Pennisetum typhoides, Avena strigosa,* and *Verbena tenuisecta.*

2. The trisomics of the second category cannot be distinguished from each other or from the disomic sibs by phenotype. Species in this group are *Triticum aestivum, Avena sativa, Nicotiana tabacum, Clarkia unguiculata,* and *Collinsia heterophylla.*

3. The species in the third category are somewhere in the middle of the extremes of the other two groups. Two of the trisomics of *Zea mays,* for example, are very distinct, three are less so but can be identified with some difficulty; the remaining five are phenotypically identical to the disomics. The trisomics of *Gossypium hirsutum* also fall into this group; some being phenotypically distinct, some less distinct, and others identical with the disomics.

The species in the first group are basic diploids with few duplications in their genomes. The imbalance caused by the presence of an extra chromosome is expressed in phenotypic deviations, lowered vigor, and reduced fertility.

In the second group, with the exception of *Clarkia unguiculata* and *Collinsia heterophylla,* the species are polyploids which carry many duplications in their genomes, evident by their tolerance for monosomy, and in the case of *Triticum aestivum* and *Avena sativa,* for nullisomy. Further duplication by the addition of a single chromosome alters the gene balance but little and there is no effect on the phenotype, vigor, or fertility of the trisomic.

However, the position of *Clarkia unguiculata* and *Collinsia heterophylla* in this group is anomalous. Both are basic diploids but their trisomics are vigorous, fertile, and phenotypically indistinguishable from each other and from the disomics. In *Clarkia unguiculata,* at least, plants with up to seven extra chromosomes have been obtained which also have normal vigor, fertility, and phenotype. The trisomics of these species have defied efforts to identify the extra chromosomes. Vasek (1961) and Garber (1964) have pointed out similarities in the behavior of the extra chromosomes of trisomic lines and the naturally occurring supernumerary chromosomes. Because of incomplete understanding of the true nature of the trisomics, these two species will be disregarded for the purposes of present discussion.

One species belonging to the third group, *Gossypium hirsutum,* is also an allopolyploid (Stephens, 1947), but a greater degree of diploidization has taken place in this species than in *Triticum aestivum, Avena sativa,* or *Nicotiana tabacum.* As discussed by Stebbins (1950), the duplication of genetic material in this species is considerably less than in the other three polyploid species mentioned. Because of this more diploidized condition, the tolerance limits for the monosomy are low (Chapter 8). When present in trisomic condition, some of the chromosomes produce enough imbalance so that the trisomics are morphologically distinct. Other chromosomes produce less imbalance and their trisomics are phenotypically indistinct. The trisomics of *Zea mays* behave like those of *Gossypium hirsutum* with regard to their phenotype. *Zea mays,* generally considered a basic diploid judging from the trisomic phenotypes, may have some duplications in its genome.

The amount of imbalance caused by the extra chromosomes can also be indicated, as discussed earlier in this book, by the transmission rates. In the diploid species, a certain proportion of $n + 1$ gametes and $2n + 1$ zygotes produced by a trisomic abort due to the imbalance caused by the extra chromosomes. This abortion is the major cause of the reduction in transmission frequency of the extra chromosome below the expected 50% level. In general, the average transmission rates of all the chromosomes of the complement of diploid species rarely exceed 25%, thereby indicating strong imbalance caused by the extra chromosomes. Data on the transmission rates of the trisomics of polyploid species though fragmentary indicate that $n + 1$ gametes and $2n + 1$ zygotes of those species are as viable as n gametes and $2n$ zygotes, and the extra chromosomes produce no imbalance. This is clear from the normal seed fertility of trisomics *Triticum aestivum* (Sears, 1954). As discussed in Chapter 4, *Zea mays* resembles the polyploid species in this respect rather than the diploids. No imbalance is caused in $n + 1$ gametes or $2n + 1$ zygotes by the extra chromosome, and the seed fertility of the trisomics is normal (Einset, 1943).

Again, because of their genetic architecture, the polyploids tolerate much higher numbers of extra chromosomes than the diploid species. From a cross of $3n \times 2n$ of *Nicotiana tabacum* ($2n = 48$), East (1933) obtained plants with chromosome numbers varying from 50 to 59. The largest class had 54 chromosomes, 6 more than the normal $2n$ complement. Kihara (1924) obtained plants with chromosome numbers ranging from 28 to 42 in the F_2 of a cross of hexaploid ($2n = 42$) and tetraploid ($2n = 28$) wheats. No data on the chromosome numbers of the progenies of triploids of *Avena sativa* and *Gossypium hirsutum,* the other two polyploid crop species, have been reported, but plants of *Gossypium hirsutum* with as many as seven to eight extra chromosomes have been reported by Brown (1966). Several other species suspected of being polyploid in origin, such as apples (Nebel, 1933), *Populus* (Johnsson, 1942), *Betula* (Johnsson, 1946), *Hyacinthus* (Darlington and Mather, 1944), and *Humulus* (Haunold, 1970), also tolerate higher numbers of extra chromosomes.

The tolerance limits for extra chromosomes in diploid species are very narrow. The gametes and zygotes with more than two to three extra chromosomes produced by the triploids of these species abort because of the imbalance. Triploid progenies mainly consist of $2n$, $2n + 1$, $2n + 2$, and $2n + 3$ individuals. For example, a progeny of triploid tomato consisted of $2n$ (37.9%), $2n + 1$ (42.7%), $2n + 2$ (16.4%), and $2n + 3$ (0.9%) plants only (Table 2.III). A similar distribution of chromosome numbers in the progenies of triploids has been obtained in *Oenothera biennis* (Van Overeem, 1921), *Frageria* sp. (Yarnell, 1931), *Nicotiana sylvestris* (Goodspeed and Avery, 1939), *Beta vulgaris* (Levan, 1942), *Sorghum vulgare* (Price and Ross, 1955, 1957), barley (Tsuchiya, 1960), *Secale cereale* (Kamanoi and Jenkins, 1962), *Antirrhinum majus* (Sampson *et al.*, 1961), *Oryza sativa* (Katayama, 1963; Watanabe *et al.*, 1969), *Lotus penduncularus* (Chen and Grant, 1968a), and *Pennisetum typhoides* (Gill *et al.*, 1970a).

However, *Zea mays,* in contrast to other diploids, has a very high tolerance for extra chromosomes. From a $3n \times 2n$ cross, Punyasingh (1947) obtained plants with chromosome numbers varying from 20 ($2n$) to 28 ($2n + 8$). Moreover, plants with 24, 25, and 26 chromosomes were as frequent as those with 21, 22, and 23.

From the foregoing discussion it is evident that the diploid species have low tolerance for aneuploidy. The number of extra chromosomes tolerated by them is low, the transmission rates of the extra chromosomes to the next generation are poor, and the trisomics have altered morphology, physiology, and anatomy, lowered fertility, and reduced vigor. The polyploid species, on the other hand, have a high tolerance for aneuploidy. They can tolerate much higher numbers of extra chromosomes, the transmission rates of the extra chromosomes to the next generation are almost normal, and the trisomics have normal morphology, fertility, and vigor. *Zea mays,* generally considered a basic diploid, has wider tolerance for extra chromosomes than other diploid species and in this respect resembles the polyploid species.

7

Sources and Cytology of Monosomics and Nullisomics

The monosomics and nullisomics have proved especially useful for the cytogenetic analysis of polyploid species although their occurrence is by no means confined to them. As we shall see later in this chapter, a few monosomics of diploid species have been reported, but their usefulness is limited because the monosomic condition is not transmitted to the progeny. In the first part of this chapter the sources and cytology of the monosomics will be reviewed. The second part will deal with the sources and cytology of the nullisomics.

Sources of Monosomics

The monosomics like trisomics can be obtained from several sources. The principal ones are: (1) normal disomics, by spontaneous occurrence or by treatment with physical and chemical agents; (2) asynaptic disomics and aneuploids; (3) polyploids, e.g., haploids and triploids; (4) intervarietal and interspecific crosses; (5) translocation heterozygotes; (6) trisomics; and (7) monosomics.

Normal Disomics

Monosomics appear spontaneously in the progenies of normal disomics. In wheat, a spontaneous monosomic with a speltoid phenotype was isolated by Nilsson-Ehle as early as 1904. However, its monosomic nature was not established until much later by Winge (1924) when he found that certain speltoids had forty-one chromosomes instead of forty-two. Since that time several workers have isolated and studied speltoid monosomics of spontaneous origin.

In the cultivated oat, *Avena sativa* ($2n = 6x = 42$), off-type plants of spontaneous origin with fatuoid phenotype have been investigated since the original report of Buchman (1857). Nilsson-Ehle (1907, 1911) tried to explain the origin of fatuoid forms on the basis of the mutation theory. Goulden's (1926) report of the association of the fatuoid phenotype with monosomy for one chromosome was confirmed by Huskins (1927) and Nishiyama (1931, 1933). Recently, efforts to isolate oat monosomics of spontaneous origin by systematic screening of varietal populations have been renewed. Riley and Kimber (1961) screened 631 seedlings of variety Sun II by root tip chromosome counts and obtained seven monosomics. They were the first to suggest the possibility of assembling a complete monosomic series by cytologically screening large populations of this species. McGinnis (1962b) determined chromosome numbers of 4023 seedlings of variety Garry and found seventeen monosomics as well as a few nullisomics and trisomics. Hacker and Riley (1963) obtained forty monosomics from a total population of 3453 seedlings of Sun II. One monosomic line of spontaneous origin was isolated by Lafever and Patterson (1964a).

One spontaneous monosomic of *Nicotiana tabacum* was reported by Clausen and Goodspeed (1926a), and Avery (1929) obtained a monosomic of *Nicotiana alata* which also appeared spontaneously. Several monosomics of spontaneous origin have been reported in *Gossypium hirsutum* by Brown and Endrizzi (1964). According to Bergner *et al.* (1940), during an 18-year period (1920–1938), fifty-five $2n - 1$ plants and chimaeras appeared spontaneously among 2,000,000 plants of *Datura stramonium*.

The results of several workers who screened varietal populations of oats indicate that a complete monosomic series may be produced in any oat variety if sufficiently large populations are screened. The series established in this manner offers the obvious advantage of being homogeneous in background. The disadvantage is that the frequency of monosomics is low and considerable time and labor are required to establish a complete series. Like trisomics, the spontaneous monosomics probably result from $n - 1$ gametes produced occasionally by normal individuals as a result of nondisjunction. The condition favoring nondisjunction, e.g., the occurrence of a few univalents at metaphase I of meiosis, has been reported by Howard (1948) in *Avena sativa*, Clausen (1931) in *Nicotiana tabacum*, and Khan (1962), Love (1941), and Riley and Kimber (1961) in *Triticum aestivum*. The frequency of the meiotic irregularity may be somewhat higher in the varieties descended from species crosses (Powers, 1932; Semeniuk, 1947).

Various physical and chemical agents can be employed in obtaining monosomics. Bergner *et al.* (1940) obtained seven $2n - 1$ plants among the 2000 plants grown from colchicine-treated seeds of *Datura stramonium*. Monosomics for at least four different chromosomes of the complement were

TABLE 7.I
Monosomics Induced by Irradiation Treatment in Different Species

Species	Authority	Irradiation treatment	Part irradiated	Remarks
Avena sativa	Costa-Rodriguez (1954)	X rays	Panicles	20 out of 279 plants or 7.2% were $2n - 1$
	Andrews and McGinnis (1964)	X rays	Panicles	18 out of 686 plants were $2n - 1$
	Chang and Sadanaga (1964b)	X rays	Panicles	6, $2n - 1$ plants
	Rajhathy and Dyck (1964)	X rays	Panicles	34 out of 200 plants were $2n - 1$
	von der Schulenburg (1965)	X rays	Seeds	13.6% of the progeny was aneuploid
	Maneephong and Sadanaga (1967)	X rays	Panicles	4.9% of the progeny was monosomic
Avena barbata	Andrews and McGinnis (1964)	X rays	Panicles	8 Plants out of 133 were $2n - 1$
Avena strigosa	Andrews and McGinnis (1964)	X rays	Panicles	2 Plants out of 484 were monosomic
Triticum aestivum	Sears (1954)	X rays	Pollen	3 Monosomics
Gossypium hirsutum	Jagathesan and Swaminathan (1961)	X rays	Seed	1 Monosomic
	Brown and Endrizzi (1964)	X rays		6 Monosomics
		Gamma rays		5 Monosomics
		Neutrons		2 Monosomics
	Galen and Endrizzi (1968)	Gamma rays	Pollen	8.9% of the progeny at 400 r and 3.7% at 1200 r levels was monosomic
Petunia	Rick (1943)	X rays	Pollen	2 Tertiary monosomics
Lycopersicon esculentum	Rick and Khush (1961) Khush and Rick (1966a, b, 1968a)	X rays and fast neutrons	Pollen	2 Primary monosomics and more than 25 tertiary monosomics
Zea mays	Baker and Morgan (1966)	X rays	Pollen	3 Monosomics for chromosome 1 and 2 tertiary monosomics

represented in the sample. Among the progeny raised from colchicine-treated germinating seeds of *Nicotiana langsdorffi* (Smith, 1943), one monosomic appeared. Swaminathan and Natarajan (1959) induced monosomy in hexaploid wheat, *Triticum aestivum,* by treating seeds with vegetable oils. Kihara and Tsunewaki (1962) obtained monosomics of tetraploid wheat, *Triticum dicoccum,* by treating florets with N_2O. Monosomics of *Avena sativa* were induced by Andrews and McGinnis (1964) by treating the florets with Myleran and EOC, and by von der Schulenburg (1965) through EMS treatment of seeds.

The application of low doses of X rays, gamma rays, and fast neutrons applied to seeds, flower organs, or pollen grains is known to generate monosomics in the X_1 or later generations. Rick (1943) was the first to obtain tertiary monosomics of Petunia by X-ray treatment of pollen. During the last 16 years, several workers have made concerted efforts to establish monosomic series in oats. Monosomics have been obtained in several other species by radiation treatments (Table 7.I).

The exact mechanism by which radiation treatment induces monosomy is not fully understood. Perhaps a chromosome is broken or damaged in the centromere region so that it is unable to show normal centromeric activity and is lost in the cytoplasm (Khush and Rick, 1966b). Costa-Rodriguez (1954) assumed that the radiation causes a break in the two chromatids, with the subsequent formation of a dicentric chromosome. The dicentric presumably goes through a breakage–fusion–bridge cycle during early cell divisions of the embryo and finally is lost.

Sometimes two nonhomologous chromosomes break in the centromere regions as a consequence of irradiation, and two arms of different chromosomes join producing a tertiary chromosome with a functional centromere and the other two arms are lost. This gives rise to a tertiary monosomic (Khush and Rick, 1966b). If a recessive marker for the chromosome is available, monosomics for a specific chromosome may be produced by design using X-ray treatment. Plants homozygous for the recessive marker can be pollinated with X-rayed pollen from plants homozygous for the normal allele. Some or all the pseudodominant plants in the progenies would be monosomic if the loss for that chromosome is tolerated.

Asynaptic Disomics and Aneuploids

In the asynaptic genotypes many univalents are present at metaphase I of meiosis. Because of irregular segregation of univalents imbalanced gametes are produced. The $n-1$ gametes are functional in the polyploid species, and when such gametes are fertilized by the n gametes, monosomic progeny is produced. Clausen and Cameron (1944) employed an asynaptic genotype of *Nicotiana tabacum* called *pale-sterile* for isolating monosomics. Many individuals in the progeny were double or triple monosomics or monosomic trisomic combinations. These complex types could be readily reduced to their elementary components by further crosses to the normal.

Using *pale-sterile,* Clausen and Cameron isolated by design specific monosomics. For example, a cross of *pale-sterile* carmine with normal white tobacco produced a small percentage of white offspring which were obviously monosomic for the chromosome carrying the white locus. This method has been used in a number of instances in *Nicotiana tabacum* (Clausen and Cameron, 1944).

Monosomics were isolated from an asynaptic strain of *Gossypium hirsutum* by Brown and Endrizzi (1964), and the progeny of an asynaptic derivative from a cross of two tetraploid species of *Avena* produced three monosomics (Thomas and Rajhathy, 1966).

Some aneuploids, partially asynaptic due to the imbalanced chromosome number, constantly yield in their progenies monosomics for different chromosomes. Thus, Sears (1954) isolated monosomics for seventeen different chromosomes in the progenies of partially asynaptic nulli-3B. Clausen and Cameron (1944) isolated several monosomics in the progenies of already established monosomics which had mildly asynaptic behavior. The fatuoid nullisomic of oats is asynaptic (Huskins, 1927) but is completely unfruitful. Hence, this source could not be utilized for obtaining monosomics of this species.

Polyploids

Among the various types of polyploids, haploids have been prolific sources of monosomics. Sears (1939, 1944, 1954) isolated monosomics for nineteen different chromosomes from the progenies of haploids of wheat. Endrizzi (1966) found three primary and one tertiary monosomic in a progeny of 121 plants of haploids of *Gossypium barbadense*. Rao and Stokes (1963) found many monosomic plants (111 or 43.6% of the progeny) from the cross of haploid and disomic *Nicotiana tabacum*. Nishiyama *et al.* (1968) isolated nine monosomic lines from the progenies of an *Avena byzantina* haploid. Generally there are double monosomic, triple monosomic, and mono-trisomic plants in the progenies of the haploids. These complex types can be reduced to their elementary components by further backcrosses to the normal disomic individuals. The various modes of origin of $n - 1$ gametes in haploids have been discussed in Chapter 2 (see Fig. 2.1).

Triploids occasionally produce $n - 1$ gametes which give rise to monosomic progeny when fertilized by n gametes. Three monosomic lines were isolated by Nishiyama *et al.* (1968) from the progeny of a triploid of *Avena byzantina*.

Intervarietal and Interspecific Crosses

Meiotic irregularities of different types occur in the intervarietal crosses of several species. The most commonly observed abnormality is the occurrence of univalents at metaphase I. Joshi and Howard (1954) observed a very high frequency of microsporocytes with univalents in the crosses of spring and winter varieties of *Avena sativa*. Out of twenty-nine F_2 plants of such an intervarietal cross, one was monosomic. Monosomic plants from intervarietal crosses of oats have also been isolated by Tagenkamp and Finkner (1954), McGinnis and Taylor (1961), and McGinnis and Andrews (1962). Brown and Endrizzi (1964) obtained five monosomics of *Gossypium hirsutum* from intervarietal hybrids.

Monosomics may also be obtained by interspecific hybridization and repeated backcrossing of selected F_2 or backcross derivatives to the species in which monosomics are to be established. Olmo (1935) obtained F_1 hybrids between Nicotiana tabacum and *Nicotiana sylvestris* and *Nicotiana tabacum* and *Nicotiana tomentosa*. *Nicotiana sylvestris* ($2n = 24$) and *Nicotiana tomentosa* ($2n = 24$) are the two putative diploid parents of the allopolyploid *Nicotiana tabacum* ($2n = 48$). The two hybrids were backcrossed as females to *Nicotiana tabacum*. The progenies obtained comprised highly variable derivatives of three general classes: (1) almost completely sterile, aberrant plants; (2) plants which resembled *tabacum* but were highly sterile and gave no progeny upon selfing; and (3) *tabacum*-like derivatives that were somewhat fertile. Cytological examination revealed that some plants belonging to the third category were monosomics. These monosomics were repeatedly backcrossed to *N. tabacum* to eliminate the genetic heterozygosity characteristic of the interspecific hybrids. Some of the partially fertile derivatives with $2n$ chromosomes, but with several univalents, were also backcrossed to *N. tabacum* to obtain monosomics for other chromosomes.

Thus, it is theoretically possible to obtain monosomics for the twelve chromosomes of the *tomentosa* genome from backcrosses of *N. sylvestris* × *N. tabacum* F_1 and the monosomics for the twelve chromosomes of the *sylvestris* genome from the backcrosses of *N. tomentosa* × *N. tabacum* F_1. Four of the former, mono-A, -B,-C, and -F, and three of the latter, mono-N, -Q, and -R, were obtained by Olmo (1935). Lammerts (1932) obtained seven monosomics of another tetraploid species of this genus, *Nicotiana rustica*, by backcrossing the hybrid between *N. rustica* and a diploid species, *N. paniculata*, to *N. rustica*. Similarly, Brown and Endrizzi (1964) isolated seven monosomic plants from the species hybrids of cotton and Rajhathy and Dyck (1964) from species hybrids of *Avena*.

Monosomics were also found in the derivatives of interspecific crosses of wheat by Love (1940). The crosses between hexaploid *Triticum aestivum* and tetraploid *Triticum durum* were made in order to transfer the rust resistance of *T. durum* to *T. aestivum*. The F_1 hybrids were allowed to self-pollinate, and selection was made for rust-resistant *aestivum*-like individuals in each generation. A plant breeder's sample from fifty different lines consisting of 336 plants was examined cytologically and 98 of the plants were found to have the monosomic chromosome number of 41. Matsumura (1940) established monosomics of the D genome by backcrossing *T. aestivum* (21 II) × *T. dicoccum* (14 II) to *T. aestivum*.

These examples show how a monosomic series may be established from interspecific crosses in a species with a high chromosome number. If monosomic series are already available in a higher polyploid, the monosomics of a lower polyploid may also be obtained by interspecific hybridization. Thus in order to obtain monosomics of *Triticum durum* ($2n = 28$, genome formula AABB), fourteen monosomic lines of *Triticum aestivum* ($2n = 42$, genome formula AABBDD) deficient for a chromosome of the A or B genome were crossed with *Triticum durum* by Mochizuki (1968a,b). Deficient F_1 plants were backcrossed with *T. durum*. The chromosome number of BC_1 individuals should theoretically vary from 27 to 35. However, no plants with twenty-seven chromosomes were found in the BC_1 but about 5% of the progeny had twenty-eight chromosomes. The plants with twenty-eight chromosomes were classified into two groups. About 30 to 50% of them showed 13 II + 2 I, and the remainder, 14 II. The former were used as the female parent for the next backcross. Of the two univalent chromosomes of such female parents, one came from *T. durum* corresponding to the monosomic chromosome of *T. aestivum*, and the other belonged to the D genome of *T. aestivum*. In the BC_2 families, monosomic plants with 13 II + 1 I were obtained. Mochizuki (1968b, 1971) was able to establish monosomics for all the fourteen chromosomes of tetraploid wheat, *T. durum*. The procedure followed is outlined in Scheme 7.1.

Triticum aestivum (20 II + 1 I) × *Triticum durum* (14 II)
↓
F_1 13 II + 1 I (A or B) + 7 I (D) × *T. durum* (14 II)
↓
BC_1 13 II + 1 I (A or B) + 1 I (D) × *T. durum* (14 II)
↓
BC_2 13 II + 1 I (A or B)

Scheme 7.1

So far the procedures for obtaining monosomics from crosses between species with different ploidy levels have been discussed. Monosomics may also be obtained from crosses of species with similar chromosome numbers. Because of reduced homology, some univalents are present at meiosis in the F_1 hybrids, and the irregular segregation gives rise to $n - 1$ gametes and $2n - 1$ zygotes. Philp (1935) obtained one monosomic from a cross of *Avena sativa gigantica*

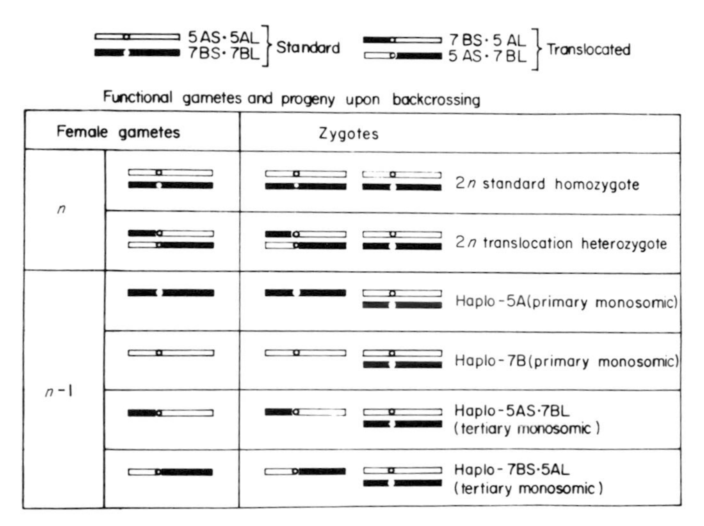

Fig. 7.1. Functional gametes and the progeny produced upon backcrossing a translocation heterozygote to normal disomic parent.

($2n = 42$) × *Avena fatua* ($2n = 42$). Another monosomic was obtained by Philp (1938) from the reciprocal cross of these two species.

It may be seen that monosomics have been obtained in several polyploid species by the interspecific hybridization technique. The main disadvantage of this technique is that the considerable heterozygosity set up in such crosses confuses the recognition and establishment of different types. Nevertheless, recurrent backcrossing eventually eliminates the troublesome segregation due to heterozygosity.

Translocation Heterozygotes

Translocation heterozygotes produce $n - 1$ and $n + 1$ gametes by nondisjunction. In the polyploid species the $n - 1$ gametes can function and yield monosomic progeny. Four different monosomics would theoretically be expected in the progeny of a translocation heterozygote, two of these, primary monosomics, and the other two, tertiary. Thus upon backcrossing to the normal $2n$, the translocation heterozygotes of wheat with the 7BS·5AL and 5AS·7BL chromosome arrangement would yield $2n$ − 5A, $2n$ − 7B, $2n$ − 7BS·5AL, and $2n$ − 5AS·7BL as shown in Fig. 7.1. No systematic search has been made for monosomics in the progenies of translocation heterozygotes of polyploid species, although Menzel and Brown (1952), Brown and Endrizzi (1964), and Brown (1966) report having obtained monosomics from the progenies of cotton plants with translocation multivalents.

Trisomics

As discussed elsewhere in this book, the three homologous chromosomes of a trisomic may occasionally fail to pair and either pass to the same pole or get lost and produce $n - 1$ gametes and $2n - 1$ progeny. Thus monosomics have been obtained among the progenies of trisomics, of maize (Einset, 1943), *Nicotiana tabacum* (Olmo, 1936; Clausen and Cameron, 1944), wheat (Sears, 1954), and cotton (Brown and Endrizzi, 1964).

Monosomics

Most of the time, the monosomics in the progeny of a monosomic are of the same type, but occasionally monosomics for other chromosomes occur. This phenomenon termed "univalent shift" by Person (1956) occurs because more than one univalent is present in some meiotic cells of the monosomic with the consequent production of $n - 1$ gametes in which the missing chromosome differs from the original monosome. Additional monosomics were isolated from progenies of existing monosomics by Clausen and Cameron (1944) in *Nicotiana tabacum* and by Brown and Endrizzi (1964) in *Gossypium hirsutum*.

Cytology of Monosomics

In a monosomic plant, $n - 1$ bivalents are usually formed during meiotic prophase leaving the monosome as a univalent. In rare instances, a trivalent

TABLE 7.II
Chromosome Pairing at Metaphase I in Eight Monosomics of *Triticum aestivum*[a]

Monosomic	No. cells examined	Metaphase I pairing		
		20 II + 1 I	19 II + 3 I	18 II + 5 I
1B	198	194	4	0
2A	77	72	5	0
3D	215	210	4	1
4A	391	385	6	0
4B	96	88	5	3
4D	131	129	2	0
5A	165	161	4	0
6D	169	168	1	0
Total	1442	1407	31	4
Percentage		97.6	2.1	0.3

[a]From Morrison, 1953.

may be formed when the univalent pairs with two homoeologous chromosomes. Olmo (1936) reported such trivalent formation in 1.5 to 2.0% of the microsporocytes of mono-N of *Nicotiana tabacum*. The other variation in chromosome pairing in the monosomics occasionally observed is the occurrence of more than one univalent. Three or five univalents may be present in some cases (Table 7.II), and these obviously result from pairing failure in one or two pairs of homologs. To be sure, one also finds occasional cells with univalents in normal disomic varieties, but the frequency of this abnormality is somewhat higher in most of the monosomics. Like trisomy, the monosomy somehow creates minor disturbances in normal meiotic pairing. According to Clausen and Cameron (1944), three univalents were observed in 5% of the microsporocytes of mono-0, 10% in mono-H and mono-I, 15% in mono-Q, about 25% in mono-U, and an even higher percentage in mono-Z. The effect of the univalent on the behavior of the other chromosomes at meiosis may also be influenced by the genotypic background. Person (1956) presented evidence to show that greater pairing disturbance occurs when the monosomic state is in the early generations of intervarietal crosses rather than in more nearly homozygous lines. Occurrence of more than one univalent in wheat monosomics has also been reported by Khan (1962) and Sasaki *et al.* (1963).

The behavior of the univalent chromosome during meiosis is of interest due to its bearing on the transmission rates. The position of the univalent during prophase in relation to paired chromosomes seems to be random (Fig. 7.2). However, at later stages the univalent shows characteristic retarded movements which first become evident at metaphase I. While all the bivalents move to the metaphase plate, the univalent lies away from the plate (Fig. 7.3). When the bivalents start undergoing disjunction, the univalent starts to move to the

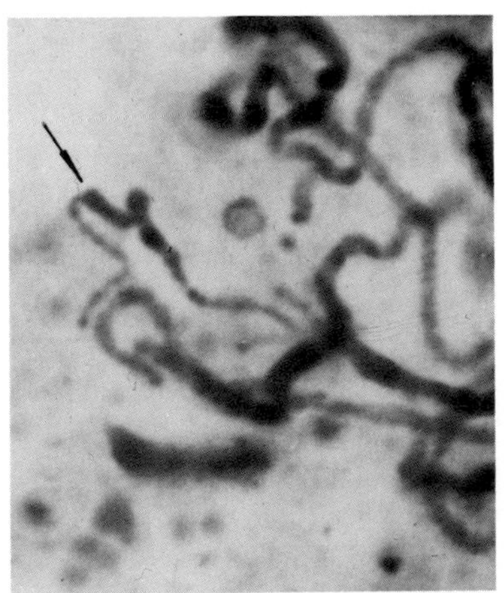

Fig. 7.2. Univalent chromosome 11 (indicated by arrow) at pachytene in mono-11 of tomato. (From Rick and Khush, 1961.)

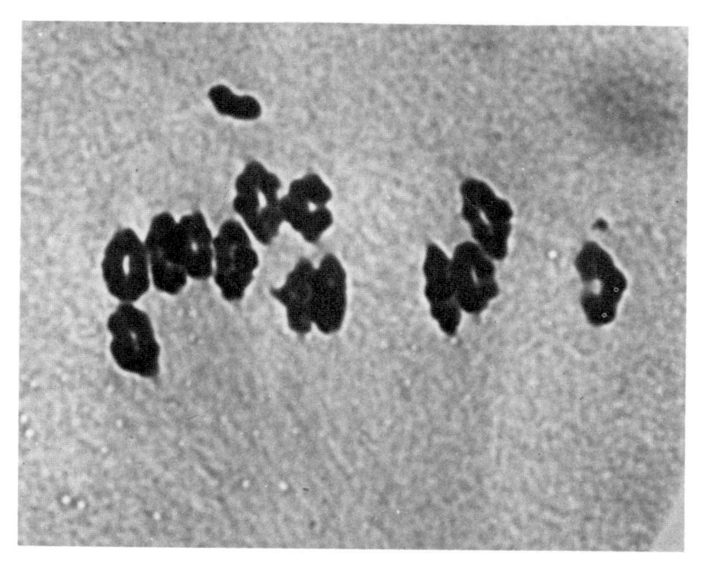

Fig. 7.3. Metaphase I in mono-1A of *Triticum durum* showing thirteen bivalents and a univalent lying away from the metaphase I plate. (From Mochizuki, 1968b.)

equatorial plate. Usually no longitudinal split of the univalent is visible during early anaphase of first division.

The later behavior of the univalent is quite variable: (1) it may pass to one of the poles undivided where it may or may not be included in the telophase I nucleus; (2) it may divide equationally with sister chromatids going to opposite poles (Fig. 7.4) or the same pole; or (3) it may misdivide.

The behavior of the univalent in mono-5A (IX) of wheat, variety Chinese Spring, was studied extensively by Sears (1952a). The univalent divided at the first division in 96% of 147 microsporocytes studied. Sanchez-Monge and MacKey (1948) also reported the division of the univalent at the first division in 97% of the microsporocytes in mono-5A of a Swedish variety of wheat. However, Morrison (1953) found that the univalent divides in slightly more than 50% of the cases in most of the monosomics. The frequency of misdivision was reported to be 39.7% by Sears in mono-5A of Chinese Spring but only 1.7% by Sanchez-Monge and MacKey in their material. The frequency of misdivision was 12% in the eleven monosomics studied by Morrison (1953). Apparently the genetic architecture of a variety influences the degree of misdivision of the univalent.

The misdivisions observed were grouped by Sears (1952a) into three classes according to the number and type of arms going to different poles: (a) one normal chromatid goes to one pole and the two arms of the other chromatid

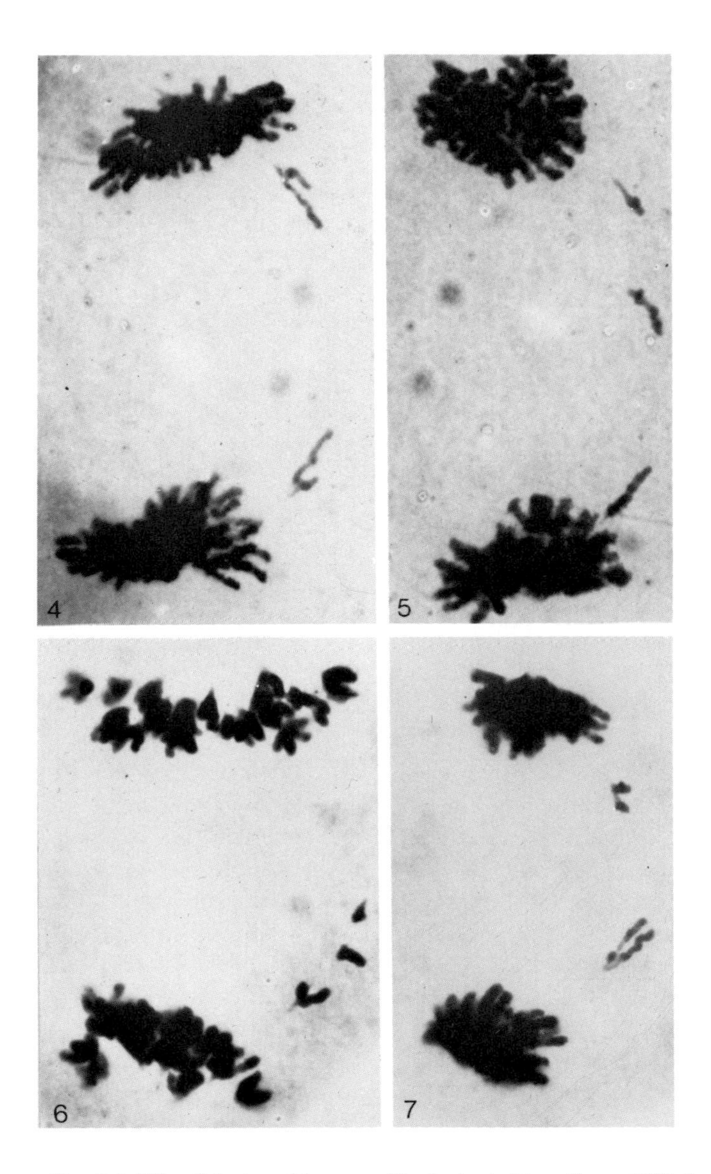

Fig. 7.4–7.7. Telophase I in mono-5A of wheat. (From Sears, 1952a.)
Fig. 7.4. Normal division of univalent.
Fig. 7.5. Two arms going separately to upper pole.
Fig. 7.6. Long arm acentric on plate.
Fig. 7.7. Isochromosome to each pole.

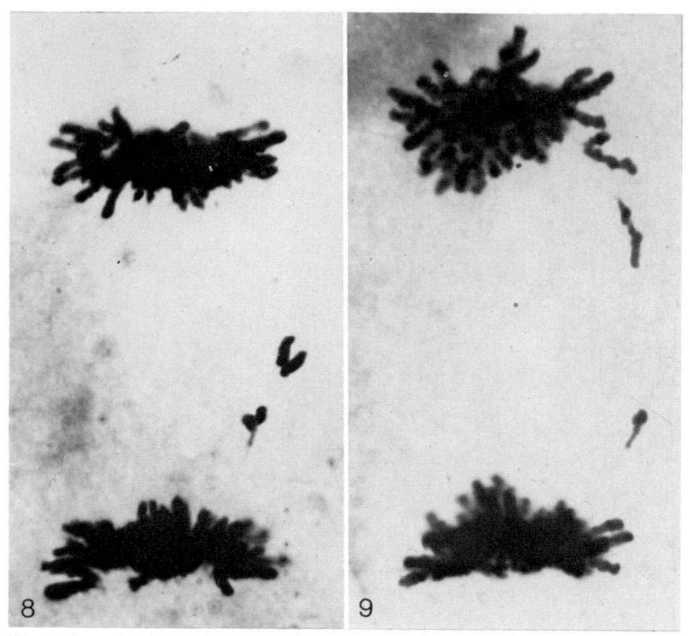

Fig. 7.8. and 7.9. Telophase I in mono-5A of wheat. (From Sears, 1952a.)
Fig. 7.8. Two long arms acentric on plate.
Fig. 7.9. Three arms to upper pole.

either pass separately to the other pole (Fig. 7.5) or one or both remain acentric on the equatorial plate (Fig. 7.6); (b) two identical arms pass to one pole and the other two arms either go to the other pole (Fig 7.7) or one or both remain acentric on the plate (Fig. 7.8); (c) three arms go to one pole and the fourth arm either passes to the other pole (Fig. 7.9) or remains on the plate. A fourth kind of division in which all four arms go separately to the poles has been observed in wheat by Li *et al.* (1948), but occurs rarely. Sears (1952a) recorded one cell with this type of division in his material. In mono-5A of Chinese Spring, type b misdivision was most frequent, 29 occurring out of 50 misdivisions seen. The frequency and types of misdivision of mono-5A observed at telophase I in Chinese Spring and in three strains in which 5A was substituted from other varieties into Chinese Spring is shown in Table 7.III.

Two other types of misdivisions have been reported in other species but were not observed in wheat monosomics. Darlington (1939) reported a misdivision in *Fritillaria* in which a single arm went to each pole, and two arms were left on the plate. Another type of division where a long and a short arm went separately to each pole was also reported by Darlington.

Since the univalent usually undergoes division much later than the separation of the bivalents, the division products usually lag between the poles. They

TABLE 7.III

Frequency and Types of Misdivision of Mono-5A at Telophase I in Four Different Varities of *Triticum aestivum*[a]

Type of misdivision	Distribution of chromosome arms at telophase I			Origin of chromosome 5A and number of microsporocytes in each class			
	Arms at one pole	Arms at other pole	Arms left on plate	Chinese	Hope	Thatcher	Red Egyptian
a	1 Long, 1 short	1 Long, 1 short	None	76	82	52	146
a	1 Long, 1 short	1 Long, 1 short	None	0	0	0	3
a	1 Long, 1 short	1 Long	1 Short	2	0	1	2
a	1 Long, 1 short	1 Short	1 Long	2	0	0	1
a	1 Long, 1 short	None	1 Long, 1 short	0	0	1	0
b	2 Long	2 Short	None	25	2	3	13
b	2 Long	1 Short	1 Short	3	1	0	0
b	2 Long	None	2 Long	1	0	0	1
c	1 Long, 2 short	1 Long	None	5	2	3	13
c	2 Long, 1 short	1 Short	None	12	8	3	11
c	2 Long, 1 short	None	1 Short	0	0	0	1
			Total	126	95	63	191
			% misdivision	39.7	13.7	17.5	23.6

[a]From Sears, 1952a.

are not included in the telophase I nuclei but form micronuclei of their own. Olmo (1936) observed 31 telophase cells with no micronuclei, 55 with one, 11 with two, and 1 with three in a monosomic of *Nicotiana tabacum*. Thus 70% of the telophase I cells had micronuclei.

During the second division, the univalent may divide normally if it has passed to one of the poles undivided during the first division. However, in those cells where it underwent equational division or misdivision during the first division, its movement is characteristically retarded. The products of the first division consisting of normal chromatids, telocentrics, and isochromosomes reach the metaphase II plate later and lag behind at telophase II (Figs. 7.10 and 7.11).

The behavior of the univalents at telophase II is again variable. (1) The normal chromatids may pass to one of the poles undivided. Two of the microspores of such quartets will have n chromosomes and the others will have $n - 1$. (2) The univalents may divide equationally a second time at telophase II. All four microspores resulting from this type of division will have n chromosomes. This type of division is apparently rare and has not been reported in wheat monosomics but was observed in four cases out of several hundred examined in mono-N and mono-B of *Nicotiana tabacum* by Olmo (1936). (3) The univalents may misdivide again at telophase II.

The misdivision of univalents at telophase II has been reported in oat monosomics (Nishiyama, 1931) and wheat monosomics (Sanchez-Monge and MacKey, 1948; Sanchez-Monge, 1950, 1951; Sears, 1952a). This misdivision generally consists of a pulling apart of the two arms of the univalent at the centromere with the short arm and long arm telocentrics going to opposite poles. Typical telophase II misdivisions are shown in Figs. 7.10 and 7.11. Here the univalent evidently divided normally at the first division, and one complete

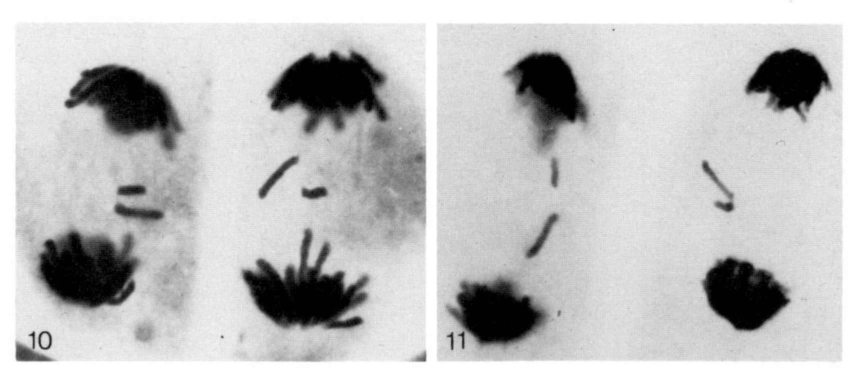

Fig. 7.10 and 7.11. Telophase II in mono-5A of wheat. (From Sears, 1952a.)
Fig. 7.10. Sister telophase II cells, with misdivision in each.
Fig. 7.11. Sister telophase II cells, with misdivision at right, possibly at left.

TABLE 7.IV
The Behavior at Telophase II of Mono-5A of *Triticum aestivum* from Various Sources[a]

Behavior of mono-5A	Source of mono-5A and % in each category				
	Chinese	Hope	Thatcher	Red Egyptian	Total
Not lagging	37.5	34.5	37.4	41.2	37.8
Passing entire	4.9	4.2	2.4	3.3	4.1
Misdivide	18.3	10.9	21.7	17.0	17.1
Lagging, unclassified	26.5	45.4	33.8	28.8	31.2
Abnormal	4.9	4.2	2.4	3.3	4.1
Telocentric	7.9	0.8	2.4	6.5	5.7
No. cells examined	328	119	83	153	683

[a]From Sears, 1952a.

chromatid was included in each telophase I nucleus. At telophase II, as at telophase I, the Chinese Spring mono-5A studied by Sears (1952a) showed a much higher rate of misdivision than did the Swedish material examined by Sanchez-Monge and MacKey (1948). Nearly three times as many lagging chromosomes misdivided in the Chinese as in the Swedish variety. Sears (1952a) compared the telophase II misdivision rates of 5A chromosomes substituted from three other wheat varieties into the genotypic background of Chinese Spring. The rates were found to be similar (Table 7.IV). According to Sears, the misdivision of univalents in other monosomics of wheat also occurs at a high frequency during both divisions of meiosis.

According to the data presented in Table 7.III, misdivision occurs in 39.7% of first division in Chinese mono-5A. Over one-half of these misdivisions are of a type which result in production of isochromosomes. However, some of these isochromosomes are converted into telocentrics as a result of misdivision at telophase II. Sears calculated that at the end of the first division 22.3% of the chromosomes would be isochromosomes and 15.3% would be telocentrics. At telophase II, misdivisions reduce the number of isochromosomes and increase the number of telocentrics.

Extensive studies of univalent behavior in monosomics of other species have not been carried out, but some misdivision of univalents probably occurs in all of them. As a result of irregular behavior of the univalent, $n - 1$ spores are produced at a much higher frequency than n spores. In addition, spores with telocentrics and isochromosomes are also produced. These include $n - 1$ + telocentric, $n - 1$ + isochromosome, $n - 1$ + telocentric + isochromosome, and $n - 1 + 2$ telocentrics. Occasionally spores with $n + 1$ and n + telocentric may also be produced. These result when the univalent divides

TABLE 7.V
Frequency of Quartets with Micronuclei in Twenty Monosomics of *Triticum aestivum*[a]

Monosomic	Quartets		No. micronuclei per quartets				
	No. analyzed	% with micronuclei	1	2	3	4	Over 4
1B	1594	39.0	26.6	10.6	1.4	0.4	0.00
1D	1695	50.1	31.7	14.4	3.2	0.6	0.20
2A	1865	65.2	30.9	27.7	5.3	1.0	0.20
2B	1602	62.9	31.0	25.8	4.5	0.9	0.60
2d	1946	63.3	27.8	29.8	4.6	1.0	0.20
3A	1832	40.4	23.3	14.7	1.9	0.4	0.00
3B	1957	47.8	26.7	17.0	3.5	0.6	0.00
3D	1698	57.4	32.2	21.8	3.2	0.3	0.00
4A	1795	64.8	29.0	28.2	5.6	1.5	0.50
4B	1651	34.3	19.4	11.3	2.9	0.4	0.10
4D	1800	40.6	24.8	13.7	1.3	0.8	0.04
5A	1641	47.7	24.9	16.2	4.9	1.3	0.50
5B	2108	55.5	27.2	22.0	4.5	1.3	0.30
5D	1777	55.1	27.2	24.3	4.0	0.8	0.60
6A	1792	49.7	32.4	15.4	1.7	0.1	0.00
6B	1946	34.1	19.8	11.0	2.2	0.1	0.00
6D	2011	60.2	35.5	19.3	4.4	0.5	0.40
7A	1879	47.4	25.8	18.5	2.6	0.5	0.00
7B	1774	46.5	25.9	16.5	3.5	0.5	0.10
7D	1819	44.9	25.9	16.4	2.4	0.2	0.00
Average	1809.1	50.3	27.4	18.7	3.38	0.66	0.18

[a]From Morrison and Unrau, 1952 reproduced by permission of the National Research Council of Canada from the *Can. J. Bot.* **30,** 371-378.

equationally and both chromatids are included in the same telophase I nucleus or when univalent is included undivided in one of the telophase I nuclei and probably divides equationally during interkinesis. Sears (1952a) observed one such cell in mono-5A of wheat. Spores with $n - 2$ and $n - 3$ occasionally result from the microsporocytes with three or five univalents as a consequence of their irregular behavior.

The proportion of n and $n - 1$ spores produced by a monosomic depends upon the frequency of lagging and misdivision of the univalent. An estimate of the proportion of n and $n - 1$ spores may be obtained by determining the frequency of spore quartets with micronuclei. Approximately 50% of the spores produced by the quartets without micronuclei will have n spores and the rest $n - 1$. About 25% of the spores resulting from the quartets with one micronucleus

TABLE 7.VI
Distribution of the Chromosome Numbers in Pollen Grains of Twenty Monosomics of *Triticum aestivum* Determined at First Pollen Mitosis[a]

Monosomic	No. pollen grains examined	18	19		20				21		22	23
			0f	1f	0f	1f	2ff	3ff	0f	1f		
1B	25	–	–	–	3	1	1	–	20	–	–	–
1D	107	–	–	–	72	5	–	–	30	–	–	–
2A	50	–	–	–	31	1	1	–	16	1	–	–
2B	72	–	–	–	35	1	1	–	34	–	–	1
2D	93	–	–	–	66	10	2	–	15	–	–	–
3A	40	–	–	–	34	–	–	–	5	–	–	1
3B	51	–	–	–	40	2	–	–	9	–	–	–
3D	25	–	1	–	21	–	–	–	3	–	–	–
4A	128	–	1	1	92	5	–	–	29	–	–	–
4B	156	1	1	–	102	8	1	–	42	–	1	–
4D	75	–	1	–	48	10	–	—	16	–	–	–
5A	141	1	1	1	78	7	1	–	51	1	–	–
5B	28	–	1	–	8	1	–	–	18	–	–	–
5D	31	–	–	–	18	2	–	–	10	–	1	–
6A	30	–	–	–	16	1	–	–	12	–	1	–
6B	83	–	–	–	39	4	3	–	35	–	1	1
6D	76	1	1	–	52	6	–	–	15	–	1	–
7A	41	–	–	–	30	–	1	–	9	1	–	–
7B	29	1	–	–	3	–	–	–	23	1	1	–
7D	28	–	1	–	24	1	–	1	1	–	–	–
Total	1309	4	8	2	812	65	11	1	393	4	6	3
% of total		0.3	0.6	0.1	62.0	5.0	0.8	0.1	30.0	0.3	0.5	0.2

[a]From Morrison, 1953.

will have n chromosomes. Morrison and Unrau (1952) determined the frequency of microspore quartets without micronuclei in twenty monosomics of wheat. The frequency varied from 34.8 to 65.9% with an average of 49.7% for all the monosomics studied (Table 7.V). They also found that 27.4% of the quartets had one micronucleus. Thus, one-half of 49.7% plus one-fourth of 27.4%, or approximately 31%, of the microspores of wheat monosomics should have n chromosomes. Morrison (1953) studied the distribution of chromosome numbers at first mitosis in the pollen grains of twenty wheat monosomics. He found that the average frequency of those with n chromosomes was 30.0%, and the frequency of pollen grains with $n - 1$ was 62.0%. The rest had one, two, or three telocentrics or isochromosomes or had $n - 2$, $n - 3$, $n + 1$, or $n + 2$ chromosomes (Table 7.VI). Bhowal (1964) also found the frequency of n and

$n - 1$ pollen grains of a substitution monosomic-3D of wheat to be 35.6 and 64.4%, respectively.

On the basis of micronuclei counts, Lafever and Patterson (1964a) calculated the frequency of gametes with *n* chromosomes produced by a monosomic line of *Avena sativa* to be only 6.03%. Similarly, Singh and Wallace (1967b) and Nishiyama *et al.* (1968) calculated the frequency of gametes with *n* chromosome number in several monosomics of *Avena byzantina* on the basis of micronuclei counts in spore quartets. In the first study, the frequency varied from 9.3 to 21.4%, and 10 to 28% in the latter.

Sources of Nullisomics

The best sources of nullisomics are the progenies of monosomics, monotelosomics, and monoisosomics. The nullisomics appear in varying frequencies in the selfed progenies of monosomics. All the possible nullisomics of hexaploid wheat have been obtained from the progenies of monosomics. Sears (1954) isolated nullisomics from the progenies of twenty-five monotelosomics and monoisosomics of wheat. Nullisomics for several chromosomes of *Avena sativa* have been obtained by various workers (See Table 8.IV in next chapter) in the selfed progenies of monosomics, but it is not known for how many different chromosomes of the complement the nullisomy is tolerated. Nullisomics were present in the selfed progenies of three of the four monosomics of *Avena byzantina* studied by Singh and Wallace (1967b). Nullisomics have also been found in varietal populations of *Triticum aestivum* (Love, 1940), *Avena sativa* (Huskins, 1927; Tagenkamp and Finkner, 1954; McGinnis and Taylor, 1961; McGinnis, 1962b; Hacker and Riley, 1963), and *Avena byzantina* (Ramage and Suneson, 1958b). However, the nullisomics thus found probably also result from the progenies of monosomics of spontaneous origin which are present in the varietal populations.

Nullisomics also appeared in the progenies of pentaploid wheat hybrids. The pentaploid F_1 plants obtained from crosses between hexaploid and tetraploid wheats have AABBD genomes and form fourteen pairs and seven univalents at prophase and metaphase I of meiosis. At anaphase I, the paired A and B genomes disjoin normally but the seven univalents segregate at random. The gametes which receive six or all the seven chromosomes of the D genome in addition to fourteen chromosomes of the A and B genomes have greater viability than those which receive fewer D chromosomes. Thus when a 20-chromosome female gamete lacking one chromosome of the D genome is fertilized by a similar male gamete, a nullisomic zygote is produced. Kihara and Wakakuwa (1930, 1935) were the first to obtain nullisomics for the chromosomes of the

D genome from this source. Nullisomics for all the seven chromosomes of the D genome were isolated by Matsumura (1952a) in this manner.

Griffiths and Thomas (1953) have described a line of *Avena sativa* with the normal chromosome number of 42 which segregated normal and 40-chromosome dwarf plants in a ratio of about 9:1. All the normal plants showed this type of segregation in their progeny. Seeds from outcrosses yielded only normal plants. According to the authors, a recessive gene in the homozygous condition brings about the loss of a particular pair of chromosomes after fertilization but before the full development of the embryo.

Cytology of Nullisomics

Only a few nullisomics of wheat and oats regularly form $n - 1$ bivalents during meiosis and yield $n - 1$ spores. However, the nullisomic condition for the majority of the chromosomes of the complement causes considerable meiotic irregularity. Thus nullisomics for chromosome 2A and 3B of wheat are partially asynaptic (Sears, 1944). Ohta and Matsumura (1961) have given a detailed account of meiotic associations in seven nullisomics of the D genome of wheat (Table 7.VII). Twenty bivalents were formed in about 90% of the microsporocytes of five nullisomics, while in the remaining two, the frequency of cells with such associations was much lower. The next most frequent class of microsporocytes had 19 II + 2 I. Occasionally cells with 18, 17, or 16 bivalents were also observed in five nullisomics, but the frequency of such associations was higher in the other two nullisomics. Similar observations were reported by Ray and Swaminathan (1959) in a few nullisomics of Chinese Spring and Redman wheat and by Mochizuki and Shigenaga (1964) in twenty nullisomics of Chinese Spring.

The fatuoid nullisomic of *Avena sativa* is also asynaptic (Huskins, 1927; Huskins and Hearne, 1933; Nishiyama, 1931, 1933), and Costa-Rodriguez (1954) reports that his oat nullisomics were partially asynaptic. Three of the nullisomics studied by Hacker and Riley (1965) were asynaptic while the rest formed twenty bivalents at meiosis. Von der Schulenburg (1965) reported still another nullisomic which showed partial asynapsis. One nullisomic studied by Lafever and Patterson (1964a) had normal meiosis and invariably formed twenty bivalents. The nullisomic of *Avena byzantina* reported by Ramage and Suneson (1958b) had normal meiosis while one of the three examined by Singh and Wallace (1967a) formed twenty pairs, one was asynaptic, and the third was desynaptic.

One nullisomic of wheat, nulli-5B, shows an effect opposite to that of asynapsis. Riley (1958) reported that the absence of 5B in haploids greatly increased the frequency of bivalents and trivalents. As many as nineteen of

TABLE 7.VII
Chromosome Associations at Metaphase I of Meiosis in Seven Different Nullisomiocs of D Genome of Wheat[a]

Nullisomic line	No. cells examined	No. cells with chromosomal associations at metaphase I						
		20 II	19 II + 2 I	18 II + 1 IV	18 II+ 1 III + 1 I	18 II + 4 I	Others	% with 20 II
a	65	59	3	0	1	2	0	90.7
b	190	120	51	0	1	11	7	63.2
c	75	70	4	0	0	0	1	93.3
d	135	124	9	0	1	0	1	91.8
e	144	130	10	1	1	2	0	90.6
f	174	162	11	1	0	0	0	93.2
g	70	31	18	5	7	4	5	44.3

[a]From Ohta and Matsumara, 1961.

the twenty chromosomes were observed in various associations, whereas in 21-chromosome haploids, the maximum number was nine. In nulli-5B, multivalent association of 3, 4, 5, and 6 chromosomes were common. More than one-half the cells had at least one multivalent, and many had several (Riley and Chapman, 1958a).

8

Breeding Behavior and Morphology of Monosomics and Nullisomics

The utility of the monosomics and nullisomics in cytogenetic studies is dependent upon their breeding behavior. The morphological differentiation, if possible, is helpful in such studies. This chapter deals with the breeding behavior and morphology of the monosomics and nullisomics. At the end of the chapter conclusions drawn from these studies are discussed.

Breeding Behavior of Monosomics

Theoretically, one expects an individual with $2n - 1$ chromosomes to produce n and $n - 1$ gametes in equal frequency. Likewise, one should be able to obtain $2n$ and $2n - 1$ individuals in equal frequency from $(2n - 1) \times 2n$ crosses. Furthermore, $2n - 1$ individuals when selfed should yield $2n$, $2n - 1$, and $2n - 2$ progeny in the ratio of 1:2:1. However, as the data in this section show, these expectations are never realized, and the departures from this idealized ratio are indeed varied. The causes of the departure are (1) production of n and $n - 1$ spores in unequal frequency; (2) reduced viability or inviability of $n - 1$ spores; (3) competition between n and $n - 1$ microspores; (4) reduced viability of $2n - 1$ zygotes; and (5) reduced viability or inviability of nullisomic zygotes. These factors which mainly determine the breeding behavior of the monosomics are reviewed below. In addition to the disomics, monosomics, and nullisomics exceptional individuals appear in monosomic progenies. The nature and the causes of occurrence of these types are also discussed.

Production of n and $n - 1$ Gametes in Unequal Frequency

From the previous discussion in Chapter 7 it was concluded that $n - 1$ spores are produced in a disproportionately high frequency due to lagging and misdivision of the univalent during meiosis. According to Clausen and Cameron (1944), in monosomics of *Nicotiana tabacum*, n and $n - 1$ spores are produced in the proportions of 20 and 80%, respectively. In wheat monosomics, as already discussed, the frequency of $n - 1$ spores is about double that of n spores (62 versus 30%). However, the monosomics of *Avena sativa* and *Avena byzantina* produce $n - 1$ spores at even higher frequencies. Various workers have calculated the frequency of $n - 1$ spores produced by the monosomics of *Avena sativa* and *Avena byzantina* on the basis of micronuclei counts (Table 8.I). On the average, 87% of the pollen grains produced by the monosomics of these two species have $n - 1$ chromosomes.

Reduced Viability or Inviability of n − 1 Spores

If the viability of the $n - 1$ megaspores is normal, then the frequency of $2n - 1$ individuals in a progeny of monosomic × normal should be comparable to the frequency of $n - 1$ microspores produced by the monosomic, assuming $n - 1$ microspores and megaspores are produced in similar frequencies. In the monosomics of wheat and oats, the $2n - 1$ megaspores have normal viability and the frequency of $2n - 1$ progeny from $(2n - 1) \times 2n$ crosses is comparable to the frequency of $n - 1$ spores. Thus Hacker (1965) calculated the average

TABLE 8.I
Frequency of $n - 1$ Pollen Grains Produced by Monosomics of *Avena* Species Determined on the Basis of Micronuclei Counts

Species	Monosomic	% $n - 1$ pollen grains	Authority
Avena sativa	Mono-C	83.3	Nishiyama (1931)
	Mono-V	93.0	Philp (1935)
	Mono-L	94.0	Philp (1938)
	Mono-14	83.4	McGinnis and Taylor (1961)
	Mono-20	91.0	Gauthier and McGinnis (1965)
	Mono-15	84.6	McGinnis and Lin (1966)
	"Clintland 60" monosomic	93.7	Lafever and Patterson (1964a)
	Mono-VI	88.0	Hacker (1965)
	Mono-VII	91.0	Hacker (1965)
	Mono-IX	87.0	Hacker (1965)
	Mono-X	84.0	Hacker (1965)
	Mono-XIII	86.0	Hacker (1965)
Avena byzantina	Mono-M 3	80.9	Singh and Wallace (1967b)
	Mono-St 7	88.3	Singh and Wallace (1967b)
	Mono-Sm 12	90.7	Singh and Wallace (1967b)
	Mono-St 17	78.6	Singh and Wallace (1967b)

frequency of $n - 1$ spores produced by five monosomics of *Avena sativa* to be 87.2% (Table 8.I). The proportion of $2n - 1$ progeny in the crosses of three monosomics with a $2n$ parent was 94, 87, and 87%, respectively. Similarly, the average frequency of monosomics in the progenies of monosomic $\times$ $2n$ crosses of oats studied by Chang and Sadanaga was 92.2%, a frequency comparable to that of $n - 1$ spores produced by the monosomics of this species.

The frequency of monosomics in $(2n - 1) \times 2n$ crosses of wheat varied from 57.3 to 81.9% and the average for all the monosomics was 72% (Tsunewaki, 1964a). This average is very close to the 75% reported by Sears (1944, 1954) and is not very far from the average frequency of 62% $n - 1$ spores produced by wheat monosomics reported by Morrison (1953). The difference becomes even narrower if we consider that Tsunewaki included plants with 40 + telocentrics, and with 40, or even 39, chromosomes in the monosomic class. Tsunewaki's statement that there is no correlation between the frequency of monosomics in the progeny and the frequency of $n - 1$ microspores in wheat monosomics is questionable. Tsunewaki compared the frequency of monosomics in the progenies of individual monosomics from his own studies with that of $n - 1$ pollen grains reported by Morrison (1953). However, the sample size studied by Morrison for each monosomic was so small that sampling error alone could have influenced the results. Under the circumstances, the comparison between the average frequencies of $n - 1$ microspores and $2n - 1$ progeny seems more justified.

The results from monosomics of wheat and of two oat species indicate that $n - 1$ spores have as good viability as n spores. This conclusion is further fortified by Tsunewaki's (1964a) observations that in the monosomics of wheat the seed fertility is only slightly below normal. These conclusions, however, do not apply to the monosomics of tetraploid species.

Although all monosomics of *Nicotiana tabacum* produce $n - 1$ spores at similar frequencies of about 80% (Olmo, 1935; Clausen and Cameron, 1944), the transmission rates through the female vary considerably. Mono-P is transmitted to only 6.5% of its progeny while 78.7% of the progeny of mono-A is monosomic (Table 8.II). In this table, the data on ovule abortion reveal that in twelve of the monosomics the percentages of ovule abortion cluster around 80. The remaining monosomics have ovule abortion values ranging from 15 to 56%. The data on ovule abortion and transmission rates show that not all of the $n - 1$ megaspores abort, even in those monosomics which approach the 80% level of abortion, shown by their ovular transmission rates of 5 to 20% or more. These results permit the conclusion that in a few monosomics some of the n ovules also abort probably due to some physiological imbalance in the monosomic caused by the missing chromosome. Clausen and Cameron (1944) cited the unpublished data of Fansler which show that the most important causes of ovule abortion in monosomics of *Nicotiana tabacum* are: (1) early

TABLE 8.II
Reproductive Features of Monosomics of *Nicotiana tabacum*[a]

Monosomic	% pollen abortion	% ovule abortion	Seed fertility (%)	Transmission through female (%)
A	6.9	17.9	116.6	78.7
B	5.1	30.7	49.9	32.3
C	24.3	32.3	83.3	45.8
D	3.2	15.6	74.0	41.2
E	6.7	29.3	62.2	81.9
F	77.4	24.4	58.2	59.8
G	75.7	84.3	18.9	6.4
H	22.3	45.4	80.7	70.4
I	78.0	76.8	19.0	7.7
J	80.7	86.9	6.4	6.0
K	79.5	56.1	40.1	48.3
L	79.1	72.9	18.9	18.6
M	83.4	32.6	33.4	59.8
N	78.7	79.2	13.3	22.1
O	85.8	16.3	89.0	77.2
P	77.0	84.4	26.1	6.5
Q	80.6	79.3	23.6	11.0
R	74.7	46.1	46.7	54.4
S	12.4	21.3	91.1	31.0
T	67.9	85.8	14.8	11.0
U	61.8	73.9	19.2	36.4
V	75.1	79.7	22.6	7.9
W	74.3	78.8	26.6	5.1
Z	82.1	82.5	5.8	8.1
Disomic	3.3	2.7	100.0	

[a]From Clausen and Cameron, 1944.

degeneration of the endosperm; (2) failure of megaspore development; and (3) absence of pollination. Greenleaf (1941) also showed that the rate of development of $n - 1$ megaspores is slower than that of n megaspores in mono-P of *Nicotiana tabacum*. As a result, most of the $n - 1$ megaspores are not ready for fertilization when the pollen tubes arrive in the ovary 3 days following pollination. Consequently only about 4% of the fertilized ovules are $n - 1$. The seed fertility values given in Table 8.II indicate that, on the whole, seed production is somewhat closely related with percentage of good ovules, especially in the group of plants with high ovule abortion. This correlation suggests that reduced viability of the $n - 1$ megaspores is the main cause of reduced female transmission of the monosomics in *Nicotiana tabacum*. Some reduction in transmission rates is probably also caused by the reduced viability of $2n - 1$ zygotes, but it is generally agreed that the tolerance for chromosomal deficiencies is higher at the sporophytic than at the gametophytic level.

Two out of seven of the *Nicotiana rustica* monosomics studied by Lammerts (1932) showed poor transmission (31%) through the female while in the rest the transmission varied from 56 to 75%. The author suggested three improbable explanations for the reduced transmission in the two monosomics: (1) selective development at the four-celled stage of megasporogenesis or megaspore replacement; (2) frequent death of $2n - 1$ zygotes; and (3) double division of the univalent chromosome to produce four spores with n chromosomes. He failed to see that the poor viability of the $n - 1$ megaspores would explain the results more logically.

The transmission rates through the female of seven monosomics of *Gossypium hirsutum* vary from 20 to 40% (Brown and Endrizzi, 1964). On the average, these are lower than those of the monosomics of *Nicotiana* species discussed above. The authors have cited the unpublished results of other workers showing that two of the monosomics of this species failed to transmit to the next generation. Most of the monosomics show considerable sterility, and it appears that an appreciable proportion of $n - 1$ megaspores abort due to the imbalance caused by the missing chromosome.

In tetraploid wheat, *Triticum durum,* the female transmission rates of the monosomics are even lower. In ten monosomic types they varied from 2.9 to 41.7% with an average of 11.14%. The fertility of the monosomics was correspondingly low and varied from 5.6 to 28.3% with an average of 16.1% (Mochizuki, 1968b). It is clear that besides a larger proportion of $n - 1$ megaspores, some n megaspores also abort.

Three monosomics and five potential monosomics ($2n - 1 + f$) of tetraploid oats, *Avena barbata,* were obtained by Andrews and McGinnis (1964) using X-irradiation treatment. Most of these plants had other chromosomal abnormalities besides the deficiency and, consequently, were highly sterile. Also, four monosomic plants isolated from an asynaptic derivative from a cross of two closely related tetraploid species, *Avena abyssinica* and *Avena barbata,* were highly sterile (Thomas and Rajhathy, 1966). The sterility was attributed to the presence of an asynaptic gene revealed by the occurrence during meiosis of several unpaired chromosomes. Information on the tolerance of aneuploidy by tetraploid species of *Avena* is only fragmentary. However, I believe transmissible monosomics of tetraploid *Avena* are obtainable. These preliminary reports should not discourage those working with oats but serve to stimulate them to work with different genotypes and techniques. At one time, the possibility of obtaining simple monosomics of tetraploid wheat seemed remote, and it was under this assumption that Longwell and Sears (1963) and Noronha-Wagner and Mello-Sampayo (1966) endeavored to produce monosomic–trisomic combinations hoping that the loss of a chromosome would be partially compensated by the homoeologous trisome. Eventually, Mochizuki (1968a, b, 1971) established all the possible, simple, transmissible monosomics of tetraploid wheat by using appropriate techniques.

The monosomics of diploid species are characterized by total inviability of the $n - 1$ spores. None of the monosomics of the diploid plant species have ever transmitted to the next generation. Avery (1929) raised two progenies of 79 plants of a monosomic of *Nicotiana alata* ($2n = 18$) and recovered only disomic plants. Branches or even whole plants of *Datura stramonium* deficient for one chromosome were obtained by Blakeslee and Belling (1924b), Blakeslee and Avery (1938), and Bergner *et al.* (1940) but the progenies of these $2n - 1$ chimaeras or plants failed to yield monosomics. The progeny of a $2n - 1$ plant of *Hyoscymus niger* ($2n = 34$) consisted only of disomic plants (Griesinger, 1937). Smith (1943) raised progenies of 12 and 75 plants from $(2n - 1) \times 2n$ and the reciprocal cross, respectively, of the monosomic of *Nicotiana langsdorfii*. Only disomic plants were obtained. The first primary monosomic of tomato, mono-11, was isolated by Rick and Khush (1961) in the progenies sired by the X-irradiated pollen. Subsequently, mono-12 and several tertiary monosomics of tomato were similarly obtained (Khush and Rick, 1966a, b). Progenies of both the primary and nine tertiary monosomics from backcrosses to the normal disomics were grown, but none of the monosomics transmitted to the progeny. Even the minute segmental deficiencies of tomato chromosomes are not transmitted to the next generation (Khush and Rick, 1967c, 1968a, 1969). It is obvious that in higher plants the gametophytic or haploid stage is more sensitive to chromosome losses than the sporophytic or diploid stage.

Even at the sporophytic level, the tolerance for monosomy in diploid species is limited to certain members of the chromosome complement or certain chromosome arms. In tomato, for example, where monosomics were sought over a 7-year period with the help of genetic markers in the irradiated progenies, simple primary monosomics were found for the two smallest members of the complement. Whole arm losses are tolerated for the fifteen short arms of the complement or for any two of these fifteen together. When mature pollen of tomato is irradiated, occasionally two chromosomes in a generative nucleus break in the centromere region, two short arms are lost, two long arms join, and give rise to an $n - 1$ nucleus. Such a pollen grain apparently takes part in fertilization and gives rise to a $2n - 1$ tertiary monosomic. Primary monosomics also arise in a similar manner. However, when these monosomics produce spores, the ones with a deficiency cannot develop to maturity, and soon after completion of meiosis they degenerate.

The longest deficiency obtained in tomato is that of chromosome 11 which is 24.8 μm or 6.8% of the total length of the chromosome complement. The total length of the two arms missing in the tertiary monosomics has never exceeded this limit. Thus, it is clear that in the diploid species, monosomy is tolerated for certain chromosomes or chromosome segments in the sporophytic stage, not in the gametophytic stage. Occasional monosomics have been reported in a few other diploid species. Kihara (1932) described one monosomic of *Pharbitis nil*. Rick (1943) obtained two tertiary monosomics of *Petunia* by X-ray treatment of pollen. Parry and Gerstel (1967) found one monosomic in a backcross progeny of (*Nicotiana forgetiana* × *N. alata*) × *N. alata*. Andrews and McGinnis (1964) obtained one monosomic and one potential monosomic ($2n - 1 + f$) in diploid *Avena strigosa* from X-irradiation of panicles. Although the breeding behavior of these monosomics has not been reported, they probably would never transmit.

Monosomics of a special type were reported by Håkansson (1945) and Hiorth (1948a, b) in diploid species of *Clarkia* (Godetia). These monosomics are transmitted to the next generation. However, these so-called "monosomics" of *Clarkia* possess all the essential chromatin of the species (Snow, 1964). As a result of translocations and loss of one centromere, the vital chromatin of two nonhomologous chromosomes is consolidated into one chromosome, and the number of chromosomes, thereby reduced. The $n - 1$ spores produced by these "pseudomonosomics" have as much chromatin as the n spores and, hence, are viable. The $n - 1$ gametes can fertilize each other and give rise to "pseudonullisomics" (Snow, 1964). Rana (1965b, 1967) has described a similar situation in *Chrysanthemum carinatum* ($2n = 18$), a diploid species. From a cross of two translocation stocks, a $2n - 1$ plant and a chimaera having $2n$ and $2n - 2$ chromosomes was discovered. F_1 progeny from the cross of these two plants yielded a nullisomic which regularly formed eight pairs of chromosomes at diakinesis. A thorough cytogenetic analysis of these monosomic and nullisomic plants is lacking. But because of their origin from the crosses of translocation stocks it appears that the situation may be comparable to that in *Clarkia*.

Finally, the position of maize in the diploid group is somewhat anomalous. McClintock (1929b) and Morgan (1956) reported $2n - 1$ chimaeras of maize. Einset (1943) obtained five monosomic plants and one potential monosomic ($2n - 1 + f$) in the progenies of trisomics. However, it is not known whether these plants were monosomic for the same chromosome or for different ones. More recently, monosomics for four different chromosomes of maize, namely, 1, 6, 9, and 10, have been obtained (Baker and Morgan, 1966; Shaver, 1965). Most noteworthy is the occurrence of monosomics for chromosome 1, the longest chromosome of the complement. Three mono-1 plants were obtained by Baker and Morgan (1966) in the progenies sired by X-rayed pollen and used on stocks which were homozygous for three recessive markers of chromosome 1. Three plants were pseudodominant for all the three markers and were found to be deficient for chromosome 1. In addition, two $2n - 1$ plants were obtained which were not hemizygous for all the three markers, but only for two. The authors concluded that the missing chromosome in these two plants was not number

1. However, it is almost certain that these two plants were tertiary monosomics in which one arm carrying the two markers of chromosome 1 and another arm of some other chromosome had been deleted and the remaining two arms had joined to give a $2n - 1$ plant.

Since, in maize, monosomics have been obtained for four chromosomes, one of which is the longest of the complement, measuring 82.4 μm or 14.9% of the total length of the complement, it appears possible to obtain monosomics for all the chromosomes. Anyone interested in this problem would do well to pollinate stocks homozygous for recessive markers of different chromosomes with X-rayed pollen and look for pseudodominants in the progeny.

The tolerance for monosomy at the gametophytic level in maize is higher than in other diploid species. In tomato, all small and large deficiencies of the euchromatic regions, including the submicroscopic ones, are lethal at the gametophytic stage. In maize, many small deficiencies of euchromatic regions are known to survive at the gametophytic stage and transmit to the next generation (McClintock, 1941, 1944; Patterson, 1952). Even large deficiencies in maize comprising one-fourth of the chromosome arm have transmitted to the next generation through the female side (Stadler, 1933, 1935). The occurrence of five monosomic plants in the trisomic progenies (Einset, 1943) suggests transmission of $n - 1$ gametes. As discussed in Chapter 4, trisomics produce $n - 1$ gametes at low frequencies which if viable give rise to monosomic individuals in the progeny. If the monosomics reported by Einset originated in this manner, then certainly some of the monosomics of maize, especially for the short chromosomes, may be transmitted to the next generation.

Mono-4 of *Drosophila melanogaster* (Bridges, 1921a) is the only example of monosomy in the animal kingdom. The monosomic condition is transmitted to about 34% of the progeny. However, chromosome 4 of this species is extremely small and probably carries a very small amount of chromatin.

Up till now the discussion has been limited to the differential viability of n and $n - 1$ megaspores. The viability of $n - 1$ microspores and megaspores is probably similarly affected. In the hexaploid species, $n - 1$ microspores have normal viability but in the tetraploid species many $n - 1$ microspores abort as in *Nicotiana tabacum* (Table 8.II). In the diploid species, all the $n - 1$ microspores abort. However, various studies have shown that like $n + 1$ gametes, the transmission of $n - 1$ gametes through the male side is generally very low or not at all. The reasons for this, form the subject of the following section.

Competition between n and n − 1 Microspores

Less is known about the transmission of $n - 1$ gametes through the male than through the female. Reliable information can be obtained only by growing the progenies of $2n \times (2n - 1)$ crosses. Sears (1944) calculated the frequency of male transmission of $n - 1$ gametes in three different monosomics of wheat in this manner and found values of 5, 9, and 19%. In addition, the frequencies of nullisomics in the self-pollinated progeny of monosomics of wheat can be used to estimate the male transmission of deficient gametes, assuming an average frequency of 75% through the female, a value widely used for illustrative purposes. On this basis, the male transmission averaged close to 4% with a range of 1 to 19% (Morris and Sears, 1967). Bhowal (1964) has reported an unusually high male transmission (61%) of $n - 1$ gametes in a monosomic for chromosome 3D, which had been substituted in a specific hybrid background. The transmission rate was influenced by environmental conditions and by the female parent in the crosses. Nishiyama (1928) also reported a male transmission rate of 37% for monosomic "f," a figure considerably higher than Sears (1944) reported for all the monosomics of wheat.

Four of the seven monosomics of *Nicotiana rustica* were transmitted through the male at a low frequency (Lammerts, 1932). Four of the five monosomics of *Nicotiana tabacum* studied by Olmo (1935) were transmitted through the male at a frequency of 0.5 to 7.3%. Clausen and Goodspeed (1926b) earlier reported a male transmission of mono-C at a frequency of 3.3%. Only one of the seven monosomics of cotton (*Gossypium hirsutum*) transmitted through the male at a very low frequency (Brown and Endrizzi, 1964).

According to Sears (1944) the low frequency of functioning $n - 1$ male gametes is due to considerable elimination of $n - 1$ pollen through competition (certation) with n pollen. The exact mechanism by which the ability of $n - 1$ pollen to compete with n pollen in affecting fertilization is impaired has not been investigated. Perhaps, like the $n + 1$ pollen of trisomics (Chapter 4), the $n - 1$ pollen (1) may mature later than n pollen, (2) may be late to germinate, (3) may produce defective pollen tubes, or (4) may produce slow growing pollen tubes. In addition, a variable proportion of $n - 1$ microspores abort in the monosomics of tetraploid species. Pollen abortion values for monosomics of *Nicotiana tabacum* are given in Table 8.II. Douglas (1968) reported pollen abortion values of 3.77 to 34.0% for six monosomics of *Gossypium hirsutum*.

The transmission rates of $n - 1$ male gametes in the monosomics of *Avena sativa* and *Avena byzantina* are quite variable. In some monosomics, there is no competition between $n - 1$ and n male gametes, while in others, the $n - 1$ male gametes are unable to take part in fertilization. In others, the transmission rates vary between these two extremes. The data of Philp (1935, 1938) showed that the transmission of $n - 1$ gametes in two different monosomics of *Avena sativa* was similar in both sexes and that no certation between $n - 1$ and n pollen grains occurred. Similar results were reported by Lafever and Patterson (1964a) for one monosomic line; McGinnis and Taylor (1961) and McGinnis *et al.* (1963) for mono-14; McGinnis and Andrews (1962) for mono-21; McGinnis and Lin (1966) for mono-15; and Hacker (1965) for one monosomic line. Two of the monosomic lines studied by Chang and Sadanaga (1964b) and one of Hacker (1965) failed to transmit any $n - 1$ gametes through the male. For the others, transmission rates through the male of 5 to 50% have been reported by Nishiyama (1933), Costa-Rodriguez (1954), McGinnis (1962b), Chang and Sadanaga (1964b), and Hacker (1966). In one monosomic line of *Avena byzantina* studied by Singh and Wallace (1967b), there was no transmission of $n - 1$ gametes through the male, in another transmission was almost as good as that of n gametes, while in the remaining two it was intermediate.

From the above discussion it is clear that the transmission of $n - 1$ gametes through the male in the monosomics of polyploid species is usually very low or does not occur at all. In hexaploid wheat, all the monosomics are able to transmit through the male but at very low frequency. The chances of $n - 1$ male gametes taking part in fertilization are on the average, one-tenth those for n gametes. In some monosomics of hexaploid oats, $n - 1$ gametes are transmitted at about the same frequency at which they are produced (80–100%), while in a few, there is no transmission. The differences in transmission rates of n and $n - 1$ gametes in the monosomics of different species are attributed to the differential ability of n and $n - 1$ pollen grains to take part in fertilization.

Reduced Viability of 2n − 1 Zygotes

As discussed earlier, in the hexaploid species, the frequency of monosomic individuals from $(2n - 1) \times 2n$ crosses is comparable to the frequency of $n - 1$ megaspores. Therefore, the viability of $n - 1$ spores and $2n - 1$ zygotes must be comparable to that of n spores and $2n$ zygotes, respectively. In the tetraploid species, the frequency of monosomic individuals is much lower than that of $n - 1$ spores. The main cause of this lower transmission probably is the reduced viability of $n - 1$ spores. However, it is possible that some of the $2n - 1$ zygotes also die in the earlier stages of development. Endosperm abnormalities in $2n - 1$ zygotes lead to early death of some embryos in *Nicotiana tabacum* as reported by Fansler (cf., Clausen and Cameron, 1944). According to Olmo (1935), selective elimination of $2n - 1$ zygotes takes place in some monosomics due to failure of a certain proportion of $2n - 1$ seeds to germinate. The average decrease in seed viability is more in those monosomics which have lower transmission rates such as mono-B and mono-N. Poor endosperm formation in $2n - 1$ individuals of tetraploid wheat has been reported by Mochizuki (1968b). The frequency of monosomics in the plants derived from shrivelled seeds was as high as 83.3% in mono-2A, the average for nine different monosomics being 55% compared to 11.14% in the random samples. Presumably some $2n - 1$ zygotes die in the earlier stage in the monosomics of tetraploid wheat also due to endosperm abnormalities. Thus extremely low fertility of $2n - 1$ of tetraploid wheat may be due to the combined effects of reduced viability of $n - 1$ spores and $2n - 1$ zygotes. On the whole, the reduced viability of $n - 1$ spores may play a greater role in lowering the transmission rates of monosomics of tetraploid species than the reduced viability of $2n - 1$ zygotes. In the monosomics of diploid species, however, absence of transmission is solely due to the inviability of $n - 1$ spores.

Reduced Viability of Nullisomic Zygotes

As discussed earlier, several monosomics of *Nicotiana rustica* and *Nicotiana tabacum* transmit the $n - 1$ gametes through the female as well as the male. Theoretically these monosomics will yield nullisomic progeny upon selfing. However, no nullisomic individuals have been found in the progenies of these monosomics (Lammerts, 1932; Clausen and Cameron, 1944). Apparently, nullisomic zygotes do not survive, and it is obvious that whole chromosome deficiencies are tolerated by the tetraploid species in hemizygous condition but not in homozygous condition. In the hexaploid species, on the other hand, where the amount of duplication of genetic material is much greater than in the tetraploid species, the nullisomics are viable. Therefore, nullisomics have been obtained in only three hexaploid species, namely, *Triticum aestivum, Avena sativa,* and *Avena byzantina*. Even in these species, especially the *Avenas,* the nullisomics are not produced at the expected frequencies. Apparently some or all of the nullisomic zygotes die in the earlier stages of development. The best way to calculate the expected frequency of nullisomics in the selfed progenies of monosomics is to determine the transmission rates of $n - 1$ male and female gametes from $2n \times (2n - 1)$ and $(2n - 1) \times 2n$ crosses. The expected fre-

TABLE 8.III
Expected Frequencies of the Different Chromosome Types in the Selfed Progeny of a Monosomic with Assumed ($n - 1$) Transmission Rates of 75 and 4% through Female and Male, Respectively[a]

	Through male		
Through female	$n = 96\%$	$n - 1 = 4\%$	Progeny:
$n = 25\%$	$2n = 24\%$	$2n - 1 = 1\%$	$2n = 24\%$ $2n - 1 = 73\%$ $2n - 2 = 3\%$
$n - 1 = 75\%$	$2n - 1 = 72\%$	$2n - 2 = 3\%$	

[a]From Sears, E. R. (1953). Nullisomic analysis in common wheat. *Amer. Natur.* **87**, 245-252. © 1953, University of Chicago Press, Chicago .

TABLE 8.IV
Frequencies of Disomics, Monosomics, and Nullisomics in the Selfed Progenies of Monosomics of *Avena sativa*

Monosomic	Progeny: Disomic (%)	Monosomic (%)	Nullisomic (%)	Authority
Mono-C	3.8	60.4	35.7	Nishiyama (1933)
Mono-9	30.0	70.0	0.0	McGinnis (1962b)
Mono-21	0.0	23.2	76.8	McGinnis (1962b)
Mono-20	3.3	32.6	64.0	McGinnis and Andrews (1962)
Mono-15	0.7	30.0	69.3	McGinnis and Lin (1966)
Mono-20	5.0	82.4	12.6	Gauthier and McGinnis (1965)
Monosomic line	0.9	16.1	82.9	Lafever and Patterson (1964a)
Mono-A	11.0	87.2	0.0	Chang and Sadanaga (1964b)
Mono-B	15.4	84.6	0.0	Chang and Sadanaga (1964b)
Mono-C	8.2	85.7	6.2	Chang and Sadanaga (1964b)
Mono-D	8.8	91.2	0.0	Chang and Sadanaga (1964b)
Mono-E	13.3	84.3	1.2	Chang and Sadanaga (1964b)
Mono-F	3.1	95.4	0.0	Chang and Sadanaga (1964b)
Mono-I	3.6	76.7	19.9	Hacker (1965)
Mono-II	9.0	62.0	29.0	Hacker (1965)
Mono-III	7.3	70.3	22.3	Hacker (1965)
Mono-IV	13.6	71.3	15.0	Hacker (1965)
Mono-V	4.6	95.3	0.0	Hacker (1965)
Mono-VI	9.0	74.0	17.0	Hacker (1965)
Mono-VII	2.0	30.3	67.6	Hacker (1965)
Mono-VIII	6.0	93.0	1.0	Hacker (1965)
Mono-IX	7.3	84.0	8.7	Hacker (1965)
Mono-X	29.0	70.0	1.0	Hacker (1965)
Mono-XI	5.6	64.7	29.7	Hacker (1965)
Mono-XII	21.0	78.0	1.0	Hacker (1965)
Mono-XIII	4.3	92.3	3.3	Hacker (1965)

quency of $2n$, $2n - 1$, and $2n - 2$ zygotes can be calculated by using the checker board square as shown in Table 8.III. The assumed transmission rates of 75% for female and 4% for male of $n - 1$ gametes for wheat monosomics shown in the table are used widely for illustrative purposes. Actual male and female transmission rates are known for only two wheat monosomics. Sears (1944) reported male and female transmission rates of 19 and 81%, respectively, for mono-3B. The frequency of nullisomics in the selfed progeny was 10.3% which is somewhat lower than the expected frequency of 15.4%. Similarly, the male and female transmission in mono-3D was 9 and 70%, respectively, and the observed frequency of 5.8% of nullisomics was quite close to the expected frequency of 6.3%. These data suggest that some nullisomics of wheat may have slightly reduced viability while the others may be normal in this respect.

The reports on the viability of nullisomics in *Avena sativa* are indeed varied. Several workers have reported very high frequencies of nullisomics in selfed progenies of some monosomics (Table 8.IV). These frequencies can be explained on the basis of normal viability of nullisomics and the absence of any selection against $n - 1$ pollen grains in these monosomics. In the progenies of some monosomics, nullisomics fail to appear. From the data presented in Table 8.IV, it cannot be decided whether the failure to get nullisomics in these progenies is due to the inviability of nullisomic zygotes or the failure of the $n - 1$ male gametes to transmit. Data of Chang and Sadanaga (1964b) on six monosomic lines and of Hacker (1965) on three monosomic lines are instructive (Table 8.V). These workers determined the frequencies of transmission of $n - 1$ gametes through both sexes and were thus able to calculate the expected frequencies of nullisomics and compare them with the observed frequencies. Their data show that nulli-VIII of Hacker and nulli-D and nulli-F of Chang and Sadanaga are inviable, and further indicate that the viability of mono-VIII and mono-C is reduced. However, nulli-VII has normal viabililty. Nishiyama (1933) demonstrated that the most fertile "fatuoid" monosomic plants had the highest frequency of nullisomics and concluded that the lower frequencies could be explained by assuming the death of nullisomic embryos or zygotes. Costa-Rodriguez (1954) and Gauthier and McGinnis (1965) also concluded that the lower frequency of nullisomics in the progenies of certain monosomics can be explained by assuming a marked lethality of 40-chromosome zygotes. The available data indicate that some nullisomics of *Avena* may have normal viability, and others reduced viability or none. Nullisomic frequency seems to be dependent upon (1) environmental conditions, (2) genetic background, and (3) genotype environment interactions. The frequency of nullisomics in the progenies of the "fatuoid" monosomic was 0.0% (Chang and Sadanaga, 1964b), 3.2% (Huskins, 1927), 39.4% (Singh and Wallace, 1967b), and 50% (Nishiyama, 1931, 1933, 1951). Since these authors investigated the breeding behavior of the fatuoid monosomic in the background of different varieties grown under different environmental conditions, the influence of genotype and environment on nullisomic frequency is evident. The absence of nullisomics in twelve of the thirteen monosomic lines of *Avena byzantina* (variety Kanota) studied by Nishiyama *et al.* (1968) and their occurrence in three of the four monosomic lines of variety Victor Grain studied by (Singh and Wallace, 1967b) supports the above conclusion.

The monosomics of Table 8.V do not appear to be representative of the complement as far as transmission rates and frequency of nullisomics is concerned. As the data of Table 8.IV show and as O'Mara (1961) pointed out, nullisomics occur much more frequently in several monosomics of *Avena sativa* than they do in wheat, and sometimes they may make up as much as 90% of the progeny.

TABLE 8.V
Transmission Frequency of $n - 1$ Gametes through the Male and Female in Monosomics of *Avena sativa,* and the Observed and Expected Frequencies of Nullisomics in the Selfed Progenies

	Transmission frequency (%)		Nullisomics (%)		
Monosomic	Female	Male	Expected	Observed	Authority
Mono-VII	97.0	73.0	70.8	77.0	Hacker (1965)
Mono-VIII	89.0	10.0	8.9	0.0	Hacker (1965)
Mono-XII	87.0	0.0	0.0	4.0	Hacker (1965)
Mono-A	96.4	17.6	17.0	0.0	Chang and Sadanaga (1964a)
Mono-B	88.9	0.0	0.0	0.0	Chang and Sadanaga (1964a)
Mono-C	88.5	25.0	22.1	6.2	Chang and Sadanaga (1964a)
Mono-D	91.8	5.6	5.1	0.0	Chang and Sadanaga (1964a)
Mono-E	88.9	0.0	0.0	1.2	Chang and Sadanaga (1964a)
Mono-F	98.7	14.3	14.1	0.0	Chang and Sadanaga (1964a)

In some of the oat monosomics studied by different authors (Tables 8.IV and 8.V), the missing chromosome may be the same. As pointed out by McGinnis (1966), monosomic series are being established in several varieties of *Avena sativa*. These authors have not employed the same nomenclature for distinguishing their monosomic lines and the correspondence between the monosomics of various workers is not known except for the ''fatuoid'' monosomic which has clear-cut traits. The nomenclature employed by McGinnis and co-workers, which identifies the monosomic by the missing chromosome as mono-15 or mono-21, seems most appropriate.

Exceptional Progeny

In addition to yielding disomic, monosomic, and nullisomic individuals, the monosomic progenies yield aberrant offspring at low frequencies. Nishiyama (1928) recorded these types in the progeny of mono-5A (Spelt monosomic) of wheat. In the progeny of the fatuoid monosomic of *Avena sativa,* Nishiyama (1933) obtained 2% aberrant individuals. Exceptional plants have been recorded and investigated in the monosomics of *Nicotiana tabacum* by Clausen (1931), Olmo (1936), and Clausen and Cameron (1944). Certain monosomics of *N. tabacum* have given rise to lines showing variegation of flower color probably through formation of ring chromosomes, which are unstable and carry the gene for color (Clausen, 1930; Stino, 1940). Corresponding trisomics are produced occasionally in the progenies of monosomics as was observed by Müntzing (1930), Uchikawa (1941), and Sears (1954) in wheat. According to Olmo (1935)

TABLE 8.VI
Frequencies of Telocentrics and Isochromosomes among the Progenies of Different Monosomics of *Triticum aestivum*[a]

Monosomic	No. plants grown	No. examined cyto-logically	No. nullisomic	No. with telocentric	No. with iso-chromosome	% examined with telo or iso
1A	381	36	8	1	2	12.0
1B	671	75	16	4	0	5.3
1D	1109	90	27	2	0	2.2
2A	962	110	48	5	2	6.4
2B	255	31	6	3	2	16.1
2D	572	78	25	1	6	11.1
3A	125	28	3	3	0	10.7
3B	2682	174	205	12	5	9.8
3D	657	75	35	1	2	4.0
4A	2159	178	138	11	4	8.4
4B	192	25	7	2	0	8.0
4D	1924	108	113	1	0	0.9
5A	832	23	28	0	3	13.0
5B	1431	80	14	2	2	5.0
5D	575	48	5	0	5	10.4
6A	598	83	15	7	7	16.9
6B	1457	74	13	5	1	8.1
6D	1084	64	30	1	3	6.2
7A	331	88	11	2	0	2.3
7B	885	55	12	2	7	16.4
7D	1002	197	2	6	0	3.0

[a]From Sears, 1954.

and Clausen and Cameron (1944), the most important recurrent aberrancy is the oscillating relation between monosomics and trisomic counterparts. Crosses of monosomic × normal give a small percentage of trisomic counterparts; those of trisomic × normal, in turn, produce occasional monosomics. Rarely, double or triple monosomics are produced in the monosomic progenies, or a monosomic for the different chromosome may appear due to "univalent shift."

The most frequent and recurrent aberrants appearing in the monosomic progenies are the plants with telocentrics or isochromosomes. The cytological basis of these aberrants and others was discussed in Chapter 7. In wheat, monotelosomics and monoisosomics have been obtained for all the chromosomes (Sears, 1954). The data of Table 8.VI show that the frequency of plants with telocentrics and isochromosomes in the progenies of wheat monosomics may vary from 0.9 to 16.9%. Similar frequencies have been reported in the monosomics of *Avena sativa* by Gauthier and McGinnis (1965), and in the monosomics of *Avena byzantina* by Singh and Wallace (1967b).

Morphology of Monosomics

The monosomics of hexaploid species of wheat and oats do not differ appreciably from the normal disomics. Under favorable conditions of environment, they are difficult to distinguish from each other and from disomic sibs. The ''speltoid'' monosomic (mono-5A) of wheat and the ''fatuoid'' monosomic of oats, which have been the subject of numerous investigations, are identified easily by the distinctive morphology of their florets. Mono-5D of wheat is of distinctly later maturity than the normal (Sears, 1944). Under less favorable conditions, there may be some departures from the normal phenotype in the direction of nullisomic traits. Monosomic phenotypes may deviate more markedly from normal in some varieties than in others. The genetic background of a variety perhaps influences the expression of the monosomic phenotypes. According to Riley and Kimber (1961), a number of unidentified monosomics of variety Holdfast of wheat had distinctive phenotypes. The monosomics of *Avena sativa* and *Avena byzantina* are difficult to identify on the basis of morphology. In addition to the ''fatuoid'' monosomic, a few may be identified by some distinctive trait such as ''narrow leaf'' monosomic of Philp (1938), a monosomic with abaxial curling of the leaves (Chang and Sadanaga, 1964b), and ''side-oat'' monosomic with one-sided panicles (McGinnis and Lin, 1966). Because of the difficulty of identifying morphologically most of the monosomics of oats, the correspondence between the monosomics isolated by different workers has not been established.

The monosomics of tetraploid species are easier to identify than those of hexaploid species. They differ from the normal disomics in a specific ensemble of morphological features similar to those by which trisomics are distinguished. Most of the monosomics of *Nicotiana tabacum* can be separated on the basis of morphological deviations from the standard normal purpurea variety. Some, of course, are more distinct from the others. Although all parts of the plant are more or less distinctively affected, Clausen and Cameron (1944) based their classifications chiefly upon flower features because of their relatively greater stability under different growing conditions. Many flower characters such as tube length, limb spread, intensity of coloration, form of limb and of corolla lobes, prominence of the infundibulum, style and filament length, time of dehiscence of the anthers, size and character of the calyx and of the capsule, and numerous other quantitative features are modified differently in different monosomics. The vegetative characters such as size of plant, rate of development, intensity of chlorophyll coloration, leaf shape and character, and degree of development of auricles are variously modified.

All the known monosomics of tetraploid cotton, *Gossypium hirsutum* (AADD), can be distinguished morphologically (Brown and Endrizzi, 1964; Brown, 1966). Most of them have smaller plant habit, smaller leaves, modified flowers, smaller or slightly distorted bolls, and reduced fertility. Among the monosomics of the A genome, boll shape and proportion rather than size is affected. The reduction in number of seeds and increase in number of motes or aborted ovules cause boll distortion and a rough boll surface. Among the monosomics of the D genome thus far identified, the characteristic effect of monosomy is an overall reduction in boll size without distortion or change in shape at maturity. Open bolls of monosomics appear slightly smaller than those of normal plants because the amount of lint is less from reduced seed count or higher mote frequency. The monosomic plants can be separated from the normal sibs on vegetative characters alone. Mono-17 has narrow, three-lobed leaves, mono-2, mono-3, and mono-10 have small leaves, mono-1 has an upright growth habit and short lateral branches, and mono-6 has long calyx lobes and bract teeth.

The monosomics of tetraploid wheat, *Triticum durum*, are also distinguishable based on morphological traits (Mochizuki, 1968b). All the monosomics have reduced vigor and fertility. They differ from normal disomics and from each other in plant height, ear (spike) traits, awn length, leaf width and angle, culm thickness, and time of maturity. Mono-5A is as tall as the normal while mono-2A is semidwarf, its height being one-half of normal. The spikes of some are modified in length, thickness, and grain density (Fig. 8.1).

The monosomics of diploid species are extremely modified. They have poor viability, extremely slow growth rates, and very high sterility. Mono-4 of *Drosophila melanogaster* was called "diminished" because of its smaller size, shorter and more slender bristles, and reduced aristae. The other distinguishing characteristics of this monosomic are paler body color, darker thorax pattern, roughish and larger eyes, and slightly blunter and spread wings. The monosomic individuals are further characterized by lower productivity, frequent sterility, delayed hatching, and heavy mortality (Bridges, 1921a). Similarly, the monosomic plants of diploid oats, *Avena strigosa* (Andrews and McGinnis, 1964), maize (Einset, 1943; Baker and Morgan, 1966), and tomatoes (Khush and Rick, 1966b) are extremely weak and reduced in size.

Tomato monosomics are probably the best known of any diploid species. Two primary monosomics (mono-11 and mono-12) and more than twenty-five different tertiary monosomics have been studied (Khush and Rick, 1966b, 1968a) in detail. All the monosomics are extremely weak as seedlings and have about one-tenth the growth rate of normals. They differ from the normals in a syndrome of morphological characters of the same nature as those by which trisomics are distinguished. However, their morphology is more drastically modified than that of the trisomics. In general, all the morphological traits are modified in

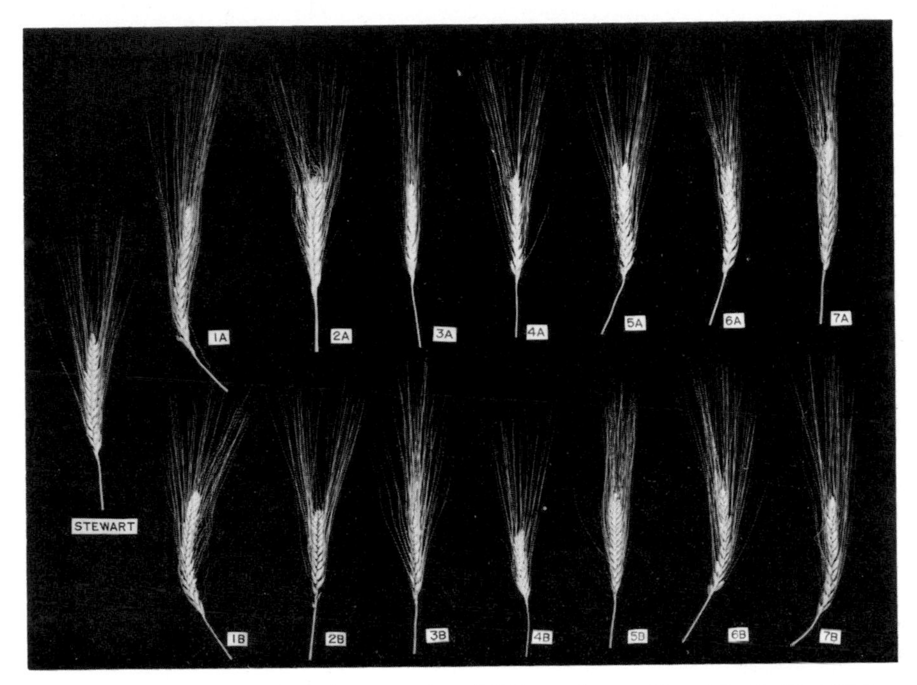

Fig. 8.1. Spikes of disomic and monosomics of *Triticum durum*. (From Mochizuki, 1971.)

the same direction in a particular monosomic. Mono-6S·11S, for example, has highly elongate leaves, and its internodes and flower parts are similarly affected. A diminutive tendency is manifest in the leaves, flowers, fruits, and compact growth habit of mono-3S·11L. The pollen fertility is extremely low (5–35% stainable pollen). Very few monosomics set fruit without artificial pollination. The seed fertility is 10–30% of the normal.

In general, the morphological features of monosomics differ from normal in the opposite direction from those of trisomics. Thus mono-F and mono-N of *Nicotiana tabacum* are small flowered, while triplo-F and triplo-N have large flowers. Mono-C on the other hand is large-flowered while triplo-C is distinctly small-flowered (Clausen and Cameron, 1944). A similar relationship is evident in tomato between mono-11 and triplo-11 as well as mono-12 and triplo-12. For example, in triplo-11, serration in the leaflets is increased, the leaf margins are curled downwards, and the density of hairs is increased. In mono-11, the serration is decreased, the leaf margins are curled upward, and the hair density is reduced (G. S. Khush, unpublished). A similar relationship is evident between mono-4 and triplo-4 of *Drosophila melanogaster* and triplo-21 and partial mono-21 ($2n - 1 + f$) of man (J. Lejeune, personal communication).

Breeding Behavior of Nullisomics

About half of the nullisomics of common wheat are either male or female sterile. The rest set some seeds upon selfing and can be maintained as nullisomic lines (Morris and Sears, 1967). Only eight are reasonably stable (Sears, 1954). Even in the progenies of these eight nullisomics, some aberrant plants are present (Table 8.VII). Out of a total of 191 plants in the progenies of these nullisomics, 17 plants had aberrant chromosomal constitution. The most common aberrants are the ones which are trisomic for one of the chromosomes besides being nullisomics (19 II + 1 III). In at least half of such nulli-trisomes, the trisome was for the homoeologous chromosome which partially compensated for the nullisome. Kihara and Wakakuwa (1935) and Kihara (1939) observed gigas plants in the progenies of nullisomics of the D genome. Such plants attracted their attention because of the greater vigor and height and improved fertility over the parental nullisomic. Matsumura (1952b, 1954) was able to confirm the earlier prediction of Kihara and co-workers that the gigas plants are nulli-trisomes having the extra chromosomes from the A or B genomes which is homoeologous to the nullisomic chromosome. The trisomes are produced as a result of irregular segregation of the unpaired chromosomes characteristic of meiosis of many nullisomics. When the $n - 1 + 1$ male gamete happens to have the extra chromosome, which compensates (homoeologous) for the missing chromosome, it is favored over $n - 1$ gametes. Hence, the nulli-trisomes for the compensating combinations are more common. The other aberrants in the selfed progenies of nullisomics include the plants which are monosomic for an additional chromosome besides being nullisomic (Table 8.VII).

TABLE 8.VII
Chromosome Constitution of Selfed Progenies of Nullisomics of *Triticum aestivum*[a]

Nullisomic	No. plants grown	No. with 20 II	No. with 19 II + 1 I	No. with 19 II + 1 III
1A	12	11	0	1
1B	68	65	1	2
1D	4	4	0	0
3A	9	9	0	0
6D	2	2	0	0
7A	15	10	3	2
7B	66	61	3	2
7D	15	13	1	1

[a]From Sears, 1954.

As mentioned in the last section, the meiosis in nulli-3B is irregular and the functional female gametes, all of which are of course deficient for chromosome 3B, may carry no additional chromosomal abnormalities or may be deficient for as many as two or three additional chromosomes, and duplicated for two or three others. Thus, 23.7% of the progeny of nulli-3B, in one season, and 44.1%, in the other, had aberrant chromosome numbers. Among 353 plants grown, 136 were aberrant (Sears, 1944).

The cross between a nullisomic and a normal plant is expected to produce only monosomic progeny. However, from a cross of Redman wheat nulli-7D × Prelude producing eight plants, only one was monosomic. Five of the others were triple monosomics, and two were double monosomics (McGinnis and Campbell, 1958, 1960).

The breeding behavior of only a few nullisomics of oats has been studied. Quite a number of them are asynaptic and sterile and yield no progenies upon selfing or backcrossing. At least one nullisomic of *Avena sativa,* which has normal meiosis, has a seed fertility of 0 to 55% depending upon the environment, the cooler temperatures being more favorable for high fertility (Lafever and Patterson, 1964b). One nullisomic of *Avena byzantina* studied by Ramage and Suneson (1958b), also with normal meiosis, sets seed upon selfing in about 10% of the florets. Manual cross-pollinations produced seed in about 25% of the florets.

Morphology of Nullisomics

Most of the nullisomics of *Triticum aestivum* variety Chinese Spring can be distinguished from the disomic and monosomic plants by seedling and mature plant traits and spike characteristics. The distinguishing traits of the corresponding monosomics, if any, are accentuated in the nullisomics. Most of the nullisomics have reduced size and vigor throughout their life cycle. In general, they have a smaller number of tillers and reduced plant height (Fig. 8.2). Some are delayed in maturity, and a few are earlier than the disomics. The leaves are broader in some nullisomics, narrower in others. The culms are similarly modified. However, the identification of specific nullisomics is usually made on the basis of spike characters (Fig. 8.3). The spike may be modified in density, length, and thickness; variations may appear in length, color, and size of awns, in stiffness of the outer and inner glumes, in size of the anthers, and in seed fertility. A thorough description of each nullisomic is beyond the scope of this volume, but interested readers are referred to two elegant publications on the subject by Sears (1944, 1954).

The nullisomics of oats also have modified morphology. The seedlings are more slender, have fewer tillers, are generally dwarfish, and less sturdy (Costa-

Fig. 8.2. Plants of wheat, variety Chinese Spring. From left to right, normal, mono-2A, nulli-2A. (From Sears, 1954.)

Rodriguez, 1954; O'Mara, 1961) . Various modifications of the panicle have been reported by Chang and Sadanaga (1964b), Hacker and Riley (1965), and Thomas (1966). These include reduction in floret number, degree of stiffness, degree of awning, compactness, size, and appearance. The nullisomics are easily identified from the disomics on the basis of vigor in the seedling stage. The differences in vigor continue to manifest at maturity (Fig. 8.4). Thus in the nullisomic of *Avena byzantina* studied by Ramage and Suneson (1958b), the height reduction was 45%, tiller reduction 50%, and florets per panicle 65% of the normal.

Conclusions

On the basis of results discussed in this chapter and summarized in Table 8.VIII, it is evident that different species do not have the same tolerance levels for monosomy and nullisomy. Nullisomy is tolerated only by the hexaploid species, *Triticum aestivum, Avena sativa,* and *Avena byzantina*. All the monosomics and nullisomics of *Triticum aestivum* are viable, and at least one-half

Fig. 8.3. Spikes of wheat, variety Chinese Spring. From left to right, normal, nulli-2A, nulli-5A, and nulli-3D. (From Sears, 1954.)

of the nullisomics can reproduce themselves. Monosomics for all the chromosomes of *Avena sativa* are probably available, and have been produced for at least twelve chromosomes of *Avena byzantina*. Nullisomics for only certain chromosomes of *Avena sativa* and *Avena byzantina* are viable, and the majority of the viable ones are sterile. The wheat monosomics are usually fully fertile, whereas oat monosomics have reduced fertility and complete sterility in the genotypic background of certain varieties. Fertility appears to be greatly influenced by minor variations in temperature and humidity and the interaction of environment and genotype. The viability of nullisomics of oats also appears to be under the influence of genotype and environment and their interaction. Therefore, it appears that monosomics and nullisomics of oats are less well buffered against the environment. This indicates that within the hexaploid group, wheat has higher tolerance for monosomy and nullisomy than oats.

The tetraploid species have no tolerance for nullisomy and as a group have lower tolerance for monosomy than hexaploids. In the hexaploid species, $n - 1$ spores and $2n - 1$ zygotes have normal viability, and the proportion of

Fig. 8.4. Plants of *Avena byzantina,* normal on left and nullisomic on right. (From *Agron. J.,* **50,** p. 52.)

monosomics in the progeny is comparable to the proportion of $n - 1$ spores produced by the monosomic (57.3–81.9% for wheat, 85–100% for oats). In the tetraploid species, on the other hand, the viability of the $n - 1$ spores and $2n - 1$ zygotes is impaired. Consequently, the proportion of monosomics in the progeny is much lower than the proportion of $n - 1$ spores produced by the monosomic. Among the tetraploid group, two species of *Nicotiana, N. tabacum* and *N. rustica,* have greater tolerance for monosomy than the other tetraploid species. All the possible monosomics of *N. tabacum* and seven of *N. rustica* have been produced; the transmission rates through the female vary from 5 to 86%. The average transmission rates and the seed fertility values of the monosomics are higher than those of monosomics of *Gossypium hirsutum* and *Triticum dicoccum*. The transmission rates of seven known monosomics

TABLE 8.VIII
Summary of Observations on Tolerance for Monsomy in Different Species

Species	Ploidy level	Chromosome number (n)	No. monosomics obtained	No. nullisomics obtained	Transmission rates (%)	Seed fertility (%)	Morphology
Triticum aestivum	Hexaploid	21	21	21	62–78	Normal	Normal
Avena sativa	Hexaploid	21	±21	±12	85–100	Variable	Normal
Avena byzantina	Hexaploid	21	12	4	90–100	28–84	Normal
Nicotiana tabacum	Tetraploid	24	24	0	5–78	6–100	Modified
Nicotiana rustica	Tetraploid	24	7	0	31–75	N.A.[a]	Modified
Gossypium hirsutum	Tetraploid	26	7 + 1[b]	0	20–40	56–80	Modified
Gossypium barbadense	Tetraploid	26	3 + 1[b]	0	N.A.	N.A.	N.A.
Triticum dicoccum	Tetraploid	14	14	0	3–41	9–28	Modified
Avena barbata	Tetraploid	14	3	0	N.A.	N.A.	Modified
Zea mays	Diploid	10	4	0	N.A.	N.A.	Modified
Datura stramonium	Diploid	12	4	0	None	Very poor	Highly modified
Lycopersicon esculentum	Diploid	12	2 + 25[b]	0	None	Very poor	Highly modified
Nicotiana alata	Diploid	12	1	0	None	Very poor	Highly modified
Nicotiana langsdorffi	Diploid	12	1	0	None	Very poor	Highly modified
Hyoscymus niger	Diploid	17	1	0	None	Very poor	Highly modified
Pharbitis nil	Diploid	15	1	0	N.A.	N.A.	Highly modified
Avena strigosa	Diploid	7	1	0	N.A.	N.A.	Highly modified
Drosophila melanogaster	Diploid	4	1	0	33.0	–	Highly modified

[a]Not ascertained.
[b]Tertiary monosomic.

of *Gossypium hirsutum* vary from 20 to 40% and are higher than those of *Triticum dicoccum* (3 to 41%). Not much is known about the monosomics of *Gossypium barbadense* and *Avena barbata*.

The diploid species have the lowest tolerance for monosomy. They tolerate monosomy only at the sporophytic level and only for certain of the smaller chromosomes. None of them transmit to the next generation because of inviability of the $n - 1$ spores. One monosomic each of *Nicotiana alata, N. langsdorfii, Hyoscymus niger, Pharbitis nil,* at least four of *Datura stramonium,* and two primary and more than twenty-five tertiary monosomics of *Lycopersicon esculentum* have been obtained, but all of them failed to transmit to the next generation. In the diploid species, even the smallest chromosomal deficiencies fail to transmit to the next generation.

Zea mays appears to be intermediate between the diploid and tetraploid groups in respect to tolerance for monosomy. Monosomics for four chromosomes of the complement, including the longest chromosomes, have been obtained. The breeding behavior of the monosomics has not been investigated, but several small and even large segmental deficiencies are known to transmit to the next generation. It appears that all the maize monosomics would likely be viable at the sporophytic stage, and some may even be viable at the gametophytic stage. This observation about the tolerance of maize for monosomy agrees well with the conclusions drawn in Chapter 6 about tolerance for trisomy.

The differential tolerance for monosomy by different species groups is also reflected in the morphology of the monosomics. The monosomics of hexaploids have unaltered morphology and normal vigor. The monosomics of tetraploid species have distinct morphology and reduced vigor. Those of diploid species have extremely reduced vigor and very distinct morphology.

9

Genetic Studies and Other Uses of Monosomics and Nullisomics

In the polyploid species, genetic analysis for quantitative and qualitative characters is somewhat difficult and less conclusive compared to similar analyses in diploid species. This difficulty is caused by the presence of gene loci, governing the same trait, on homoeologous chromosomes belonging to different genomes. These homoeologous gene loci can interact additively or nonadditively and greatly influence the phenotypic expression of a trait. The feasibility of monosomic analysis in tetraploids and monosomic and nullisomic analysis in hexaploids greatly facilitates genetic study of these species. In addition, monosomics can be employed in the solution of certain practical problems. The first part of this chapter deals with monosomic and nullisomic analysis; in the second part, other uses of monosomics and nullisomics are reviewed.

Monosomic Analysis

The most efficient method of associating genes with their respective chromosomes is by monosomic analysis. In its essential features, the method is similar to sex-linked inheritance where the heterogametic sex is monosomic for the sex chromosome and, hence, hemizygous for genes located in it. Thus, if a complete set of monosomics is available, some of the well known advantages of sex linkage in the study of gene-chromosome relationships become applicable to all chromosomes of the set. The number of generations studied and the procedures followed for monosomic analysis vary slightly according to the gene action involved. Monotelodisomics, which are hemizygous for one chromosome arm, can be employed for locating genes to chromosome arms and for estimating the map distance of the marker from the centromere.

Locating Recessive Genes

The method consists of crossing genetic marker stocks with each of the monosomics. Part of the F_1 population is monosomic, and if the gene is recessive, the monosomic plants of the critical F_1 family are all of the recessive phenotype. In the other families, no plants with the recessive phenotype will appear. Thus, even if it is difficult to distinguish between monosomic and disomic fractions of the F_1 progenies, the appearance of recessives in a particular family is clear evidence that the gene is located in the chromosome which is monosomic in that family. The crosses may be represented as follows:

Marker and the monosomic associated:

P_1 monosomic (dominant) × P_2 disomic (recessive)

F_1 disomic (dominant) + monosomic (recessive)

Marker and the monosomic not associated:

P_1 monosomic (dominant) × P_2 disomic (recessive)

F_1 disomic (dominant) + monosomic (dominant)

This is the simplest type of monosomic analysis—the critical information is obtained simply by raising the F_1 progeny. The simplicity and usefulness of the method was first demonstrated by Bridges (1921a). He crossed mono-4 *Drosophila melanogaster* with the recessive mutants *bent* and *eyeless*. All the F_1 mono-4 individuals were pseudodominant, i.e., expressed the recessive phenotype. Similarly, Clausen and Goodspeed (1926b) crossed mono-C of *Nicotiana tabacum* with the recessive white flower color mutant. All the disomic plants of the F_1 family had carmine flowers (normal) while all the monosomic plants had white flowers. This information led to the conclusion that the gene for flower color was located on chromosome C.

Monosomic analysis is feasible only in those species in which the monosomic condition is transmitted to the next generation. Consequently, it cannot be applied to the diploid species (except for mono-4 of *Drosophila melanogaster*). However, in the diploids, it is possible to associate a few genes with their respective chromosomes through radiation-induced monosomy. As an example, the *a–hl* linkage group of tomato had not been associated with its chromosome. A marker stock homozygous for these two recessive genes was pollinated with radiated pollen carrying the normal alleles. Nine out of 2357 plants of this cross were pseudodominant for *a* and *hl*. Upon cytological examination these nine plants turned out to be monosomic for chromosome 11. This permitted the localization of *a–hl* linkage group to chromosome 11 (Rick and Khush, 1961). However, this is not monosomic analysis in the strict sense of the word. This technique is limited to those members of the complement which tolerate monosomy at the sporophytic level. Compared to simple monosomic analysis, the method is extremely laborious and very large progenies must be grown to obtain a few monosomics. The use of special types of trisomics may permit monosomic analysis in diploid species, as discussed in Chapter 5, but due to the rarity of such trisomics, the technique appears to be of limited value.

Locating Dominant Genes

If the marker to be located is dominant, it is crossed with each monosomic. All the F_1 offspring of all the families express this trait. The F_2 populations are grown from the monosomic F_1 plants and in the critical F_2 family no recessive segregates appear. The cross may be represented as follows:

P_1 monosomic (recessive) × P_2 disomic (dominant)

F_1 disomic (dominant) + monosomic (dominant)

F_2 disomic (dominant) + monosomic (dominant)—critical family

F_2 disomic (3 dominants:1 recessive) + monosomic (3 dominants:1 recessive)—noncritical family

BC disomic (dominant) + monosomic (recessive)—critical family

All the monosomic F_1's, except one, are heterozygous for the dominant allele and segregate three dominants to one recessive in the F_2. The F_1 monosomics in the critical family are hemizygous for the dominant allele, and no recessives appear in their progenies. Thus, an F_1 monosomic plant from a cross of mono-6 of cotton with Lc_2 marker was selfed and an F_2 population of seventeen plants

TABLE 9.I
Segregation for Red Glume Color in F_2 Families from Crosses of Red Glumed Variety (Federation 41) with Seventeen Different Monosomics of Chinese Spring Wheat (White Glumed)[a]

Chromosome tested	Red glumed plants	White glumed plants		Total plants	X^2 for 3:1 ratio
		No.	Percentage		
1B	528	38	6.7	566	100.93[b]
1D	1308	417	23.6	1725	0.63
2A	637	193	23.2	830	1.35
2D	314	99	23.9	413	0.23
3B	628	219	25.8	847	0.33
3D	814	264	24.5	1078	0.15
4A	600	196	24.5	796	0.06
4B	690	213	23.5	903	0.96
4D	708	224	24.2	932	0.47
5A	248	73	22.7	321	0.87
5B	652	202	23.6	854	0.83
5D	578	174	23.0	752	1.39
6A	748	228	23.3	976	1.45
6B	797	248	23.8	1045	0.89
6D	679	212	23.7	891	0.69
7A	915	292	24.2	1207	0.42
7B	659	208	23.9	867	0.47

[a]From Unrau, 1950.
[b]Significant at 1% level.

was grown (Endrizzi, 1963). There were eight disomics and nine monosomics and all were Lc_2. The absence of recessives in the family indicated that Lc_2 was located on chromosome 6. This association was further confirmed by crossing the F_1 mono-6 as female to a recessive tester stock. The backcross family consisted of eleven disomics, all dominant, and eight monosomics, all recessive.

In the above discussion it was assumed that nullisomics do not appear in the progenies of monosomics, which is indeed the case in tobacco and cotton. If nullisomics do appear, as in wheat and oats, all the F_2 families are characterized by the presence of some recessive segregates. However, all F_2 families except one segregate 25% recessives, while the critical F_2 has a much lower proportion of recessives—generally less than 10%. All the recessives in the critical family are nullisomics which carry neither the dominant allele nor the recessive but express the recessive phenotype. Since the proportion of nullisomics in the progenies of wheat monosomics rarely exceeds 10% and generally is much lower, the critical family is easily detected by the marked deficiency of the recessives. As an example, Unrau (1950) crossed the stock homozygous for the dominant marker gene red glume color with seventeen different monosomic lines. As the data of Table 9.I show, sixteen families gave excellent fit to the 3:1 ratio expected for disomic segregation. In the F_2 of mono-1B, however, only 6.7% recessives appeared, and the deviation from 3:1 was highly significant. The gene for red glume color was thus located on chromosome 1B. This conclusion was further confirmed by the fact that all the recessives in the mono-1B family were nullisomics. Actually it is possible to identify the critical family by growing F_2 populations with as few as twenty plants each. Sears (1953) grew F_2 populations of twenty to twenty-four plants each, from crosses to the monosomics of the pubescent-glumed variety Indian of common wheat. In twenty of the twenty-one F_2 populations, there was no significant deviation from the ratio 3 pubescent: 1 nonpubescent, and the nonpubescent plants in each family included some nonnullisomics. In the F_2 population, involving chromosome 1A, however, there were nineteen nonnullisomics, all pubescent, and one nullisomic nonpubescent. From these data the pubescent gene was placed on chromosome 1A.

If the nullisomics appear with a frequency of about 25%, then the proportion of the recessives in all the families will be similar. It would still be possible, however, to identify the critical family by checking the nullisomic fraction of each family. All the nullisomics in the critical family will be recessive, in the other families only 25% of the nullisomics will show the recessive phenotype. The frequency of nullisomics in the progenies of monosomics of oats varies from 0 to 86% (Table 8.IV). The proportion of recessives in the critical F_2 families of the crosses of oat monosomics would thus be quite variable. Some would have no recessives, while in others the recessives would form the preponderant class.

Locating Complementary Genes

In the polyploid species, several traits are governed by duplicate loci. It is feasible to locate these loci to specific chromosomes by means of monosomic analysis. For example, the yellow burley trait of *Nicotiana tabacum* was known to be under the control of two mutant loci. Crosses of normal $2n$ and yellow burley varieties segregate 15 normal:1 yellow burley. One of the factors, yb_1, was located in chromosome B, the other, yb_2, in chromosome O by Clausen and Cameron (1944, 1950) using monosomic analysis. Some of the critical crosses for locating yellow burley on specific chromosomes are summarized in the following tabulation:

	Phenotypes		Genotypes	
P_1	mono-B (green) × yellow burley		$yb_1^+ - yb_2^+ yb_2^+ \times yb_1 yb_1 yb_2 yb_2$	
F_1	disomic (green) + mono-B (green)		$yb_1^+ yb_1 yb_2^+ yb_2 + yb_1 - yb_2^+ yb_2$	
F_2	76 green:6 yellow	66 green:23 yellow	↓	↓
Ratios	15:1	3:1	15:1	3:1

The results show beyond doubt that yb_1 is located on chromosome B, which belongs to the *tomentosa* genome. It was assumed that yb_2 might be located in a member of the *sylvestris* genome. Consequently, further tests were confined to the nine monosomics of the *sylvestris* genome available at that time. These monosomics were made heterozygous for yellow burley and backcrossed to yellow burley. Eight monosomics segregated approximately 3 normals to 1 yellow burley, a disomic ratio expected for a digenic segregation. The progeny of mono-O, however, segregated 50% yellow burley. These results implicated yb_2 to chromosome O of the *sylvestris* genome. The crosses with mono-O are shown in the tabulation below:

	Phenotypes	Genotypes
P_1	mono-0 (green) × yellow burley	$yb_1^+ yb_1^+ yb_2^+ - \times yb_1 yb_1 yb_2 yb_2$
F_1	disomic (green) + mono-0 (green)	$yb_1^+ yb_1 yb_2^+ yb_2 + yb_1^+ yb_1 yb_2 -$
BC[a]	19 green : 17 yellow	
Ratios	1:1	1:1

[a]BC, backcross.

Thus recessive traits governed by two genes segregate 15:1 in the F_2's of all heterozygous monosomics except two. In the two critical families the ratio is modified to 3:1.

Locating Genes by Absence of Expression in Nullisomics

Certain dominant genes can be located simply by observing the absence of their effect in the appropriate nullisomic. The variety Chinese Spring of common wheat has a dominant gene for red seeds. Nulli-3D has white seeds. Therefore, the gene for red seed color of Chinese Spring must be on chromosome 3D. Similarly, nulli-4B and nulli-6B have increased awn development, and each of the chromosomes 4B and 6B must carry a dominant gene for awn inhibition (Sears, 1953). Several genes have been located in *Avena sativa* by this technique. The nullisomic studied by Philp (1935) was albino. He concluded that a gene for chlorophyll production was located on the chromosome pair missing in the nullisomic but was unable to identify the chromosome involved. McGinnis and Taylor (1961) and McGinnis and Andrews (1962) associated two loci for chlorophyll production with chromosome 21 and 15, respectively, by nullisomic analysis.

In wheat there is a class of dominant genes which cannot be located in this manner because the critical nullisomic still shows the dominant phenotype. For example, variety Chinese Spring carries the dominant, *nonsphaerococcum* gene on chromosome 3D. Nulli-3D's are *nonsphaerococcum,* however. The recessive alleles of such genes are hemizygous ineffective and can be located by studying the F_2 populations.

Locating Hemizygous Ineffective Genes

The hemizygous ineffective recessives referred to above cannot be located simply by observation of F_2 ratios; closer study of the F_2 segregates is required. In the critical F_2 families from crosses with the monosomics or nullisomics, these genes are expressed only in the disomic segregates. Both the monosomics and the nullisomics express the dominant phenotype. Since, in wheat, the disomics appear at a frequency of approximately 25% (Table 8.III) , the critical F_2 also segregates 3 dominants:1 recessive and does not differ significantly from the other F_2 families in this respect. The critical family can, however, be identified by cytological examination of the recessive segregates, all of which would be disomic. In the other families, only about 25% of the recessives would be disomic. Thus, as soon as a monosomic or nullisomic with recessive phenotype is identified, examination of the family can be discontinued.

The recessive gene *s* (*sphaerococcum*) mentioned earlier was located in this manner by Sears (1947). The *s* stock was crossed with seventeen nullisomics.

All the F_1 monosomics were phenotypically normal, and all the F_2 families segregated 3 dominants:1 recessive. In the F_2 family involving chromosome 3D, there were fourteen dominants (*nonsphaerococcum*), including one nullisomic, and six recessives (*sphaerococcum*). Upon cytological examination, all the recessives were found to be disomics. These results were conclusive in locating *s* in chromosome 3D. Two other hemizygous ineffective genes, *v (Neatby's virescent)* and *q (speltoid)*, were located in this manner in chromosomes 3B and 5A, respectively.

Locating Genes by Studying Selected F_3 Populations

If the scoring of an F_2 population is difficult due to environmental interaction or if several genes are to be located requiring test procedures, it may be advisable to study F_3 progenies from four or five disomic individuals selected by cytological examination in the F_2 generation. All the disomics from the critical family will be homozygous for the gene under study.

Hemizygous ineffective genes may also be located by studying F_3 populations. The critical family can be distinguished in the F_3 since the dominant F_2 phenotypes from the F_1 monosomic plants are all monosomic and all will segregate in the F_3.

Locating Genes by Making Chromosome Substitutions

Some traits such as certain types of disease resistance are difficult to classify in the segregating populations. This difficulty is further exaggerated if modifying genes are also segregating along with major genes. Under these conditions single chromosome substitutions can be used for locating genes. This method involves transferring each chromosome of the variety carrying the trait to the control variety by means of backcrosses to the respective monosomics; the effect of the substituted chromosome is then accurately ascertained. Several genes in wheat have been located by this technique.

Since substitution lines have been used in the genetic analyses of qualitative as well as quantitative traits, various methods of producing substitution lines and the kinds of genetic analyses possible with them are discussed in the next chapter.

Locating Genes in the Parental Species

Monosomics of a polyploid species may also be utilized in determining gene–chromosome relationships in the parental species. For example, a recessive gene in *Nicotiana tomentosa,* a diploid progenitor of tetraploid *N. tabacum,*

TABLE 9.II
Differences in Days to Flowering between Monosomics and Disomics in F_1 Families of Crosses between Redman Monosomics and Caid Eleize Wheat[a]

	Monosomics		Disomics		Delayed flowering of monosomics in days
F_1 family	No. of plants	Average days to flowering	No. of plants	Average days to flowering	
1A	11	66.8	5	69.4	−2.6
2A	10	71.2	5	72.0	−0.8
3A	9	73.4	5	72.6	0.8
4A	11	71.0	5	72.4	−1.4
5A	12	80.9	7	69.3	11.6[b]
6A	10	71.8	6	7.5	1.3
7A	14	70.6	6	69.8	0.8
1B	9	72.1	6	72.3	−0.2
2B	10	71.8	6	69.2	2.6
3B	12	70.8	6	70.5	0.3
4B	11	72.9	7	71.0	1.9
5B	13	72.2	7	69.4	2.8[c]
6B	11	68.6	6	69.5	−0.9
7B	11	71.9	6	71.5	0.4
Redman × Caid Eleize			17	70.6	

[a] From Kuspira and Millis, 1967.
[b] Significant at 1% level using t test.
[c] Significant at 5% level using t test.

may be located by crossing the marker stock as a male to the monosomics of the *tomentosa* genome of *N. tabacum*. The monosomic progenies will consist of disomic (12 II + 12 I) and monosomic (11 II + 13 I) plants. All the monosomic plants in the critical family will be recessive, but no recessives will appear in the other families. The monosomics of the *sylvestris* genome may be similarly employed for locating recessive genes in *N. sylvestris*.

The above type of cytogenetic analysis has not been applied in either *Nicotiana* or *Gossypium,* but the genes for winter growth habit in tetraploid wheat were located by Kuspira and Millis (1967) using monosomics of hexaploid wheat. Variety Caid Eleize of *Triticum durum* (AABB) with winter growth habit was crossed as male parent with monosomics for the A and B genomes of hexaploid wheat, variety Redman, which has spring growth habit. The F_1 progenies were analyzed cytologically to distinguish the disomic (14 II + 7 I) plants from the monosomics (13 II + 8 I). There was clear-cut segregation for growth habit, expressed in terms of delayed flowering in the mono-5A family. All the monosomic plants of this family were distinctly late in flowering (Table 9.II), indicating that they were hemizygous for winter growth habit (a recessive trait). The gene for winter growth habit was therefore located to chromosome 5A of tetraploid wheat. The data indicated that chromosome 5B of Caid Eleize may also carry a gene or genes for growth habit, but the segregation was not clear-cut. Actually, cytological analysis of the F_1 is not essential for the establishment of gene–chromosome associations. For example, the gene or genes located on chromosome 5A could have been associated with this chromosome because of obvious segregation for growth habit in the F_1 family of this monosomic.

It should be noted that only recessive genes or incompletely dominant genes can be located by this method and that hemizygous ineffective genes or dominant genes cannot.

Locating Genes to Chromosome Arms

After a gene has been located to a specific chromosome, its arm location can be determined by the use of monotelodisomics for either of the two arms. The monotelodisomic as the female is crossed to the stock homozygous for the gene to be located. The recessive gene is located simply by observing the F_1 monotelodisomics. If the gene is located on the arm missing in the telocentric chromosome, all the F_1 monotelodisomics express the recessive phenotype, and all the disomics are normal. However, if the gene is located on the telocentric arm, all the progeny are normal. Thus, the *cluster flower* (cl_1), a a recessive mutant of cotton, was known to be on chromosome 16. Endrizzi and Kohel (1966) crossed cl_1d_1 stock with mono 16L. The F_1 monotelodisomics were normal flowered and cl_1 was concluded to be in 16S.

If the marker to be located is dominant, F_2 or backcross populations are grown. For the markers situated on the arm missing in the telocentric, no normal (recessive) phenotypes appear in the F_2 or the backcross generation. The cross may be represented as follows:

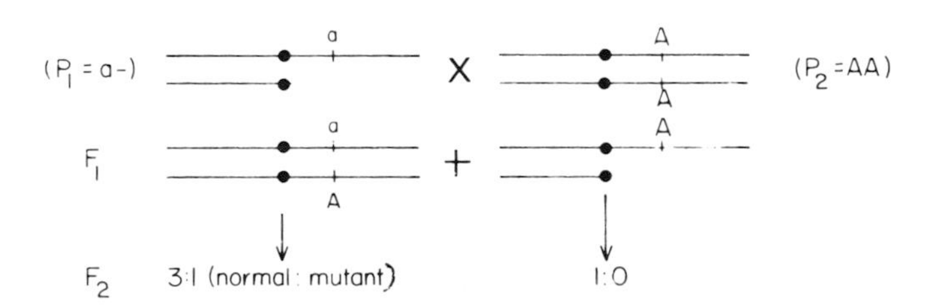

Similarly, in the backcross population all the disomics will have the mutant (dominant) phenotype, and the monotelodisomics, if any, will have the normal (recessive) phenotype.

It may be noted, however, that if the marker is located on the telocentric arm but is tightly linked to the centromere, similar results may be obtained. Tests with the second monotelodisomic or the segregation of a linked marker would usually be helpful in differentiating between the two possibilities. Thus two dominant mutants of cotton, H_2 and *Lc,* were known to be on chromosome 6. In order to determine their arm location, mono-6S and mono-6L were made heterozygous for these markers, and the F_1 monotelodisomics were crossed as males to the homozygous recessive stock (Endrizzi and Kohel, 1966). The data on their backcross progenies as well as the control backcross are given in Table 9.III. In the backcross population of mono-6S, 53 of the 405 plants were mono-6S, which resulted from the functioning of $n-6S$ gametes. All of these were *lc*. The remaining 352 plants were disomic and all were *Lc*. These results indicated that *Lc* should be either in the short arm or in the long arm closely linked to the centromere. Four plants (3 disomic and 1 monotelodisomic) were crossovers, giving a crossover value of 1.0% between H_2 and the centromere; this placed H_2 in the long arm. Since the crossover distance between H_2 and *Lc* is 22.12%, *Lc* cannot be between H_2 and the centromere and must, therefore, be on the short arm.

Though conclusive for placing *Lc* on the short arm, these results were verified from the backcross family of mono-6L. Only disomic plants were obtained and all were H_2, again showing that H_2 is in the long arm. Sixteen of the 360 plants were *lc,* which were crossovers giving a crossover value of 4.4% between the centromere and *Lc*. This also demonstrated that *Lc* is in the short arm.

One aspect of the genetic data presented in Table 9.III is noteworthy. In mono-6S, only 1.0% recombination was observed between H_2 and the centromere. In the study with mono-6L, 4.4% recombination was obtained between the centromere and *Lc*. The combined percentages give 5.4 units between H_2 and *Lc,* which is much less than the 22.1 units obtained between H_2 and *Lc* in the disomic control family. The absence of an arm caused reduction in recombination in the centromere region. According to the data presented by Endrizzi and Kohel (1966), nonpairing of the heteromorphic pairs occurred in less than 1.0% of the cells in cotton monotelodisomics. Therefore, asynapsis could account for only a very small fraction of this reduction. Thus in the monotelodisomics there seems to be a suppression of recombination in the centromere region. The application of the techniques described above in determining the arm location of the genes has been limited to cotton. Endrizzi and Kohel (1966), White (1966), and Endrizzi and Taylor (1968) using six monotelodisomics ascertained the arm locations of thirteen genes of cotton belonging to five linkage groups.

TABLE 9.III
Genetic Segregation in Backcrosses of Disome, Mono-6S, and Mono-6L F_1's and Crossover Values between the Linked Markers and between the Markers and Centromere in Cotton[a]

Family[b]	Backcross	$H_2\,lc$	$H_2\,Lc$	$h_2\,lc$	$h_2\,Lc$	Total
Diplo	$\frac{h_2 \quad lc}{h_2 \quad lc} \times \frac{h_2 \quad Lc}{H_2 \quad lc}$	82	21	27	87	217
Mono-6S	$\frac{h_2 \quad lc}{h_2 \quad lc} \times \frac{h_2 \quad Lc}{H_2}$	(52)[c]	3	(1)	349	405
Mono-6L	$\frac{h_2 \quad lc}{h_2 \quad lc} \times \frac{H_2 \quad Lc}{lc}$	16	344	0	0	360

[a]From Endrizzi and Kobel, 1966.
[b]Percentage crossover values: diplo, H_2–Lc = 22.12 ± 2.81; mono-6S, H_2–*centromere* = 0.99 ± 0.49; mono-6L, centromere –Lc = 4.44 ± 1.09.
[c]Numbers in parentheses in the two classes are individuals resulting from functioning of $n - 6S$ pollen.

In the hexaploid species where nullisomy is tolerated, monotelosomics, ditelosomics, or monoisosomics have been employed for determining the arm location of the marker. These chromosomal variants result from misdivision of univalents in monosomics and are homozygous deficient for one chromosome arm. They can be used for locating markers to specific arms by absence of expression as the nullisomics. Sears (1954) located most of the known genes of the Chinese Spring wheat to chromosome arms by observing whether the character concerned was expressed in the monotelosomic or the ditelosomic. Thus the genes for speltoid suppression, pubescent node, and spring habit of growth are located on chromosome 5A. Monotelosomics and monoisosomics for 5AL are normal, i.e., have nonspeltoid spikes, nonpubescent nodes, and spring growth habit. These three genes must therefore be located on 5AL. Love (1938, 1943) examined the off-type plants with white chaff occurring in variety Dawson which has golden chaff. The off-types were found to be ditelosomic, and the white chaff trait was associated with homozygous deficiency for one chromosome arm. However, at that time the individual chromosomes of wheat had not been identified, and the arm carrying the gene or genes for golden chaff of Dawson was not identified.

Nullisomics for chromosome 14 of *Avena sativa* are albino. Therefore, it is evident that chromosome 14 carries a gene for chlorophyll production. Monotelosomics for the short arm of chromosome 14 are also albino. The gene for chlorophyll production must, therefore, be located on 14L (McGinnis *et al.*, 1963) . Nullisomics for chromosome 20 have kinky neck and show abaxial curling of leaves. Ditelosomics for 20L have normal neck and abaxially curled leaves. Gene(s) for kinky neck must therefore be on 20L and those for abaxial curling on 20S (Dubuc and McGinnis, 1970). Genes for regular synapsis and fatuoid suppression were similarly located on the short arm and long arm, respectively, of the same chromosome by Thomas and Mytton (1970).

Chromosome Mapping

Sears (1962) pointed out the use of monotelodisomics in chromosome mapping and later presented detailed data on the technique and discussed its advantages. For example, chromosome 6B of common wheat carries several genes whose arm locations were known. The marker *co* was known to be on the short arm and *B*2, *Sr*11, *Ki*, and *Lr*9, on the long arm. However, the crossover distances of these genes from the centromere and their order on the long arm were not known. Sears (1966a) determined these using monotelodisomics.

In one experiment, telo 6BL carrying *B*2 and *Sr*11 was combined with the normal chromosome carrying *b*2, *sr*11, and *Co*. Pollen from these heterozygous monotelodisomics was used for pollinating plants hemizygous for 6BS and homozygous deficient for 6BL. The short arm telocentric carried *Co*. All the

TABLE 9.IV

Crossing over between *B*2, *Sr*11, and the Centromere of Chromosome 6B of *Triticum aestivum*[a]

Co × Sr 11 B2 / Sr 11 b2 Co

No. of plants	Constitution of male gamete	Region of crossover
126	*sr*11 *b*2 *Co*	None
99	*Sr*11 *b*2 *Co*	*Sr*11–*B*2
1	*Sr*11 *B*2 –	None
1	*sr*11 *B*2 –	*Sr*11–*B*2
3	*Sr*11 *B*2 –/*sr b Co*	None
1	*Sr*11 *B*2 *Co*	*B*2–centromere
231		

[a]From Sears, 1966a.

female gametes were deficient for *B*2 and *Sr*11 and were, therefore, effectively *b*2 and *sr*11. About 75% were completely deficient for the short arm and were effectively *co*, and 25% of the female gametes received the short arm telocentric and were *Co*. As the data of Table 9.IV show, only six *B*2 plants appeared. Two of these had the paternal telocentric, as evidenced by their being *co*. Among the remaining four, three had both paternal chromosomes and thus were noncross-overs, and one had the normal paternal chromosome and was, therefore, a crossover. The crossover value for the centromere-*B*2 region was found to be 1/231 or 0.44%. There were 100 crossovers between *Sr*11 and *B*2 and one between the centromere and *B*2. Thus there were 101/231 crossovers giving a crossover value of 43.7% between *Sr*11 and the centromere.

The crossover distance between *Ki* and *Sr*11 was known to be about 9%. In order to determine the gene order of *B*2, *Sr*11, and *Ki* on the long arm, another experiment was carried out. Mono-6BS plants heterozygous for *Ki* and *Sr*11 were used as the females in crosses with *ki ki*, *sr*11 *sr*11, and *co co* testers. The data of Table 9.V show there were 40/99 crossovers between *Ki* and the centromere giving a map distance of 41.4%, 9/99 crossovers between *Ki* and *Sr*11 giving a map distance of 9.1% and 50/99 crossovers between *Sr*11 and the centromere giving a map distance of 50.5%. These data indicate the gene order is *co*–centromere–*B*2–*Ki*–*Sr*11. The fact that only one of the nine crossovers in the *Sr*11–*Ki* region was also a crossover between *Ki* and *B*2 confirms the above gene order. If the order of *Sr*11 and *Ki* were reversed, eight of the nine would be double crossovers, an obvious impossibility.

TABLE 9.V
Crossing over in the Regions *Sr* 11–*Ki* and *Ki*–Centromere of Chromosome 6B of *Triticum aestivum*[a]

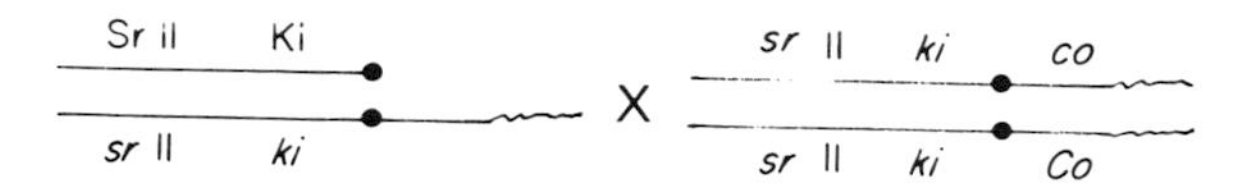

No. of plants	Constitution of female gametes	Region of crossover
21	*Sr*11 *Ki* –	None
29	*sr*11 *ki Co*	None
22	*Sr*11 *Ki Co*	*Ki*–centromere
18	*sr*11 *ki* –	*Ki*–centromere
6	*Sr*11 *ki Co*	*Sr*11–*Ki*
2	*sr*11 *Ki* –	*Sr*11–*Ki*
1	*sr*11 *Ki Co*	Both
99		

[a]From Sears, 1966a.

The map distance between the centromere and *v* (*Neatby's virescent*) on 3BS was determined to be 0.28% through the use of mono-3BL (Sears, 1966a). Thus two genes were located within one crossover unit of the centromere in wheat. Sears suspected that crossing over may be reduced near the centromere in heteromorphic bivalents. As discussed in the previous section, Endrizzi and Kohel (1966) showed convincingly that, in the heteromorphic bivalents of cotton, crossing over in the regions adjacent to the centromere is reduced to as low as 25% of the normal. However, no serious reduction in crossing over was encountered in the distal regions of heteromorphic bivalents of cotton.

Several genes in wheat have been mapped using this technique. Law and Wolfe (1966) mapped *ml* and *e*2 on the long arm of chromosome 7B. Briggle and Sears (1966) mapped the *Pm*3 and *Hg* genes on the short arm of chromosome 1A. Sears and Loegering (1968) mapped *Sr*9a and *Sr*16 on the long arm of chromosome 2B. Sears and Briggle (1969) mapped *pml* on the long arm of chromosome 7A. In cotton, nine genes belonging to three linkage groups were mapped by using this technique (Endrizzi and Kohel, 1966).

Genes Located through Monosomic Analysis

According to Cameron (1966), forty mutant loci have been associated in *Nicotiana tabacum* through monosomic analysis. These markers were associated with seventeen of the twenty-four chromosomes of the complement. As yet no markers have been found for the remaining seven. In cotton (*Gossypium hirsutum*), eight monosomics have been obtained. Mutant loci have been associated with six of the chromosomes through the use of these monosomics (Endrizzi, 1963; Endrizzi and Brown, 1964; White and Endrizzi, 1965; White, 1966; Endrizzi and Taylor, 1968).

Several authors have isolated monosomics of oats (*Avena sativa*). Chang and Sadanaga (1964a) and Sadanaga (1970) associated six mutant loci and four genes for disease resistance with five chromosomes using the monosomic technique. In addition, associations were found between a monosomic and chlorophyll production (Philp, 1935), a monosomic and broad leaves (Philp, 1938), three monosomics, each associated with a different gene for chlorophyll production (McGinnis and Andrews, 1962; McGinnis and Taylor, 1961; McGinnis *et al.*, 1968), a monosomic and panicle shape (McGinnis and Lin, 1966), and a monosomic and kinky neck (Gauthier and McGinnis, 1965).

A large number of mutant genes have been located in common wheat by monosomic and nullisomic analyses as well as by the substitution method (Table 9.VI). The arm locations of several of these markers have also been determined and a few have been mapped using monotelodisomics.

Other Uses of Monosomics and Nullisomics

Besides their use in gene location work, the monosomics and nullisomics have been employed in several other cytogenetic investigations and in practical plant breeding projects.

Identification of Chromosomes Belonging to the Different Genomes

Since polyploid species have two or more distinct genomes, the missing chromosome of the monosomic can be identified in relation to the basic genomes. Goodspeed and Clausen (1928) reported that *N. tabacum* ($2n = 24$) has the genomes of *N. tomentosa* ($2n = 12$) and *N. sylvestris* ($2n = 12$). Crosses of the various *N. tabacum* monosomics with two parental species were used to identify the chromosomes of each genome. The crosses of *N. tabacum* with both of the parental diploid species exhibit 12 II + 12 I as the modal condition for chromosome pairing. The crosses of monosomics of *tabacum* and the diploid progenitors are of two classes: one with a modal association of 12 II + 11 I, and the other, 11 II + 13 I; the former indicates that the missing chromosome does not belong to the species used in the cross, but the later does. Clausen and Cameron (1944) were able to assign all the individual chromosomes of *N. tabacum* either to the *tomentosa* genome or the *sylvestris* genome by this technique.

TABLE 9.VI
List of Major Genes for Wheat Characters Located by Means of Monosomic and Nullisomic Analyses and by Means of Substitution Method[a]

Variety source	Gene	Character	Location		Method[b]	Authority
			Chr.	Arm		
Indian, Sonora, Jones Fife, Prelude	*Hg*	Pubescent glume	1A	S	Nulli. Mono.	Sears (1953); Kuspira and Unrau (1960); Tsunewaki (1966a); Briggle and Sears (1966)
Parker	*Lr* 10	Leaf rust resistance	1A	S	Mono.	Reddy (1969)
NP 790		Stem rust resistance	1A		Mono.	Singh and Swaminathan (1960)
Marquis, Dirk	*Rf* 1	Fertility restorer	1A		Mono.	Robertson and Curtis (1967); Yen *et al.* (1969)
Indian	*Pm* 3c	Mildew resistance	1A	S	Mono.	Briggle (1966); Briggle and Sears (1966)
Asosan	*Pm* 3a	Mildew resistance	1A		Mono.	McIntosh and Baker (1969a)
Federation 41		Brown glume color	1B		Mono.	Unrau (1950)
Turkey 3055	*Bt* 4	Bunt resistance	1B			Schmidt *et al.* (1969)
Rio	*Bt* 6	Bunt resistance				Metzger *et al.* (1963)
T. spelta	*Pt*	Procumbent stem	1D		Nulli.	Matsumura (1947)
Hope, Gabo, Mentana	*Sr* 18	Stem rust resistance	1D	S	Subs.	Sears *et al.* (1957); Baker *et al.* (1970)
NP 790, Yaqui		Stem rust resistance	2A		Mono.	Singh and Swaminathan (1960); Singh (1967b)
Khapli × Chancellor[8]	*Pm* 4	Powdery mildew resistance	2A		Mono.	R. A. McIntosh (personal communication)
Chinese 166	*Yr* 1	Stripe rust resistance	2A			Macer (1966)
Elgin, Jones Fife, etc.	*a* 1	Awn promoter	2A		Mono.	Tsunewaki (1966b)
Marquis Kenya 338		Stem rust resistance	2A		Subs.	Sheen and Snyder (1964; 1965)
Kharkov, Jones Fife	*Ne* 2	Complementary semilethal	2B	S	Mono.	Tsunewaki (1960)
Redman, Kenya W 744	*D* 2	Complementary dwarf	2B	L	Mono.	Hurd and McGinnis (1958); McIntosh and Baker (1969a)
Marquillo-Timstein		Dwarf stature	2B		Mono.	Piech (1968)
	W 1	Waxy leaf bloom	2B		Mono.	Allan (1959); Tsunewaki (1966b)

Selection 5075		Wax inhibition (allelic with W 1)	2B	S	Mono.	Driscoll and Jensen (1964a)
Red Egyptian	*Sr* 9a	Stem rust resistance	2B	L	Subs.	Sears *et al.* (1957)
Kenya Farmer	*Sr* 9b	Stem rust resistance	2B	L	Subs.	Sears and Loegering (1968)
Redman	*Sr* 1	Stem rust resistance	2B	L	Mono.	Campbell and McGinnis (1958)
Thatcher	*Sr* 16	Stem rust resistance	2B	L	Subs.	Sears *et al.* (1957); Sears and Loegering (1968)
C.I. 12632, C.I. 12633	*SrTt* 1	Stem rust resistance	2B	L	Mono.	Nyquist (1957)
Vernstein	*Srd* 1v	Stem rust resistance	2B	L	Mono.	McIntosh and Baker (1970)
Eligulate	*lg* 1	Liguleless	2B	L		McIntosh and Baker (1968)
Martin	*Bt* 1	Bunt resistance	2B		Mono.	Sears *et al.* (1960)
Chinese Spring		Hollow stem	2B		Mono.	Larson (1959)
Thatcher, Kharkov, Jones Fife	*Sg* 3	Winter growth habit	2B		Mono.	Kuspira and Unrau (1957); Tsunewaki (1966b)
Thatcher	*Yr* 7	Stripe rust resistance	2B			Johnson *et al.* (1969)
Eligulate W 1342	*lg* 2	Liguleless	2D		Mono.	McIntosh and Baker (1968)
Hymar, Little Club	*C*	Compactum (club) head type	2D		Mono.	Unrau (1950)
Red Egyptian, Kenya Farmer, Eureka	*Sr* 6	Stem rust resistance	2D	L	Subs.	Sears *et al.* (1957); Sheen and Snyder (1965); Driscoll and Baker (1965)
S 615		Hollow stem (inhibitor of pith)	2D		Mono.	Larson and MacDonald (1959)
Yaqui 53	*a* 2	Awn promoter	2D		Mono.	Singh (1967a)
Marquillo-Timstein Timstein	*D* 1	Dwarf stature	2D	L	Mono.	Hermsen (1963); Piech (1968); McIntosh and Baker (1968)
	l_2-W	Waxlessness	2D		Mono.	Tsunewaki (1966b)
Compair	*Yr* 8	Stripe rust resistance	2D			Riley *et al.* (1968)
Kenya W 1483	*Lr* 15	Leaf rust resistance	2D		Mono.	Luig and McIntosh (1968)
Kenya 338		Stem rust resistance	3A		Subs.	Sheen and Snyder (1965)
Chinese Spring	Viri-dis *508*	Chlorophyll mutant	3A		Mono.	Washington and Sears (1970)
		Hermsen's virescent	3A		Mono.	Sears and Sears (1968)
Yaqui 53		Stem rust resistance	3A		Mono.	Singh (1967b)

TABLE 9.VI (Continued)

Variety source	Gene	Character	Location		Method[b]	Authority
			Chr.	Arm		
Red Bobs	*R2*	Red grain color	3A		Mono.	Metzger (personal communication)
	v	Neatby's virescent	3B	S	Mono.	Sears (1953, 1954, 1956a)
Hope		Brown necrosis (pseudo-black chaff)	3B		Subs.	Kuspira and Unrau (1958)
P.I. 116301	*R* 3	Grain color	3B			Metzger (personal communication)
Thatcher, Redman Yaqui	*R*	Stem rust resistance	3B		Subs.	Sears *et al.* (1957); Campbell and McGinnis (1958); Singh (1967b)
Hope		Adult plant stem rust resistance	3B		Subs.	Kuspira and Unrau (1958)
S-615		Solid stem, top internode	3B		Mono.	Larson and MacDonald (1959, 1966)
		Hermsen's virescent	3B		Mono.	Sears and Sears (1968)
T. sphaerococcum	*s*	Sphaerococcum	3D		Mono.	Sears (1947)
Chinese	*Ch* 2	Complementary, semilethal	3D		Mono.	Tsunewaki and Kihara (1961)
NP 790		Stem rust resistance	3D		Mono.	Singh and Swaminathan (1960)
S-615		Solid stem, top internode	3D		Mono.	Larson and MacDonald (1959)
Pawnee, Chinese Spring	*R* 1	Red seed color	3D		Mono.	Heyne and Livers (1953); Sears (1953)
Federation 41		Modifier of Martin gene for bunt resistance	3D		Mono.	Unrau (1950)
Selection 82al-2-4-7	*Hp* 1	Hairy peduncle (derived from Rye)	4A		Mono.	Driscoll and Sears (1965)
Klein Cometa		Stripe-rust resistance	4A		Mono.	Singh and Swaminathan (1959)
Chinese Spring		Adult plant leaf rust resistance	4A		Nulli.	McIntosh and Baker (1966)

Chinese	*Hd*	Hooded	4B	S	Nulli.	Sears (1944)
K. Farmer, Sapporo, Kenya 117 A	*Sr* 7a	Stem rust resistance	4B	L	Mono.	Knott (1959); Loegering and Sears (1966)
Hope, Marquis, Redman, Spica	*Sr* 7b	Stem rust resistance	4B	L	Subs.	Sears *et al.* (1957); R. A. McIntosh (personal communication)
Timstein, Gabo, Spica, K. Farmer	*D* 3	Dwarf stature	4B	L	Mono.	Hermsen (1963); R. A. McIntosh (personal communication)
		Alkaline phosphatase	4D			Brewer *et al.* (1969)
Chinese	*Q*	Speltoid suppression, square head	5A	L	Mono.	Sears (1944)
Chinese		Pubescent nodes	5A	L	Nulli.	Sears (1944)
Hymar	*sg*2	Winter growth habit	5A	L	Mono.	Unrau (1950)
Klein Cometa		Stripe rust resistance	5A		Mono.	Singh and Swaminathan (1959)
S-615 Rescue		Solid culm, lower internode	5A		Mono.	Larson and MacDonald (1959; 1962)
Elgin, Jones Fife, Yaqui 53, Redman	B_1	Awnlettedness	5A		Mono.	Tsunewaki and Jenkins (1961); Singh (1967a); Campbell and McGinnis (1958); McGinnis and Campbell (1961)
Prelude, *T. macha*	Ne_1	Complementary, semi-lethal (Necrosis)	5B		Mono.	Tsunewaki (1960)
Chinese, Holdfast		Suppression of homoeologous pairing	5B	L	Mono. hybrid	Sears and Okamoto (1958); Riley (1958)
S-615		Solid stem, lower internodes	5B		Mono.	Larson and MacDonald (1959)
Chinese Spring		Awn inhibitor	5B		Mono.	Okamoto (1960)
Yaqui 53	*Sg* 5	Spring growth habit	5B		Mono.	Singh (1967a)
Thatcher	sg_1	Winter growth habit	5D		Subs.	Kuspira and Unrau (1957)
Yaqui 53	Sg_1	Spring growth habit	5D		Mono.	Singh (1967a)
S-615		Solid culm, lower internode	5D		Mono.	Larson and MacDonald (1959)
Uruguay, Malakoff	*lr* 1	Leaf rust resistance	5D	L		McIntosh *et al.* (1965)
Ulka, C.I. 12632, S 2303	*Pm* 2	Powdery mildew resistance	5D	S		R. A. McIntosh (personal communication)

TABLE 9.VI (Continued)

Variety source	Gene	Character	Location		Method[b]	Authority
			Chr.	Arm		
Red Egyptian, Kenya 338, Frontana, Mentana	*Sr* 8	Stem rust resistance	6A		Mono.	Sears *et al.* (1957); Sheen and Snyder (1965)
Khapstein	*Sr* 13	Stem rust resistance	6A		Mono.	R. A. McIntosh (personal communication)
Klein Cometa		Stripe rust resistance	6A		Mono.	Singh and Swaminathan (1959)
S-615, Rescue		Hollow stem (inhibitor of pith)	6A		Mono.	Larson and MacDonald (1959; 1962)
Chinese Spring	*B* 2	Awn inhibitor	6B	L	Nulli.	Sears (1944; 1966a)
Chinese Spring	*Co*	Corroded (leaf necrosis)	6B	S	Mono.	Sears (1966a)
Timstein, K. Farmer	*Sr* 11	Stem rust resistance	6B	L	Mono.	Sears and Rodenhiser (1948); Loegering and Sears (1966)
Pawnee, Mentana Mediterranean	*lr* 3	Leaf rust resistnce	6B	L	Mono.	Livers (1949); Heyne and Livers (1953); R. A. McIntosh (personal communication)
Canthatch	*Rf* 2	Fertility restorer	6B		Mono.	Yen *et al.* (1969)
"Transfer"	*Lr* 9	Leaf rust resistance	6B	L		Sears (1966a)
Thatcher, "Marquis," Yaqui, Kanred	*Sr* 5	Stem rust resistance	6D		Subs.	Sears *et al.* (1957); Sheen and Snyder (1964); Singh (1967b)

S-615		Hollow stem	6D		Mono.	Larson and MacDonald (1959)
Canthatch	*Rf* 3	Fertility restorer	6D		Mono.	Yen *et al.* (1969)
Hope	Re_1	Red coleoptile	7A		Mono.	Sears (1954)
Axminster	*Pm* 1	Mildew resistance	7A	L	Mono.	Sears (1954); Sears and Briggle (1969)
S-615		Hollow stem	7A		Mono.	Larson and MacDonald (1959)
Chinese Spring	Chlorina-1	Chlorophyll mutant	7A		Mono.	Sears and Sears (1968)
Chinese Spring	Chlorina-448	Chlorophyll mutant	7A		Mono.	Pettigrew and Driscoll (personal communication)
Hope	*Pc*	Purple culm	7B	S	Subs.	Kuspira and Unrau (1960); Law and Wolfe (1966)
Hope	*el*	Earliness	7B	S	Subs.	Kuspira and Unrau (1958); Law and Wolfe (1966)
Yaqui 53		Stem rust resistance	7B		Mono.	Singh (1967b)
Hope, Renown, Redman, Spica	*Pm* 5	Mildew resistance	7B	L	Subs.	Law and Wolfe (1966); McIntosh *et al.* (1967)
	Lr 14a	Leaf rust resistance	7B			McIntosh *et al.* (1967)
	Lr 17	Leaf rust resistance	7B			McIntosh *et al.* (1967)
S-615		Hollow stem	7D		Mono.	Larson and MacDonald (1959)
Dirk	*Rf* 4	Fertility restorer	7D		Mono.	Yen *et al.* (1969)
Synthetic spelta		Purple coleoptile	7D		Mono.	Jha (1964)
Chinese Spring	Chlorina-214	Chlorophyll mutant	7D		Mono.	Washington and Sears (1970)
Agatha	*Lr* 19	Leaf rust resistance from *Agr. elongatum*	7D			Sharma and Knott (1966)

[a]Modified from Burnham, 1962.
[b]Nulli., nullisonic; Mono., monosomic; Subs., substitution.

The twenty-one chromosomes of hexaploid wheat (AABBDD) were similarly assigned to the three genomes. Each of the monosomics of *Triticum aestivum* was crossed with *Triticum dicoccum* (AABB). Seven of the monosomics formed 13 II + 8 I, and the rest formed 14 II + 6 I. The former were thus placed in the D genome. Sears (1944) numbered chromosomes of the A and B genomes as I to XIV and those of the D genome as XV to XXI. Monosomics for the D genome were established and identified also by Matsumura (1952a, b, 1954).

Larson (1952) tried to identify chromosomes of the A and B genomes mainly on the basis of studies of stem solidness of the aneuploids, but her classification was far from satisfactory. Okamoto (1957, 1962) distinguished the chromosomes of the A and B genomes by crossing plants having the 20 II + telocentric for chromosomes of the A and B genome with the synthetic amphidiploid (AADD) of *T. aegilopoides* × *Aegilops squarrosa*. In the F_1, if the telocentric was from an A genome chromosome, it paired with an A genome chromosome from the male parent and formed an unequal bivalent at a high frequency. If the telocentric was from a B genome chromosome, it had no pairing partner and either formed no unequal bivalent or formed it at a very low frequency. By using these techniques the twenty-one chromosomes of hexaploid wheat were divided into three groups of seven each, corresponding to the A, B, and D genomes.

Chapman and Riley (1966) verified the classification of Okamoto by crossing ditelocentric lines for chromosomes of the A and B genomes (chromosome I–XIV) of *Triticum aestivum* with *Triticum thaoudar* (AA). The participation of the telocentric in bivalent formation could be scored readily. There were two distinct classes of behavior, each displayed by seven hybrids. In the first class, the telocentric chromosomes paired in 48 to 70% of the cells, in the second class, in less than 2%. The seven chromosomes marked by telocentrics that paired frequently were allocated to the A genome, the other seven, to the B genome. The resulting classification confirmed Okamoto's identifications for twelve chromosomes, but chromosomes XIII and II, which were placed in the A and B genomes, respectively, by Okamoto, were shown to belong to the B and A genomes, respectively, by Chapman and Riley (1966).

Identification of Homoeologous Sets of Chromosomes

Since the genomes of the polyploid species are related phylogenetically, they are partially homologous (homoeologous). In other words, each member of one genome has a homoeologous partner in the other genome or genomes of the polyploid species. In hexaploid wheat, which has three genomes of seven chromosomes each, there should be seven sets of homoeologous chromosomes with three members each. There are several methods of identifying the homoeologous sets. In wheat, a successful method was the test of the ability

Fig. 9.1. Spikes of wheat showing ability of tetra-2B to compenstate for nulli-2D. From left to right: nulli-2D, normal, nulli-2D-tetra-2B, tetra-2B. (From Sears, 1944.)

of a particular tetrasome to compensate for a particular nullisome (Sears, 1952b, 1954). As mentioned in the last chapter, the presence of an extra chromosome may compensate phenotypically for the absence of a particular chromosome. Thus nullisomics of wheat are reduced in size, vigor, and fertility, and most are either male or female sterile. However, if a homoeologous chromosome is added, the morphology and fertility is altered toward the direction of the normal disomic. If a pair of a homoeolog is added there is greater improvement in morphology, vigor, and fertility; many of such nulli-tetra combinations are practically indistinguishable from the disomics. For example, plants nullisomic for chromosome 2D in Chinese Spring wheat are dwarfish with shorter and narrower leaves and are female sterile. Plants nullisomic for 2D but also tetrasomic for 2B are nearly normal in all respects and are fertile. Not only does the tetrasomy for one homoeologous pair eliminate the modifications caused by nullisomy, but the abnormalities caused by the tetrasomy are also rectified in the compensating nulli-tetra combinations (Fig. 9.1). In other words, the imbalance caused by the absence of one pair and duplication of the other is neutralized in these combinations due to similar gene contents of homoeologs.

Some of the nulli-trisomes arise spontaneously in the progenies of nullisomics (See Chapter 8), and nulli-tetras can be obtained in the selfed progenies of nulli-trisomes. Sears (1966b) has outlined several methods for producing nulli-tetra combinations experimentally. In the compensating nulli-tetra combinations, the morphology, vigor, and fertility is almost normal. In the noncompensating combinations, the chromosomes involved are nonhomoeologous, and there is no improvement in morphology, vigor, and fertility. Such combinations are inferior to the nullisomic concerned, and many are extremely abnormal. Since six different nulli-tetra combinations are possible within each homoeologous group of three, a total of 42 compensating combinations are possible in wheat, all of which have been tested by Sears (1954, 1966b) and Sears and Okamoto (1956). A total of 378 noncompensating combinations are theoretically possible in wheat, and 61 of these were tested by Sears (1966b).

It is clear that the twenty-one chromosomes of common wheat can be classified into three genomes of seven each (Sears, 1944, 1954; Okamoto, 1957, 1962; Chapman and Riley, 1966). Previous to the allocation of all the chromosomes to their respective genomes, Sears (1952b, 1954) had shown that the complement could also be divided into seven homoeologous groups, each with three pairs of chromosomes with similar gene contents. The two classifications were then fitted together so that each genome had one chromosome pair in each homoeologous group, while each homoeologous group had one pair in each genome. On the basis of this two-way classification, Sears (1958) suggested that the Roman numerals formerly used should be superceded by designations indicating the homoeologous group and the genome to which each chromosome belongs. These designations 1A, 1B, 1D, 2A, 2B, 2D, . . . 7D were consequently introduced (Table 9.VII) and have received universal acceptance by wheat workers. These designations show the evolutionary relationships of the different chromosomes of the complement. Thus 1A is in homoeologous group 1 and in A genome, and so on to 7D. Supporting evidence for these designations was provided by the findings of Okamoto and Sears (1962) that the pairing in haploids of wheat is largely between chromosomes belonging to the same homoeologous group and by studies of Riley and Kempanna (1963) who found that pairing in the 5B deficient wheat was mainly between homoeologous chromosomes.

The genome analysis in wheat, which has been carried to its logical conclusion, proceeded in four more or less distinct steps. First, Sakamura (1918) and Sax (1922) discovered that the common wheat was hexaploid with three distinct genomes. Second, the putative donors of these genomes were identified by MacFadden and Sears (1945), Sarkar and Stebbins (1956), and Riley *et al.* (1958). Third, the homoeologous chromosomes were identified by Sears (1952b, 1954); and finally the chromosomes belonging to the different genomes were identified by Sears (1944), Okamoto (1957, 1962), and Chapman and Riley

TABLE 9.VII
Classification and Designation of the Chromosomes of *Triticum aestivum*[a]

Homoeologous group	Genome					
	A		B		D	
	New	Old	New	Old	New	Old
1	1A	XIV	1B	I	1D	XVII
2	2A	II	2B	XIII	2D	XX
3	3A	XII	3B	III	3D	XVI
4	4A	IV	4B	VIII	4D	XV
5	5A	IX	5B	V	5D	XVIII
6	6A	VI	6B	X	6D	XIX
7	7A	XI	7B	VII	7D	XXI

[a]From Chapman and Riley, 1966.

(1966). The last two steps in the analysis were possible only with the availability of aneuploids.

Studies on Genetic Control of Chromosome Pairing

As discussed in the two previous sections, hexaploid wheat has three closely related genomes. These three genomes have been combined from three closely related but distinct diploid species. The genetic content of the homoeologous chromosomes of the different genomes is similar as indicated by compensation studies. Furthermore, when the diploid progenitors are crossed with each other, their chromosomes pair and recombine to a limited extent. However, common wheat behaves cytologically as a diploid, as there is regular bivalent formation during meiosis. Usually pairing and recombination between homoeologous chromosomes does not occur. For a long time wheat cytogeneticists were unanimous in thinking that the three genomes of common wheat became differentiated by small chromosomal rearrangements during their evolution and that this differentiation prevented intergenomic pairing and recombination. However, independent observations of Riley and Chapman (1958a) and Sears and Okamoto (1958) showed conclusively that absence of pairing and recombination between homoeologous chromosomes in wheat is under genetic control, and when this control is broken down, the homoeologous chromosomes pair and recombine with each other. Riley and Chapman (1958a) reported that the absence of one particular chromosome in haploids (nulli-haploids) greatly increased the frequency of bivalents and trivalents. As many as nineteen of the twenty chromo-

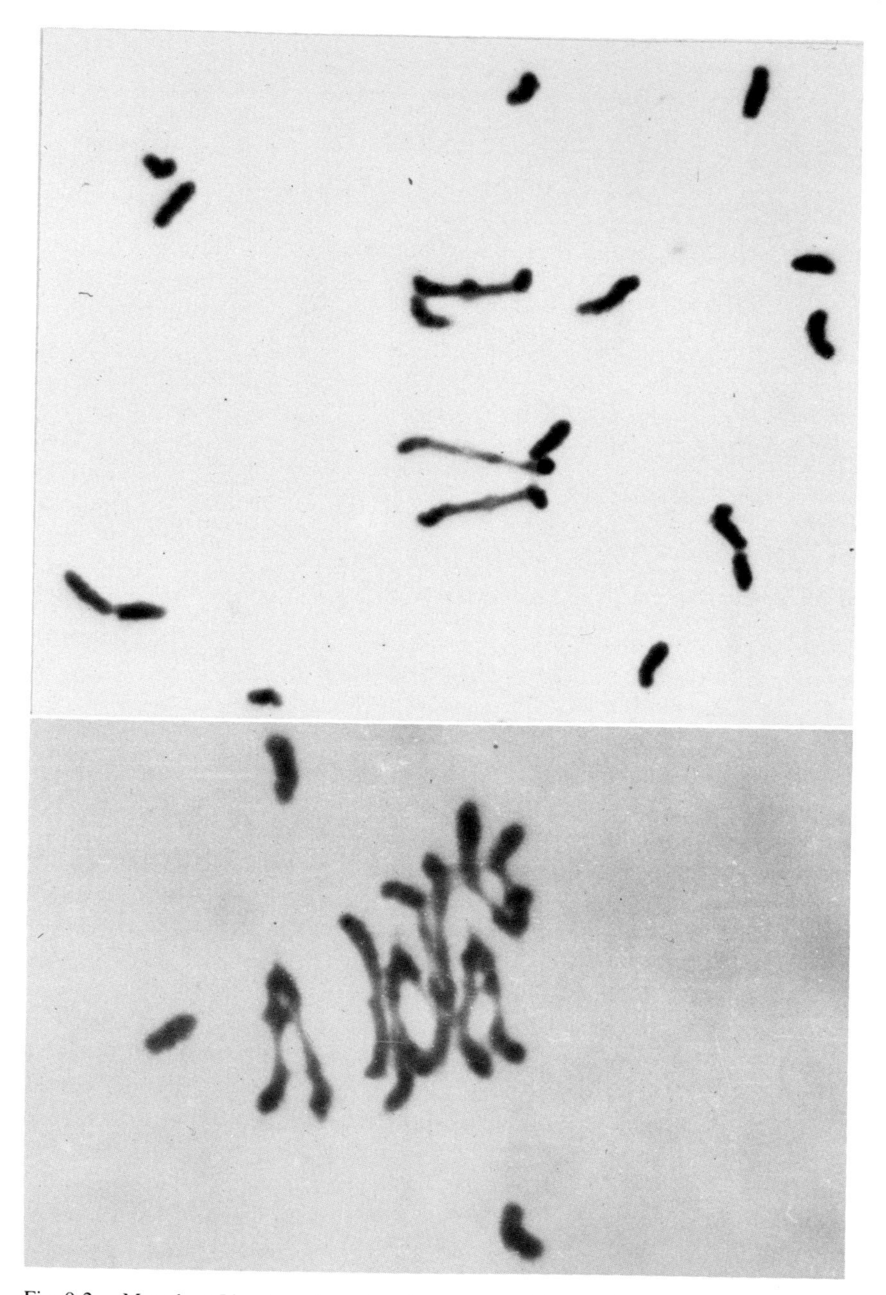

Fig. 9.2. Metaphase I in wheat. Top, three bivalents and fifteen univalents in euhaploid. Bottom, five trivalents, one bivalent, and three univalents in nulli–haploid. (From Riley and Chapman, 1958a.)

TABLE 9.VIII
Mean Pairing at First Metaphase of Meiosis in Euploid and Nulli-5B *Triticum aestivum*[a]

Plant	Cells examined	Univalents	Bivalents	Trivalents	Quadri-valents	Penta-valents	Hexa-valents
Euploid	200	0.14	20.93	–	–	–	–
Nulli-5B	60	1.41	17.22	0.47	0.59	0.02	0.05

[a]From Riley, 1960b.

somes were observed to have taken part in pairing associations whereas in euhaploids, the maximum number was 9. In the euhaploids, trivalents are rarely formed, but in the nulli-haploids trivalents were common (Fig. 9.2). In nullisomics with the same chromosome missing, multivalent associations of 3, 4, 5, and even 6 chromosomes were common (Table 9.VIII). More than one-half the cells had at least one multivalent, and many had several.

Similarly, Sears and Okamoto (1958) reported that when chromosome 5B was missing there were many more bivalents in hybrids between hexaploid wheat (AABBDD) and the amphidiploid AADD from *T. aegilopoides* × *Aegilops squarrosa* and also in hybrids of AABBDD with AA. Thus the latter hybrid forms 2–7 bivalents in the presence of 5B, but in its absence it forms 5–13 bivalents (average about 10). The authors hypothesized that a mutant locus on chromosome 5B was responsible for inhibiting pairing thus helping the hexaploid wheat achieve diploid pairing. Later studies showed that the chromosome missing in Riley's material was also 5B. Further studies on localization of the gene suppressing homoeologous pairing and investigations of the control mechanism were carried out by Riley and his colleagues at Cambridge and Sears and his co-workers at Columbia. The important conclusions drawn from these investigation are as follows:

1. The effect of 5B is expressed only in its complete absence, i.e., in nullisomic condition. The meiosis in mono-5B is normal (Riley and Chapman, 1958a).
2. The gene for suppression of homoeologous pairing is located on the long arm of 5B (5BL). In the presence of 5BL, as in the case of monotelodisomics for 5BL or 5BL monoisosomics, the pairing is normal. In the absence of 5BL and presence of 5BS, homoeologous pairing takes place (Riley and Chapman, 1958a; Riley, 1960b; Riley and Chapman, 1964).
3. The gene locus is probably located in the distal half of the long arm of 5B as there was homoeologous pairing in a wheat × rye hybrid with a large terminal deficiency of 5BL (Riley *et al.*, 1966b).
4. The increased dosage of 5BL (in the absence of 5BS), as in plants with two and three 5BL isochromosomes, reduces the frequency of chiasmata formed

between homologs and also increases the abnormalities of synapsis (Feldman, 1966).

5. However, in the tetrasomics for the complete 5B, which also have increased dosage of 5BL, the synapsis is normal (Riley *et al.*, 1966c). Therefore, it seems that presence of 5BS influences synapsis in the reverse direction from that of 5BL. This hypothesis was further confirmed by Riley and Chapman (1967) from studies of plants with various dosages of 5BS and 5BL.

6. None of the diploid putative progenitors of wheat from which 5B may have been derived produces a genetic effect on pairing like that of 5B of hexaploid wheat. This suggests that the diploidizing function of 5B arose by mutation after the origin of polyploidy in wheat, probably at the tetraploid level (Riley, 1960b; Riley *et al.*, 1961).

7. A hypothesis was advanced by Feldman (1966) to explain the operational aspects of the 5BL system. He observed that the pairing was reduced in plants with six doses of 5BL, there were interlocking bivalents, and there was heteromorphic bivalent formation, presumably because of pairing between homoeologs. To explain increased homoeologous pairing but decreased pairing of homologs, Feldman assumed that in nulli-5B, homoeologs as well as homologs are associated premeiotically. All six homoeologs enter meiosis lying near each other and both homologous and homeologous pairing can then ensue. In the presence of 5B, however, the tendency toward somatic association is reduced to the point that only homologs, not homoeologs, are associated, and at meiosis only pairing of homologs occurs. Extra dosage of 5B reduces somatic association still further until at six doses not only homoeologs but also homologs are situated at random with respect to each other. When meiosis starts in this material with extra 5B chromosomes, some homoeologs are closer to each other than to their homologs, and, therefore, they pair; and since pairing begins at the ends of the chromosomes, some bivalents, as they come together, trap one or more other chromosomes between them, causing interlocking. The confirmation of the tendency for somatic association in root tip cells of seedlings was provided by Feldman *et al.* (1966) using telocentric chromosomes as cytological markers.

8. It is possible to induce mutations which suppress the effect of 5B (Riley *et al.*, 1966b).

9. The homoeologous chromosomes 5D and 5A also exercise a regulatory influence over meiotic pairing. The effect of the deficiency of 5D is only fully revealed at temperatures of 15°C or lower, so that there is genotype-temperature interaction since meiosis is unaffected in disomics at this temperature. The absence of 5D results in almost total failure of pairing at low temperatures. Therefore, one function of 5D is to stabilize meiotic pairing under such conditions (Riley, 1966a). Meiosis in plants nullisomic for 5A is normal at low temperatures.

However, meiosis in plants that are nulli-5D–tetra-5A is also normal at these temperatures. The pairing failure expected from the absence of 5D does not occur, and this is interpreted to mean that increased dosage of 5A compensates for the deficiency of 5D in this respect (Riley, 1967).

10. The practical applications of 5BL system in wheat improvement have been discussed by Riley and Chapman (1958a, 1963), Riley (1966b, 1967), and Sears (1967a). This topic is considered in Chapter 11.

Studies on Gene Action in Wheat

One type of recessive gene whose action has been investigated with the aid of aneuploids deserves discussion as its occurrence seems to be restricted to wheat. This is the recessive gene that is ineffective in hemizygous condition. As discussed in the early part of this chapter, such genes are expressed only in the disomics. Monosomics with one dose of such genes and the nullisomics with no dose express the dominant phenotypes. The dominant alleles of these genes show no dosage effect. This is in contrast to the gene action observed in diploid species like *Drosophila,* maize, and tomato where all the recessive genes tested with the help of deficiencies express themselves in hemizygous condition, and there is no dosage effect.

According to Sears (1953), the likely reason for this unusual type of gene in wheat is the polyploid nature of this species. In a diploid species, a gene with positive effect is almost always dominant because one dose of the gene is adequate to ensure full expression. In a polyploid species, however, the gene expression is subject to the action of modifiers not only from its own genome but also from the homoeologous genome or genomes. Under these conditions some genes in a single dose would be unable to express themselves. They would be recessive and would be ineffective when hemizygous. The action of one such hemizygous gene, *v* (*Neatby's virescent*), has been investigated thoroughly by Sears (1956a, 1957). This gene is located on chromosome 3B. Homoeologous chromosomes 3A and 3D carry genes which compete with *v*. Null mutations have been obtained at the *v* locus on 3B and the homoeologous locus on 3A which segregate conventionally (Sears and Sears, 1968).

It may be noted, however, that this type of gene action has been encountered only in hexaploid wheat. Three genes of spontaneous origin—*Neatby's virescent,* speltoid suppressing gene, *nonsphaerococcum* gene—and, four EMS-induced chlorophyll mutants—chlorina-1, chlorina-448, chlorina-214, and viridis-508 (Washington and Sears, 1970)—behave in this manner. In the tetraploid species, *Gossypium hirsutum* and *Nicotiana tabacum,* several recessive genes located with the help of monosomics have behaved conventionally.

Identifying the Chromosomes Involved in Translocations

The chromosomes involved in a reciprocal translocation can be identified by crossing the translocation stock with each of the monosomics. In two of the monosomic F_1 families, the monosomics would exhibit $n - 2$ bivalents and a trivalent, and the chromosomes involved would thus be identified. In the remaining families, the monosomics would exhibit $n - 3$ bivalents plus a translocation quadrivalent and a univalent. The chromosomes involved in several translocations in *Nicotiana tabacum* were identified in this manner (Cameron, 1966).

Monosomics in Practical Breeding

Monosomics have been employed in practical breeding programs for transferring a whole chromosome or a chromosome segment from one variety into another or from one species into another. This subject is reviewed thoroughly in the following chapters. Since several nullisomics are male sterile, Lafever and Patterson (1964b) have suggested using them as male sterile female parents for producing F_1 hybrids for commercial use.

10

Cytogenetics of Substitution Lines

When one chromosome pair of a variety is replaced by a homologous pair from another variety of the same species, it is referred to as an intervarietal substitution line or simply a substitution line. The variety which receives the chromosome pair is called the recipient, the one which contributes, the donor. As mentioned in the previous chapter, substitution lines have proved useful in associating genes for disease resistance with respective chromosomes, and they are becoming increasingly popular for the analysis of quantitative traits. This chapter describes different methods of producing substitution lines and their utility in genetic analysis and practical plant breeding programs. Since substitution lines have mainly been studied in wheat, the following discussion is confined to this species.

Production of Substitution Lines

Two different methods of producing substitution lines have been discussed by Sears (1953), Unrau *et al.* (1956), and Kuspira (1966). They differ primarily in the type of aneuploid employed as the recurrent female parent, the basic procedure being the same for both methods.

In method 1, the monosomics are used as recurrent parents. The monosomics of the recipient variety are crossed as female with the donor. The F_1 monosomics are selected and backcrossed on the monosomic recurrent parent. The monosomic offspring from the first backcross are again backcrossed to the same monosomic stock, and this procedure is repeated five or six times. After the final backcross, the monosomic plants are selfed, and the disomic individuals which represent the substitution line are selected.

The major drawback of this method is that the monosomic plants selected in different backcross generations may not have the univalent from the donor parent. It is well known that occasional $n - 1$ pollen grains (about 4% on the average) function. If such a pollen grain fertilizes a 21-chromosome egg then the univalent chromosome in monosomic individuals sired by such fertilizations will not be from the donor parent but from the recipient. One way of avoiding this would be to self-pollinate the monosomics and select disomic progeny. These disomics would be used as males for the next backcross to the original monosomic. This method requires twice as many generations and consequently takes twice as much time and labor. If, however, the donor chromosome carries a recessive gene for a trait that is easily recognized, the selfing is not necessary, since one can then distinguish phenotypically for the presence of the desired monosome.

Method 2 is the most satisfactory and is used when monotelosomics or monoisosomics are available. It involves using a monotelosomic or monoisosomic as the recurrent female parent in the backcrosses. The monotelosomic or monoisosomic stocks are crossed as the female with the donor variety. If the monotelosomic is used as female, the F_1's include monosomics and monotelodisomics. The monosomic F_1 plants are selected and backcrossed on the same monotelosomic stock. The procedure is repeated at least five or six times. The monosomic plants are then selfed and disomic progeny selected. These disomics represent the substitution line. The advantages of this method are: (1) it is as rapid as method 1, (2) there is little danger of an undetected shift to a chromosome that is different from the one represented by the telocentric, and (3) there is no chance of the univalent coming from the recipient parent as in method 1.

The major disadvantage of the method is that at least one monotelosomic or monoisosomic line for each of the chromosomes of the complement must be available. Such lines have been isolated for all the chromosomes of the wheat variety Chinese Spring. Because of the availability of these types in Chinese Spring, chromosomes from other varieties such as Kenya Farmer, Mida, Marquis, Thatcher, Hope, and Timstein have been substituted into this variety.

Development of Chromosome Deficient Lines in Varieties other than Chinese Spring

If the other varieties are to serve as recipient or if reciprocal substitutions are desired, it is necessary first to establish a monosomic series of these varieties (Sears, 1953). This is done by crossing the existing monosomics as females to a variety in which monosomics are to be established. The monosomic F_1's are selected and again crossed as females with the same variety (recurrent parent). The process is repeated until all the chromosomes of the original variety are replaced by those from the recurrent parent. Usually five or six backcrosses are considered adequate. At the end of the backcrossing program, a complete set of monosomics is available in the desired varieties.

Since a univalent shift (see Chapter 7) may occur during the backcrossing program, it is necessary to check the identity of the monosomics at the termination of the backcrossing program. This may be done by crossing each derived monosomic line with its corresponding ditelosomic. If the univalent in the derived monosomic line is the expected one, the monotelosomic offspring from these crosses will have a telocentric univalent. If there had been a univalent shift, the telocentric will form a heteromorphic bivalent with its normal homolog and such offspring will be monosomic for a different chromosome. By using these techniques, complete monosomic series have been established in several wheat varieties such as Rescue, Cadet, Red Bobs, Redman, Prelude, Thatcher, Lehmi, Kharkov, and Witchia.

A new method of avoiding errors associated with a univalent shift was proposed by Anderson and Driscoll (1967). It involves the use of monosomic alien substitution lines (MAS line). A MAS line possesses twenty pairs of wheat chromosomes and one alien monosome. An alien monosome is a single chromosome derived from a species related to wheat (see Chapter 11). The main value of a MAS line is that cases of univalent shift would be readily detected by routine analysis of monosomics. In the absence of univalent shift, following pollination of a MAS line with the recurrent parent, F_1's involving 20-chromosome eggs will possess twenty wheat bivalents and one wheat univalent (20 II + 1 I). However, if a univalent shift has occurred, these F_1 individuals will possess nineteen wheat bivalents, two wheat univalents, and one alien univalent (19 II + 3 I). In this way, all individuals that undergo a univalent shift can be detected and discarded. The monosomic F_1's retained are identical to the monosomic F_1's obtained by using conventional monosomics because the alien chromosome has now been eliminated. However, in this case, one is sure of the identity of the univalent. Bielig and Driscoll (1971) have established a number of MAS lines for this purpose.

The monosomic series having been thus established, these varieties can serve as recipients for substituting chromosomes from other varieties. Method 1 of the substitution will have to be followed until monotelosomics or monoisosomics of these varieties are isolated, then method 2 can be employed. In practice, a combination of methods 1 and 2 may be used for establishing a complete set of substitution lines.

Genetic Analyses Using Substitution Lines

The use of substitution lines in associating genes with their respective chromosomes was first suggested by Sears (1953). Since then substitution lines have been employed in determining the number and dominance relationships of the genes governing a trait, for the genetic analysis of quantitative traits, for determining the number of genes controlling a quantitative trait on a single chromosome, and for mapping these genes.

Locating Major Genes

Some traits such as certain types of disease resistance are difficult to classify in segregating populations. This difficulty is further exaggerated if modifying genes are also segregating along with major genes. Substitution lines are of great value for revealing the presence of such genes on the substituted chromosomes. The substitution lines are normal with respect to chromosome constitution, show no meiotic abnormalities, are fully fertile, and breed true like the standard varieties. A large amount of seed of each line can be obtained. These lines can be tested for disease and insect resistance, grain quality, and many other desirable agronomic traits, and then compared with the recipient and donor varieties. The genes governing these traits can be assigned to the respective chromosomes by such comparison.

Wheat varieties, such as Hope, Thatcher, Red Egyptian, and Timstein, are resistant to stem rust and have been used as sources of resistance in breeding programs. In order to locate the genes for stem rust resistance of these varieties to the respective chromosomes, Sears *et al.* (1957) substituted each of the twenty-one chromosomes of these four varieties to Chinese Spring (susceptible to stem rust). The substitution lines were then compared for stem rust resistance with Chinese Spring and the donor resistant varieties. A total of nine different substitution lines (chromosomes 4B and 1D from Hope, 3B, 2B, and 6D from Thatcher, 6A, 2B, and 2D from Red Egyptian, and 6B from Timstein) were found to be resistant to stem rust. Genes for stem rust resistance were thus located on eight different chromosomes of the four varieties. Similarly, major genes for stem rust resistance were located on chromosomes 2A, 3B, and 6D of variety Marquis (Sheen and Snyder, 1964) and chromosomes 2A, 3A, 6A, 3B, 4B, 6B, and 2D of variety Kenya Farmer (Sheen and Snyder, 1965). Several genes for disease resistance and a number of morphological traits have been located by various workers by the substitution method (Table 8.VI).

Determining the Number and Dominance Relationships of Genes

Before they are assigned to the respective chromosomes, the number of genes governing a trait and their dominance relationships are usually known from the conventional genetic analyses. If, however, this information is not available, the critical substitution lines can be crossed with the recipient variety, and

TABLE 10.I
Segregation for Awning, Culm Color, and Reaction to Pseudoblack Chaff in F_2 Populations from Crosses of Substitution Lines with the Recipient Variety Chinese Spring[a]

Cross	Awning		Culm color		Reaction to pseudoblack chaff		X^2 for
	Awnless	Apically awned	Purple	White	Resistant	Susceptible	3:1
Chinese × Chinese (Thatcher-3B)	147	43					0.57
Chinese × Chinese (Thatcher-4A)	126	37					0.46
Chinese × Chinese (Thatcher-3A)	111	43					0.70
Chinese × Chinese (Thatcher-7D)	119	42					0.10
Chinese × Chinese (Hope-7B)			184	68			0.53
Chinese × Chinese (Hope-3B)					138	40	0.61

[a]From Kuspira and Unrau, 1958.

the inheritance of the traits can be studied in the F_1 and F_2 generations. In this way Kuspira and Unrau (1958) found that the substitution line Chinese Spring (Hope-3B) which is Chinese Spring with 3B substituted from Hope, was susceptible to pseudoblack chaff like the donor variety Hope. It was concluded that the gene or genes for susceptibility to pseudoblack chaff must be located on 3B. However, the number of genes governing the trait and the dominance relationships were not known. Kuspira and Unrau crossed Chinese Spring (Hope-3B) with Chinese Spring. The F_1 was resistant to pseudoblack chaff, and the F_2 segregated 3:1 for resistance and susceptibility (Table 10.I). This information was diagnostic in concluding that variety Hope carries a single recessive gene on chromosome 3B for susceptibility to pseudoblack chaff. Each of the Thatcher chromosomes 3A, 3B, 4A, and 7D, which were known to control awn development, were found to have a single recessive gene for this trait, and control of purple culm color on 7B of Hope was also found to be due to the activity of a single dominant gene (Table 10.I). Dominance relationships and number of genes controlling stem rust resistance on several chromosomes of varieties Marquis and Kenya Farmer were determined by Sheen and Snyder (1964, 1965) using similar techniques.

Associating Genes for Quantitative Traits with Chromosomes

The substitution lines are highly uniform, and each can be treated as a variety in replicated experiments along with the recipient and donor varieties. Since each line represents a genotype that differs from the recipient parent only in the genes carried on the substituted chromosome, significant differences that appear may be ascribed to genes on this chromosome.

Clear-cut effects of certain substituted chromosomes on such quantitative traits as height, yield, and lodging were observed by Kuspira and Unrau (1957). Variety Thatcher has significantly higher yield potential than Chinese Spring. Kuspira and Unrau compared the yield of Thatcher, Chinese Spring, and the nineteen substitution lines in which a pair of Chinese Spring chromosomes had been replaced by a pair of Thatcher chromosomes. As the data of Table 10.II indicate, four of the substitution lines yielded significantly higher than the Chinese Spring. It is evident that chromosomes 1B, 3A, 3B, and 4B of Thatcher carry genes which are responsible for its higher yield. Kuspira and Unrau also found that a large number of chromosomes of Thatcher, Timstein, and Hope carry minor genes for lodging resistance. More recently, the genetic control of solid stem has been investigated using substitution lines (Larson, 1966; Larson and MacDonald, 1966), and substitutions of Cheyenne chromosomes into Chinese Spring have been used to study the genetics of flour quality (Morris *et al.*, 1966). By the substitution method, Halloran and Boydell (1967) have identified the chromosomes carrying genes for vernalization response, and Sasaki *et al.* (1968) have located genes for several quantitative characters on

TABLE 10.II
Yield in Grams of Chinese Spring, Thatcher, and Thatcher Substitutions into Chinese Spring[a]

Substitution lines and varieties	Significantly higher than Chinese Spring	Same as or similar to Chinese Spring
1B	830.7	—
2A	—	682.0
2B	—	426.3
2D	—	590.3
3A	721.3	—
3B	704.9	—
4A	—	502.2
4B	763.5	—
4D	—	580.8
5A	—	637.8
5B	—	678.3
5D	—	445.0
6A	—	631.4
6B	—	625.7
6D	—	647.5
7A	—	492.9
7B	—	630.9
7D	—	688.7
Chinese	—	523.7
Thatcher	1013.8	—

[a]From Kuspira and Unrau, 1957.

their respective chromosomes. Law (1966a, 1967) has applied techniques of biometrical analysis for studying such quantitative traits as time to ear emergence, tiller number, fertile tiller number, height, and spike density.

Determining the Number of Genes Controlling a Quantitative Trait and Their Mapping

The substitution lines can identify the chromosomes carrying genes which influence a quantitative trait, but the number and kind of genes located on the chromosome cannot be determined. To undertake this type of analysis, further crossing procedures are employed in which the substituted chromosomes are broken down by recombination into components which are closer to the conventional units of inheritance, the genes. A technique for this type of analysis was first suggested by Unrau (1958) and employed with minor modifications by Law (1966b, 1967) and Law and Wolfe (1966) to locate some of the factors controlling time to ear emergence and mildew resistance on chromosome 7B and for mapping these genes.

As outlined by Unrau (1958) the technique consists of three steps. In the first step, the substitution line is crossed with the recipient variety. The F_1 produced is heterozygous only for the genes located on the substituted chromosome. The crossing over results in recombinations of the genes located on this chromosome only. With one locus, there are two gametic types; with three loci, eight, and with n loci, there are 2^n possible combinations.

In the second step, the F_1 is crossed as the male with the corresponding monosomic of the recipient variety. For example, if chromosome 7B has been substituted, the F_1 is crossed with mono-7B of the recipient variety. The monosomic offspring of this backcross are isolated, all of them having the univalent chromosome which is the recombinant or nonrecombinant chromosome of the F_1. There are as many genetically different univalents as there are recombinations. If three loci are segregating, there are eight genetically different univalents.

In the third step, these monosomic backcross plants are kept separate and permitted to self. The disomic progeny obtained upon selfing (approximately 24%) each monosomic constitute a distinct line. These lines represent different recombinant and nonrecombinant classes. Estimation of the number of loci involved and mapping of loci can be done directly from the relative frequency of different recombination classes.

Law (1966b) employed this technique to determine the number of loci governing time to ear emergence. Chromosome 7B of Hope was known to influence the time of ear emergence. The substitution line, Chinese Spring (Hope-7B) was crossed with recipient variety Chinese Spring. The F_1 was crossed as male to the Chinese Spring mono-7B. The monosomic plants from this backcross progeny were selfed, and disomic offspring were isolated in each line. Eighty-one such lines were established, each of which originated from a different monosomic plant of the backcross progeny. These lines were scored for time to ear emergence. The bimodal distribution of these lines in respect to this trait indicated that one locus with large effect on time of ear emergence was present on chromosome 7B. This locus was designated as *e*1. Presence of another locus (*e*2) with small effect was also revealed by the analysis of genetic variance. Linkage of *e*1 with another dominant gene for purple color of culm (*Pc*) was also detected.

Law and Wolfe (1966) used the same technique to determine the number of loci governing mildew resistance on chromosome 7B of variety Hope. After following the procedures enumerated above, seventy lines were established and tested for mildew resistance. Forty were found resistant, and thirty, susceptible. This agreed with the 1:1 ratio, expected if only one gene determines the mildew resistance. This gene for mildew resistance was designated *ml*. These seventy families were also segregating for *Pc* and *e*1 besides *ml* and were classified into eight phenotypic classes with respect to these three loci. Recombination values were calculated from these data, and the gene order was determined.

The arm locations of these genes were determined with the aid of a telocentric for the long arm of 7B. Further analysis by Law (1967) indicated the presence of four factors on chromosome 7B of Hope influencing four quantitative traits such as plant height, grain weight, grain number, and tiller number.

Realizing the potential value of substitution lines in genetic studies, a European Wheat Aneuploids Cooperative has been organized. Composed of wheat cytogeneticists from almost every country of Europe, their primary objective is the development and exploitation of substitution lines (Sears, 1969). Four widely grown varieties (Cappelle-Desprez, Fanal, Besostaya I, and Mara) have been selected, and work on the production of all the 252 substitution lines possible among them has been started. Analysis of these and other substitution lines will enhance our understanding of the quantitative inheritance in wheat.

Other Uses of Substitution Lines

As pointed out by Kuspira and Unrau (1957), the substitution method may be employed to improve existing varieties. To improve a variety in characters controlled by one or several genes located on the same chromosome, the substitution method is quicker and more efficient than the backcross method regardless of whether the genes in the donor are dominant or recessive. The necessity of interbackcross selfing or progeny testing for the presence of characters like disease or insect resistance is eliminated. Only the final line is evaluated for these traits in contrast to the backcross method which requires testing every backcross generation.

The use of substitution lines to study hybrid vigor was suggested by Sears (1969). This requires reciprocal substitutions for all the chromosomes of two varieties whose hybrid displays heterosis. Each chromosome can then be tested for its effect when in two doses, in one dose (monosomic), or in one dose accompanied by its homolog from the other variety.

11

Cytogenetics of Alien Additions and Substitutions

Ever since the advent of modern plant breeding, researchers have been interested in exploiting the variability in the wild relatives of cultivated crop species. Efforts have been made to transfer hardiness and disease and insect resistance from the wild species of a given genus or of related genera to the cultivated species. There are several examples of crop improvement accomplished through gene transfers from related species. The feasibility of exploiting alien variation in conventional hybridization programs is dependent upon the occurrence of synapsis and recombination between the chromosomes of the cultivated species and the alien chromosomes. In many interspecific and intergeneric crosses, the genomes are so well differentiated that synapsis and recombination does not occur. Cytogeneticists, however, have devised ways to exploit alien variations. These techniques involve the addition of a single alien chromosome, or an alien chromosome pair, to the full chromosome complement of the recipient species (alien addition lines), or the substitution of a single chromosome, or a chromosome pair, of the recipient species with a single alien chromosome, or chromosome pair (alien substitutions). This chapter deals with the production, cytology, breeding behavior, and agronomic values of alien additions and substitutions.

Addition Lines

The addition of a single alien chromosome to the full complement of the recipient species is called monosomic addition, while the addition of a pair of homologs is called a disomic addition. Although the addition of a single chromosome of *Secale cereale* to common wheat was first accomplished by

Leighty and Taylor in 1924, systematic investigations on this subject were not started till 1940 when O'Mara recognized the potential value of the technique and undertook a program of producing the alien additions experimentally. Since then both monosomic and disomic addition lines of wheat having alien chromosomes of different species have been produced. Except for one or two instances in *Nicotiana* and *Avena,* addition lines have been produced entirely in wheat.

Production of Addition Lines

The technique of O'Mara (1940) has been used widely with minor modifications. It involves the production of an amphiploid between two species, for example, wheat and rye. If individual chromosomes of rye are to be added to the wheat chromosome complement, the amphiploid is crossed to wheat, producing progeny with two complete sets of chromosomes of wheat and one set of rye. During meiosis in these individuals, the two wheat sets pair and segregate normally while the chromosomes of rye remain unpaired and segregate at random. Thus the resulting gametes usually, but not always, contain a complete set of wheat chromosomes and a varying number (from zero to seven) of chromosomes of rye. The progeny of these plants thus have two sets of chromosomes of wheat and a varying number of chromosomes of rye. After further backcrosses to wheat, individuals with a full chromosome complement of wheat and one chromosome of rye can be selected. The method is represented in Scheme 11.1.

The monosomic addition plants thus produced are selfed to obtain disomic addition lines. O'Mara (1940) added three different rye chromosomes to wheat, and disomic additions were obtained for each of the three chromosomes. Riley and Chapman (1958b) obtained four disomic additions of rye chromosomes to wheat, while Evans and Jenkins (1960) obtained seven. Several other workers have produced addition lines of wheat having one chromosome or a pair of alien chromosomes from species of related genera (Table 11.I).

A slight variation of the O'Mara technique was used by Sears (1956b) for producing an addition line of *Aegilops umbellulata* to wheat. Intergeneric hybrids between wheat *(Triticum aestivum)* and *Aegilops umbellulata* are difficult to produce. To overcome this difficulty, Sears first crossed *Triticum dicoccoides,* which has the genomic constitution AABB, with *Aegilops umbellulata.* The amphiploid produced from this hybrid was backcrossed to wheat. The product of this cross was backcrossed to wheat several times while selecting for rust resistance. With repeated backcrossing, the genotype of wheat was reconstituted, and, because of rigid screening for rust resistance, only those individuals which possessed the rust-resistance carrying chromosome of *Aegilops umbellulata* were

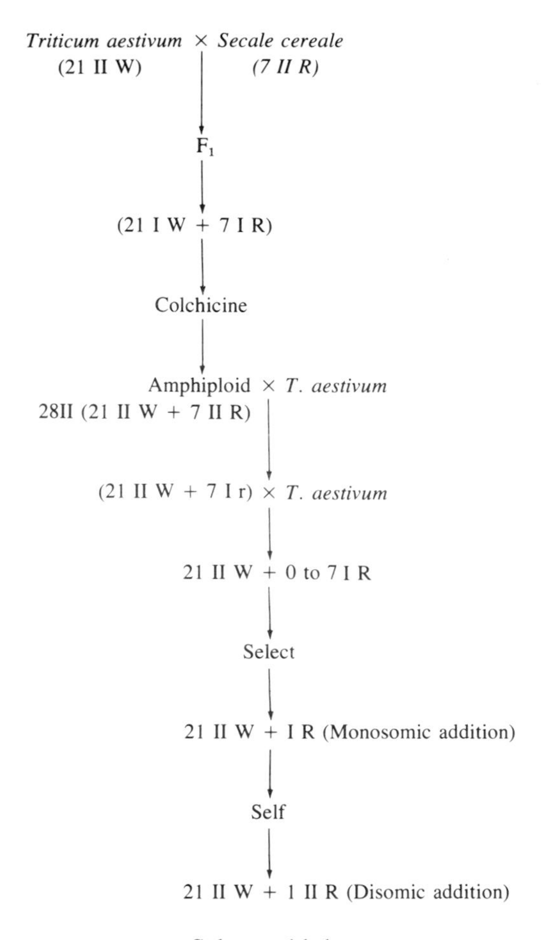

Scheme 11.1

saved after each backcross. In this manner, Sears isolated two monosomic addition lines and one disomic addition line. Somewhat later, Kimber (1967) was able to produce the F_1 hybrid between wheat and *Aegilops umbellulata* by embryo culture and obtained six monosomic and six disomic addition lines using the conventional technique. He also obtained several telocentric and isochromosome additions (G. Kimber, personal communication). Hyde (1953) used procedures similar to those of Sears to produce six of the seven possible additions of *Haynaldia villosa* chromosomes to wheat.

If the fertility of the initial hybrid is good, it may not be necessary to produce the amphiploid. The backcrossing to the recipient species can be started directly thus bypassing one step of the O'Mara technique. In their program, Riley *et al.* (1968) started backcrossing the F_1 *Triticum aestivum* × *Aegilops comosa* to *T. aestivum* to produce additions of *Aegilops comosa* chromosomes to wheat.

TABLE 11.I
Alien Additions of Different Species Produced to Date

Recipient species		Donor species		Number of addition lines		
Name	Chromosome number	Name	Chromosome number	Monosomic addition	Disomic addition	Authority
Triticum aestivum	$n = 21$	*Secale cereale*	$n = 7$	1	1	Leighty and Taylor (1924); Taylor (1934)
				1	1	Florell (1931)
				3	3	O'Mara (1940)
				1	1	Chapman and Riley (1955)
				4	4	Riley and Chapman (1958b)
				7	7	Evans and Jenkins (1960)
				7	7	Riley and Kimber (1966)
		Aegilops umbellulata	$n = 7$	2	1	Sears (1956b)
				6	6	Kimber (1967)
		Aegilops comosa	$n = 7$	1	1	Riley *et al.* (1966a)
		Agropyron elongatum	$n = 7$		1	Konzak and Heiner (1959)
				1		Knott (1961)
		Agropyron intermedium	$n = 21$	1	1	Wienhues (1966)
		Haynaldia villosa	$n = 7$	6	5	Hyde (1953)
Triticum durum	$n = 14$	*Secale cereale*	$n = 7$	2		Sakanaga (1956)
				2		Sadanaga (1957)
		Agropyron elongatum	$n = 7$	7	6	Mochizuki (1962)
Avena sativa	$n = 21$	*Avena strigosa*	$n = 14$	1	1	Dyck and Rajhathy (1963)
		Avena hirtula	$n = 7$	6	6	Thomas (1968)
Nicotiana tabacum	$n = 24$	*Nicotiana glutinosa*	$n = 12$	1	1	Gerstel (1945)
		Nicotiana plumbaginifolia	$n = 10$	3	3	Cameron and Moav (1957)
		Nicotiana paniculata	$n = 10$	1	1	Lucov *et al.* (1970)

Sometimes it may be difficult to obtain disomic additions from the selfed progeny of the monosomic additions. The extra alien chromosome may not transmit through the male due to competition between normal (n) pollen and the pollen with the alien chromosome ($n + 1$). This difficulty may be overcome by obtaining monosomic addition plants of such chromosomal constitution that they produce fewer normal pollen grains than those produced by strictly monosomic additions resulting in less competition for the $n + 1$ pollen. For example, plants having two extra alien chromosomes (double addition monosomic) or plants monosomic for one chromosome of the recipient species and also having an extra alien chromosome would produce far less normal pollen than those having only one extra alien chromosome. Translocation heterozygosity in the basic complement of the monosomic addition line would also reduce the proportion of normal pollen. Hyde (1953) made use of such techniques to obtain disomic additions of *Haynaldia villosa* to wheat from monosomic additions.

Morphology of Addition Lines

The monosomic and disomic addition lines differ from the recipient species in a specific ensemble of morphological features similar to those by which trisomics and monosomics are distinguished from the disomics. Generally the differences are quantitative in nature. Occasionally, qualitative differences may be present. For example, hairy neck, a qualitative trait, is introduced when chromosome 5R of rye is added to wheat.

From the morphological descriptions of rye additions to wheat presented by O'Mara (1940), Riley and Chapman (1958b), and Evans and Jenkins (1960), it is evident that most of the plant parts are affected by the addition of an alien chromosome. The leaves of the addition lines are modified in intensity of color, length, width, and erectness. The spikes differ in length, width, spikelet density, and degree of awning. Some of the addition lines are shorter in height while others are taller (Fig. 11.1). The tillering is also affected in both directions Fertility of the addition lines is generally reduced (Table 11.II).

The additions of *Aegilops umbellulata* chromosomes to wheat show similar differences. Sears (1956b) noted the effect of one chromosome in reducing the length of the culm and the spikelet and in lowering the fertility. Kimber (1967) described the spike traits of addition lines which were diagnostic in identifying the different chromosomes (Fig. 11.2). The addition lines differed from wheat also in plant height and in time to flowering.

Hyde (1953) described the morphology of monosomic and disomic additions of *Haynaldia villosa* chromosomes to wheat. Six monosomic and five disomic additions were identified by spike traits and fertility, by plant vigor and height, and by length, width, and color of leaves. Disomic addition lines of *Agropyron*

Fig. 11.1. Plants illustrating the series of seven disomic addition lines in which single pairs of chromosomes of rye are added to the full complement of wheat chromosomes. (a, far left) *Triticum aestivum* var. Holdfast. (b–h) rye additions III, II, VI, V, I, IV, and VII, respectively. (From Report of the Plant Breeding Institute, Cambridge, 1964/1965, by permission of the director.)

elongatum to tetraploid wheat, *Triticum durum* var. Stewart (Mochizuki, 1962) differed from durum wheat by leaf traits, spike density, culm thickness, presence or absence of wax, tillering ability, and plant vigor, and all the addition lines had very low fertility.

An addition line of a *Nicotiana glutinosa* chromosome to *N. tabacum* was considerably shorter in plant height, and later in flowering and had crowded inflorescences and shorter internodes and leaves (Gerstel, 1945). Addition lines of *Avena hirtula* chromosomes to *Avena sativa* differed from the recipient species in several panicle traits such as compactness, degree of awning, and size of lemma and palea (Thomas, 1968).

TABLE 11.II
Data on Morphological Traits of Three Monosomic and Disomic Rye Additions to Wheat Variety Kharkov[a]

Line	No. of plants	Height (inches)	Density spikelets/cm	Fertility seeds/spikelets	Tillering capacity
Kharkov	10	50.5	1.60	2.87	40.8
V monosomic	9	50.1	1.47	2.25	41.8
V disomic	10	45.2	1.37	2.08	26.7
VI monosomic	9	45.6	1.56	1.50	47.6
VI disomic	11	36.8	1.47	0.75	32.8
VII monosomic	3	46.0	1.80	2.70	56.3
VII disomic	9	44.6	1.83	2.76	35.2

[a]From Evans and Jenkins, 1960.

Cytology of Addition Lines

In the monosomic additions, the alien chromosome remains unpaired during prophase, and its behavior during later stages of meiosis is comparable to the univalent chromosome of wheat monosomics (Chapter 7). Some of the univalent alien chromosomes also influence the synapsis of the chromosomes of the recipient species. Thus Riley (1960a) observed that monosomic additions of chromosomes III and IV of rye to wheat (variety Holdfast) had a higher proportion of cells with 20 II + 3 I and 19 II + 5 I compared to Holdfast and to other monosomic additions.

In the disomic additions, the alien chromosomes pair to form a bivalent but may occasionally fail to synapse with each other. Riley (1960a) compared the univalent frequency in four disomic additions of rye chromosomes to wheat. The univalent frequency was greater than that of the wheat variety Holdfast in chromosome III and IV additions (Table 11.III). Cells with two univalents constituted the majority of those in which there was some pairing failure, but it was difficult to determine whether the univalents were rye or wheat chromosomes. The paired rye chromosomes could be recognized in occasional cells with univalents. The frequency of cells with four or more univalents was also higher than in Holdfast. Thus, these alien chromosomes caused some asynapsis of the wheat chromosomes. On the other hand, additions of chromosomes II and V had regular meiosis as in Holdfast.

Some of the disomic additions of *Agropyron elongatum* chromosomes to wheat had regular bivalent formation while, in others, univalents were observed in 5 to 30% of the cells (Mochizuki, 1962). The presence of an alien rye chromosome

Fig. 11.2. Spikes illustrating the series of six disomic addition lines in which single pairs of chromosomes of *Aegilops umbellulata* are added to the full complement of wheat chromosomes. From left to right. *Triticum aestivum* var. Chinese Spring, disomic additions of chromosomes A, B, C, D, E, and G, of *Aegilops umbellulata*, respectively. (From Kimber, 1967.)

in tetraploid wheat increased the frequency of asynapsis of one or two pairs of the wheat chromosomes (Sadanaga, 1957).

Cytological Stability of Addition Lines

The stability of the addition lines is affected by the irregular behavior of the alien chromosomes in the somatic tissues as well as during meiosis. The alien chromosomes in the addition lines of *Nicotiana tabacum* show somatic instability. Cameron and Moav (1957) produced monosomic addition lines of *N. plumbaginifolia* to *N. tabacum*. In these lines, when the two homologous chromosomes of the *tabacum* genome carried the recessive marker and the alien chromosome carried the dominant marker, the plants were frequently variegated. For example, plants of *wswsWs* genotype having *Ws* on the alien chromosome had colored flowers with white streaks. Variegation was found to be the consequence of somatic loss of the alien chromosome, and the cell lineages from cells that had lost the alien chromosome with dominant marker *Ws* (normal flower color) exhibited the recessive phenotype (white flower). Another alien chromosome which carried the *Kl* locus (pollen killer) behaved in a similar manner. When it was eliminated occasionally from a somatic cell and this cell gave rise to sporogenous tissue, the pollen grains developing from this cell

TABLE 11.III
Meiotic Chromosome Pairing in Disomic Additions of Rye Chromosomes to Wheat[a]

Line	No. plants	No. cells examined	% cells with bivalents only	% cells with univalents		
				2	4	6
II	10	850	91	9	0	0
III	9	470	71	21	6	2
IV	14	730	85	13	1	1
V	4	210	91	8	1	0
Holdfast (wheat)	4	200	94	5	1	0

[a]From Riley, 1960a.

lineage were normal. Further investigations into the mechanism of somatic elimination of alien chromosomes in *Nicotiana* were carried out by Moav (1961) and S. B. Gupta (1968). The stability of the alien chromosome may be affected by adverse environmental conditions. Lucov *et al.* (1970) found that exposing the heterozygous alien addition plants of *N. tabacum* having an added chromosome of *N. paniculata* to lower temperatures greatly increased the amount of variegation.

The relationship of the meiotic behavior of the disomic additions to their cytological stability has been investigated by several workers. From his observations on the study of the addition of rye chromosome I to wheat, O'Mara (1953) concluded that stable addition lines of rye chromosomes to wheat would be improbable because of irregularities in meiotic pairing of alien chromosomes. This conclusion has been verified by subsequent investigators. Riley (1960a) found that 10–20% of the progeny of four disomic additions of rye to wheat were not disomic (Table 11.IV). Most of the aberrant individuals were monosomic additions. Since $n + 1$ pollen is at competitive disadvantage in monosomic additions, about 75% of the progeny of these would be euploid (i.e., revert to the chromosome number of the recipient species). With this type of instability, in only seven or eight generations, the frequency of disomic additions in a population grown without selection would be less than 20%. With cross-pollination, which must occur under field conditions, reversion to the euploid chromosome number would be even faster. This behavior has been termed "addition decay" by Riley and Kimber (1966). Similar results have been reported by O'Mara (1951) and Evans and Jenkins (1960) with rye additions and by Hyde (1953) with *Haynaldia* additions to wheat. Much higher rates of addition decay were reported for three *Agropyron* additions to durum wheat (Mochizuki, 1962).

TABLE 11.IV
Chromosome Numbers of the Progeny of Disomic Additions of Rye Chromosomes to Wheat[a]

Line	No. plants	% with various chromosome numbers				
		41	42	43	44	45
II	64	–	–	6	91	3
III	23	4	–	9	83	4
IV	51	–	–	8	92	–
V	61	1	–	18	79	2

[a]From Riley, 1960a.

Agronomic Value of Addition Lines

To date several alien chromosomes have been added to wheat from related genera which carry genes for resistance to important diseases of wheat, such as, leaf and stem rust, brown and yellow rust, and powdery mildew. For example, rye chromosome III additions are immune to powdery mildew. The *Agropyron elongatum* chromosome added by Knott (1964) conveys resistance to brown rust and stem rust. The chromosome of *Aegilops umbellulata* carrying leaf rust resistance was added to wheat by Sears (1956b), and an addition line of wheat was produced by Riley *et al.* (1968) which carried the yellow rust resistance of *Aegilops comosa*. Other traits of economic importance by which some of the addition lines differ from the recipient species are earliness (Sears, 1956b), winter hardiness (Evans and Jenkins, 1960), and protein content (Riley and Ewart, 1970). In spite of the useful traits, particularly disease resistance, which are added to the recipient species by alien chromosome, none of the addition lines have found a place in agriculture. The factors limiting the success of these lines are: (1) poor stability, (2) reduced fertility, and (3) addition of undesirable traits by the alien chromosomes.

As discussed in the last section, in the absence of selection the addition line would revert to the euploid condition within a few generations due to addition decay. Even if the addition lines were relatively stable, the effect of the alien chromosome on fertility and agronomic traits is generally adverse. Riley (1960a) reported that the four disomic additions of rye to wheat were partially sterile. The most fertile of the four had fertility of 80% of Holdfast (the recipient wheat variety), and one of them had a fertility of less than 2%. Knott (1964) reported a similar effect when an *Agropyron elongatum* chromosome was added to wheat. Low fertility of the addition lines has also been reported by Evans and Jenkins (1960), Hyde (1953), and Mochizuki (1962).

The effect of the alien addition chromosome on other characters of agronomic importance has been pointed out by several workers. The quantitative traits such as tillering, spike conformation, and straw strength and length are affected adversely (Riley and Kimber, 1966). Thus the commercial value of addition lines has been limited by cytological instability, reduced fertility, and adverse effects on agronomic traits.

The alien additions, although unacceptable commercially, have nevertheless been utilized in incorporating small segments of alien chromosomes into the basic chromosome complement of the recipient species by radiation-induced translocations. In the first successful attempt, Sears (1956b) transferred to a wheat chromosome complement, a segment of *Aegilops umbellulata* chromosome carrying rust resistance by irradiating a monosomic addition line. Similar transfers were later made by Knott (1961) and Sharma and Knott (1966) from *Agropyron elongatum* to wheat, by Wienhues (1966) from *Agropyron intermedium* to wheat, and by Driscoll and Jensen (1964b) and by Sears (1967b) from rye to wheat. Such transfer products do not suffer from some of the disadvantages of the alien additions. However, the replacement of chromosome segments of the recipient species by the alien chromosome segments is at random and such transfers generally suffer from being deficient for one chromosome segment and duplicated for the other. It is not surprising that none of these transfers have become commercial varieties.

Alien Substitutions

When a single chromosome pair of a species is replaced by a homoeologous pair of a related species of the same genus or a related genus, the line is referred to as an alien substitution. Alien substitutions of rye chromosomes to wheat have been known since 1938. Systematic investigations on this subject were not started till much later when Unrau *et al.* (1956) outlined a method for producing the alien substitutions experimentally. Since then a great deal of knowledge on this subject has accumulated.

Production of Alien Substitutions

The alien substitutions may arise spontaneously during a backcrossing program following interspecific hybridization. The first known alien substitution, that of rye chromosome 1 (hairy neck chromosome) for one of the wheat pairs (Kattermann, 1938) originated in this manner. Similarly, an alien monosomic substitution of a rye chromosome for 5A of wheat arose spontaneously among the offspring of a disomic addition line of a rye chromosome to wheat (O'Mara, 1947). The disomic addition line was slightly asynaptic. Presumably, as a result of slight asynapsis and segregation of a resultant univalent, it produced some twenty-one chromosome gametes, which had twenty wheat and one rye chromosome (20 I W + 1 I R). When a gamete of this constitution was fertilized by a gamete having twenty-one wheat chromosomes, an alien monosomic substitution plant (20 II W + 1 I W + 1 I R) was produced. Upon selfing, these individuals produced a plant with 21 II consisting of twenty pairs from wheat and one pair from rye.

Several other alien substitutions of spontaneous origin have been obtained in the breeding programs aimed at incorporating disease resistance from an alien species into the cultivated species. The genomes of the parental species involved are characterized by the absence of chromosome pairing and recombination in the hybrids. In these programs, an amphiploid is produced and backcrossed several times to the cultivated variety. After each backcross, the progeny is screened for disease resistance and only the resistant individuals are used for the next backcross. Since the gene or genes for resistance cannot be transferred across recombination barriers, the resistant segregates selected in the backcross progenies prove to be either alien addition lines or alien substitutions. Thus a rust-resistant line of wheat produced by backcrossing an amphidiploid of wheat and *Agropyron elongatum* to the rust-susceptible wheat variety Thatcher had twenty-one pairs of chromosomes and upon cytogenetic analysis proved to be an alien substitution. One chromosome pair of *Agropyron elongatum* had been substituted for chromosomes 6A of wheat (Knott, 1958, 1964). Alien substitutions of *Agropyron* chromosomes to wheat, which originated in the manner described above, have also been analyzed by Bakshi and Schlehuber (1958), Wienhues (1965, 1966), Quinn and Driscoll (1967), and Larson and Atkinson (1970). A mosaic resistant variety of *Nicotiana tabacum* which was produced by repeatedly backcrossing the amphiploid of *Nicotiana tabacum* and *N. glutinosa* to *N. Tabacum* was found to be an alien substitution in which a pair of *N. glutinosa* chromosomes had replaced a specific pair of *N. tabacum* chromosomes (Gerstel, 1943).

A method of producing alien substitutions experimentally was outlined by Unrau *et al.* (1956) and Riley and Kimber (1966). In the first step, the disomic addition line of the alien chromosome to be introduced into the recipient species is crossed as a male to the monosomic of the recipient species. From the progeny, plants are selected which are simultaneously monosomic for the alien chromosome and a specific chromosome of the recipient species. These plants are either self-pollinated or crossed with the same disomic addition line, or its telocentric derivative, as an addition.

If the plant is self-pollinated, individuals with several different chromosomal constitutions are produced in the progeny. Riley *et al.* (1966a) have enumerated the different types theoretically expected. In addition, several misdivision products of the univalents may be present. Plants having *n* bivalents are isolated.

Some of these are disomics like the recipient species, and the rest are the desired disomic alien substitutions. Discrimination between the two is based on the morphology of the plant or by diagnoses of somatic chromosome preparations. However, the ultimate test of the validity of a substitution can only be made by crossing to the recipient species and obtaining a progeny that is exclusively made up of plants with $n - 1$ bivalents and two univalents at meiosis. If the alien chromosome carries some phenotypic markers, such as disease resistance, the identification of the substitutions is facilitated. An alien substitution of a chromosome of *Aegilops comosa* into wheat was obtained by Riley *et al.* (1966a) using the technique described above.

When a plant simultaneously monosomic for the alien chromosome and a chromosome of the recipient species (monosomic substitution) is pollinated by the corresponding alien disomic addition, the progeny contains some plants with n bivalents plus one univalent. These plants are of two constitutions. In one, the univalent is the alien chromosome and all the bivalents belong to the recipient species. In the second group, the univalent and $n - 1$ bivalents belong to the recipient species, and the other bivalent is formed from the alien chromosomes. Disomic alien substitutions are recovered from the selfed progeny of such plants. This procedure was employed for substituting specific rye chromosomes for specific wheat chromosomes by Jenkins (1966), Anderson and Driscoll (1967), Sears (1968), Lee *et al.* (1969), and Bielig and Driscoll (1970, 1971).

If the alien chromosome is not karyotypically recognizable, it may be difficult to distinguish between the two types of progeny. To overcome this difficulty, Riley and Kimber (1966) have suggested the use of telocentrics of alien chromosomes. If telocentric misdivision products of alien chromosomes are available, then a precise distinction is possible. Instead of crossing a line with the disomic addition of the complete alien chromosome to the wheat monosomic, a plant with the disomic addition of an alien telocentric is used. In the progeny of this cross, plants are isolated with forty-two chromosomes, one of which is telocentric. These plants are crossed as the female with a disomic addition of the complete alien chromosome. Individuals will be isolated with forty-three chromosomes, one of which is telocentric. These plants will have a meiotic configuration of twenty-one bivalents—one of which is heteromorphic—and one wheat univalent. Alien substitution would be derived from the selfed progeny of such plants.

Driscoll (c.f., Sears, 1969) has proposed a method for producing alien substitution which bypasses the production of addition lines. The method involves production of amphiploids between wheat monosomics and a donor species, such as *Secale*, *Agropyron*, or *Aegilops*. In the pollen grains produced by such amphiploids, the homoeologous chromosomes of the donor species would compensate for the absence of a particular wheat chromosome. Such pollen grains would be at a selective advantage and when they fertilize the female gametes of similar chromosomal constitution, an alien substitution would be the result.

Identifying the Chromosomes Replaced by the Alien Chromosomes

In the alien substitutions produced experimentally, the identity of the chromosome pair of the recipient species which has been replaced by the alien chromosome pair is known. However, in the alien substitutions of spontaneous origin, the identity of this pair is unknown. As mentioned earlier, the validity of the supposed substitution is verified by crossing the suspected alien substitution with the euploid of the recipient species. The presence of $n - 1$ bivalents and two univalents confirms the supposed nature of alien substitution. The alien substitution is then crossed with all the possible monosomics of the recipient species, and the meiotic metaphase I configurations are analyzed. The F_1 plants either have $n - 2$ bivalents and three univalents or $n - 1$ bivalents and one univalent. The latter configurations would occur only in crosses involving the monosomic line for the chromosome which had been replaced in the alien substitution. Thus Knott (1964) crossed a rust-resistant alien substitution of wheat, in which one pair of *Agropyron elongatum* chromosomes had been substituted with twenty-one monosomics of wheat. A configuration of 20 II + 1 I occurred only in the cross involving mono 6A—evidence that the *Agropyron* pair had replaced chromosome 6A in the alien substitution. Bakshi and Schlehuber (1958) similarly found that in a leaf rust-resistant alien substitution of wheat, a pair of *Agropyron elongatum* chromosomes had replaced the wheat pair 3D.

If available, the ditelosomic lines may be employed instead of the monosomic. These would serve as cytological markers for the appropriate chromosomes. Quinn and Driscoll (1967) crossed an alien substitution of wheat which had a pair of *Agropyron elongatum* chromosomes with wheat lines ditelosomic for 7A, 7B, and 7D. In the F_1 of ditelo-7D, there were twenty pairs and two univalents, one of which was telocentric—proof that the alien chromosome pair had replaced the 7D pair in the substitution. In the F_1 of the ditelo-7A, on the other hand, the telocentric was paired with an entire chromosome showing that 7A was not involved in the substitution. Larson and Atkinson (1970), in a similar way, identified the three wheat chromosomes replaced by *Agropyron* chromosomes in a triple alien substitution line by crossing with ditelosomic lines of several wheat chromosomes.

Cytology, Fertility, and Phenotype of the Alien Substitutions

Reports on the cytology of the alien substitutions are rather meager. Three alien substitutions in which rye chromosome II replaced the chromosomes of the homoeologous group 6 of wheat were reported to be cytologically stable and to regularly form twenty-one bivalents (Riley, 1965). Most of the *Agropyron* substitutions to wheat also form twenty-one bivalents regularly and breed true for chromosome number. A few of them have normal fertility, such as the rust-resistant alien substitution of an *Agropyron intermedium* chromosome to wheat, which is commercially grown under the name Weique in Germany (Wienhues, 1965). An alien substitution of wheat in which one *Agropyron elongatum* chromosome pair replaces the 6D pair was reported to be fully fertile and vigorous by Anderson and Driscoll (1967). The alien substitution studied by Knott (1964) was barely distinguishable from the recipient variety, even in such specialized traits as grain yield and flour quality. The alien substitution lines in which an *Aegilops comosa* chromosome replaced the chromosomes of homoeologous group 2 of wheat were all fertile and of good vegetative vigor and differed only quantitatively in spike conformation and grain shape from the recipient variety Chinese Spring (Riley *et al.*, 1966a).

The three alien substitutions of rye chromosome VI for the wheat chromosomes of homoeologous group 3 are slightly less vigorous than the recipient variety and have reduced fertility (Lee *et al.*, 1969). Similarly, the alien substitutions of rye chromosome III for the wheat chromosomes of the homoeologous group 2 are partially sterile and differ slightly from the recipient variety in spikelet traits (Sears, 1968).

Relationships of Chromosomes Participating in Alien Substitutions

Of interest are the relationships between the alien chromosomes taking part in the alien substitutions and chromosomes of the recipient species being replaced. The alien substitutions are really nullisomic for one member of the chromosome complement. Nullisomics of wheat are known to be phenotypically abnormal and generally of poor vigor and low fertility, and nullisomics of *Nicotiana tabacum* are not even viable. Yet plants with disomic substitutions of alien chromosome pairs show little evidence of the abnormalities normally associated with nullisomy for the substituted chromosomes. Consequently, it appears that the genetic activity of the alien chromosome pair must compensate for the defects normally associated with deficiency, such as poor vigor and sterility. The situation is comparable to the compensating nulli-tetra combinations of wheat discussed in Chapter 9. The inescapable conclusion is that the substituting alien chromosomes are phylogenetically related to chromosomes of the recipient species they are replacing and have similar gene contents. Strong evidence in favor of this hypothesis is also provided by the fact that an alien chromosome substitutes successfully only for the members of a specific homoeologous group but not for others. Thus Knott (1964) found that a specific chromosome of *Agropyron elongatum* tended to substitute spontaneously for chromosome 6A of wheat. The alien substitution line was barely distinguishable from the recipient wheat variety. Johnson (1966) and Anderson and Driscoll (1967) succeeded in substituting the same *Agropyron* chromosome for 6D of wheat, and this alien substitution line was also fully vigorous and fertile. Two other chromosomes of *Agropyron*

Fig. 11.3. Spikes illustrating the disomic addition and alien substitutions of chromosome 2M of *Aegilops comosa* to wheat. From left to right, wheat var. Chinese Spring, 2M disomic addition line, and the three disomic alien substitutions 2M for 2A, 2M for 2B, and 2M for 2D. (From Riley *et al.*, 1968.)

elongatum, each conditioning leaf rust resistance, substitute very well for 3D (Bakshi and Schlehuber, 1958) and 7D (Quinn and Driscoll, 1967), respectively.

Two chromosomes of *Agropyron intermedium* have been substituted into wheat, and the compensation is so good in both cases that the alien substitution lines are fully fertile and vigorous (Wienhues, 1965, 1966). The three alien substitution lines of chromosome 2M of *Aegilops comosa* for the chromosomes of homoeologous group 2 of wheat are also vigorous and have relatively normal spikes (Fig. 11.3) and fertility. The compensation in the alien substitutions of rye chromosomes for those of wheat is not as good as in those of substitutions of *Agropyron* and *Aegilops* chromosomes. This is evident from the low fertility of the rye substitutions (Sears, 1968; Bielig and Driscoll, 1970, 1971). However, there is definite improvement over the phenotype, vigor, and fertility of the corresponding nullisomics (O'Mara, 1947; Riley, 1965; Sears, 1968; Lee *et al.*, 1969).

Convincing evidence for genetic affinity (homoeology) of the chromosomes of *Agropyron*, *Aegilops*, and rye with specific chromosomes of wheat also comes from compensation effect in the pollen grains. According to Knott (1964), pollen grains with twenty wheat chromosomes and one *Agropyron* chromosome function just as well as pollen grains with twenty-one wheat chromosomes. Likewise, Smith (1963) found that rye chromosome I compensated for wheat chromosome 5A in the pollen grains.

The problem of compensation in the pollen grains has been studied thoroughly by P. K. Gupta (1968, 1969) in rye-wheat substitutions. Wheat gametes with twenty chromosomes have very poor male transmission. In the absence of compensation, 20 W + 1 R gametes will have equally poor transmission. However, if the rye chromosome compensates for the missing wheat chromosome, the transmission will be about as good as with twenty-one wheat chromosomes. P. K. Gupta (1969) produced twenty-one monosomic alien substitutions using three chromosomes of rye and seven chromosomes of the D genome of wheat. Each one of these lines had a chromosomal constitution of 20 II W + 1 I W + 1 I R and was used as a male parent in crosses with disomic wheat. The transmission frequency of different kinds of male gametes of each line was inferred from the chromosome counts of the progeny. The monosomic alien substitutions would produce four kinds of gametes, e.g., 20 W + 0 R, 20 W + 1 R, 21 W + 0 R, and 21 W + 1 R. In the compensating combinations, 20 W + 1 R gametes would be functional, and a substantial number of progeny (29% according to Gupta's calculations) have 41 W + 1 R chromosomal constitution. In the noncompensating combinations, on the other hand, very few or no individuals with this constitution will appear. Analyzed on this basis, the data of Table 11.V show that rye chromosome VI (I according to Bhattacharyya and Jenkins, 1960) is homoeologous to chromosomes 1D and 3D of wheat, and rye chromosome II (IV of Bhattacharyya and Jenkins) to 6D of wheat. Chromosome V of rye did not show compensation for any of the chromosomes of the D genome of wheat. The conclusions regarding genetic affinity of rye chromosome VI with chromosomes 1D and 3D of wheat were confirmed by Lee *et al.* (1969) from the study of alien substitutions of this chromosome. The relationship of rye chromosome II to chromosomes of homoeologous group 6 were already known (Riley, 1965). Sears (1968) showed that rye chromosome III is homoeologous to chromosomes of group 2.

The affinity of rye chromosome VI with the chromosomes of two groups (1 and 3) of wheat appears to be anomalous. Theoretically, a chromosome of alien genome should show homoeology with one homoeologous group. However, reciprocal translocation could alter the chromosome arrangements so that partial compensation by one alien chromosome for chromosomes of two groups may be possible. It is indeed known that the chromosomal arrangement of *Secale cereale,* the cultivated rye, differs from the ancestral chromosome arrangement of *Secale montanum,* a wild ancestor, by two translocations involving three chromosome pairs (Riley, 1955; Khush and Stebbins, 1961; Khush, 1962).

TABLE 11.V
Distribution of Progenies from (21 II W) × (20 II W + 1 I W + 1 I R) Crosses Showing Male Transmission of Different Kinds of Gametes[a]

Rye chromosome	Wheat monosome	Chromosome number of progeny			Total
		41W + 1R	42W	42W + 1R	
I[b] (VI)[c]	1D	9	28	4	41
	2D	–	49	4	53
	3D	13	15	3	31
	4D	1	49	10	60
	5D	–	51	6	57
	6D	–	27	6	33
IV[b] (II)[c]	1D	–	20	3	23
	2D	–	22	2	24
	3D	1	48	10	59
	4D	–	40	9	49
	5D	–	20	4	24
	6D	11	36	6	53
	7D	1	21	7	29
V[b]	1D	–	28	3	31
	2D	–	29	1	30
	3D	–	24	5	29
	4D	1	36	6	43
	5D	–	37	6	43
	6D	1	47	3	51
	7D	–	29	2	31

[a]From Gupta, 1969.
[b]According to nomenclature of Bhattacharyya and Jenkins (1960).
[c]According to nomenclature of Riley and Chapman (1958b).

Future research should identify two more chromosomes of rye having cross homoeologies with wheat chromosomes. It is possible that the chromosome involved in both the translocations may have been altered so much that it may not show genetic affinity (compensation) with any of the seven homoeologous groups of wheat. In fact, chromosome V failed to compensate for any of the chromosomes of the D genome of wheat in the pollen grains (Gupta, 1969) and may be the chromosome involved in two translocations.

Several designations for the same rye chromosome have been suggested. Jenkins (1966) proposed that a letter be given to the rye genome and Arabic numbers of 1 to 7 assigned to individual chromosomes to correspond with those of the wheat homoeologous groups. Sears (1968) gave letter designation of R to the rye genome and on the basis of compensation data proposed that

TABLE 11.VI
Homoeologous Relationships of Rye Chromosomes to Those of Wheat Determined from Alien Substitutions and Compensation Studies in Pollen Grains

Substituted rye chromosome[a]	Suggested new designation	Homoeologous relationships determined by
I (VI)	5R	5A (Kattermann, 1938; O'Mara, 1947; Smith, 1963; Jenkins, 1966)
		5A, 5B, and 5D (Bielig and Driscoll, 1970)
		5D (Jenkins, 1966; Muramatsu, 1968)
II (IV)	6R	6A, 6B, and 6D (Riley, 1965)
		6D (Jenkins, 1966; Gupta, 1969)
III	2R	2B and 2D (Sears, 1968)
VI (I)	1R/3R	1A, 1B, 1D, 3A, and 3D (Jenkins, 1966; Lee *et al.*, 1969)
		1D and 3D (Gupta, 1969)
?	3R	3A, 3B, and 3D (Bielig and Driscoll, 1971)

[a]Designations of Riley and Chapman (1958b), those of Bhattacharyya and Jenkins (1960) in parentheses.

rye chromosome I, II, and III be renumbered as 5R, 6R, and 2R, respectively. Lee *et al.* (1969) numbered VI as 3R although this chromosome is also partially homoeologous with chromosomes of group 1 of wheat. Gupta (1971) has rightly assigned a designation of 1R/3R to this chromosome to indicate its cross-homoeology. Bielig and Driscoll (1971) have identified a chromosome which is homoeologous to chromosomes of group 3, but, it is not known whether this chromosome is the same as 1R/3R of Lee *et al.* (1969) and Gupta (1971). The two older systems and the newer and more acceptable system of numbering rye chromosomes are shown in Table 11.VI.

Agronomic Value of Alien Substitutions

Before it can become a commercial variety an alien substitution must meet certain criteria. For example, the substituted chromosome (1) must integrate with the remaining chromosomes of the recipient species so that the meiotic stability is not upset; (2) must compensate for the missing chromosome; (3) provide some additional function, such as disease resistance, not available in the recipient species; and (4) should not introduce any undesirable traits into the recipient species.

Most of the alien substitutions of rye chromosomes to wheat studied to date have reduced fertility and, as such, are of little commercial value. However, the *Agropyron* chromosomes seem to integrate better when substituted for wheat chromosomes. The substitutions of *Agropyron elongatum* chromosomes into wheat studied by Bakshi and Schlehuber (1958), Knott (1964), and Quinn and Driscoll (1967) produced lines comparable to wheat in vigor, fertility, and yield and were meiotically stable. In addition, the substituted chromosomes contributed the leaf rust resistance from *Agropyron*. Yet none of these lines has become a commercial variety for, as pointed out by Bakshi and Schlehuber (1958), they have inferior grain quality.

There is only one example of an alien substitution line becoming a commercial variety. In this case, one chromosome of *Agropyron intermedium* compensates so well for a particular wheat chromosome that the rust-resistant alien substitution line has been released in Germany for commercial production under the name of Weique (Wienhues, 1965). However, the general performance of the variety relative to other wheat varieties is not known. Although reduced seed fertility is not a handicap in crop production in tobacco, the substitutions of the chromosomes of *Nicotiana glutinosa* and *N. plumbaginifolia* for those of *N. tabacum* have also not been very successful. Many disease-resistant varieties of *N. tabacum* have been developed in which resistance has been incorporated from the alien species. Cytogenetic analyses reveal that these varieties do not possess the intact alien substituted chromosomes but only segments (Gerstel and Burk, 1960; Mann *et al.*, 1963; D. U. Gerstel, personal communication). Occasional synapsis and recombination take place between the alien chromosome and the homoeologous chromosome of the recipient species in the monosomic alien substitutions, and segments of the alien chromosome with traits of economic importance are substituted.

It is evident that alien additions and substitutions have not been very useful in grain crops because the overall modifications resulting from additions or substitutions of alien chromosomes are usually too gross and too arbitrary. In *Nicotiana,* however, lines with substituted alien segments have become important cultivated varieties. As mentioned earlier, pairing and recombination between wheat chromosomes and the alien chromosomes from rye, *Agropyron,* and *Aegilops,* do not occur, and the alien chromosome segments cannot be substituted as in *Nicotiana*. Radiation-induced transfers have been attempted, but such transfers are deficient for one chromosome segment and duplicated for another. These disadvantages may be overcome if the alien chromosome can be induced to pair and recombine with the homoeologous chromosomes of the recipient species. The recombination products would incorporate alien chromosome segments smaller than the entire chromosome and in which the replacement by the alien segment would not be at random. The alien chromosome segment in such transfers would compensate for the deficiency of the chromosome segment being replaced. Sears (1967a) outlined several procedures for inducing homoeologous recombination by removing the restriction to homoeologous pairing imposed by the 5B system (see Chapter 9 for discussion of the 5B system). The first successful attempt to transfer a useful trait from a wild species to wheat by suppressing the 5B effect was by Riley *et al.* (1968). An alien chromosome segment carrying a gene for yellow rust resistance of *Aegilops comosa* was substituted by induced recombination between homoeologous chromosomes 2D of wheat and 2M of *Aegilops comosa*. The activity of 5B was inhibited by introducing a genome of *Aegilops speltoides*. The line with the substituted alien chromosome segment is fully fertile and, in the hybrids with other wheat varieties, chromosome pairing and segregation are normal so that the line can be treated as any other varietal parent. It has been given the varietal name Compair. The usefulness of Compair in agriculture either as a commercial variety or as a parental material in breeding programs remains to be seen. However, it is the result of application of a sophisticated cytogenetic technique in wheat improvement. Due to the availability of aneuploids and knowledge of homoeologous relationships, it now appears possible to substitute either whole chromosomes, or chromosome segments, carrying traits of economic importance from alien species into wheat.

12

Aneuploidy in Animals and Man

Much work has been done with the aneuploids of plant species. Because of the economic importance of crop species, cytogeneticists have thoroughly investigated aneuploids of several crops. However, little is known about the aneuploids of animal species. Until 1959 the only known aneuploids in the animal kingdom were those for sex chromosomes and chromosome 4 of *Drosophila melanogaster*.

One reason for the few known aneuploids in the animal kingdom is the problem of detecting the aneuploid condition in organisms that have a large number of chromosomes as do man and domesticated animal species. As mentioned in Chapter 1, Langdon-Down (1866) first described the anomalous phenotype in man now called Down's syndrome. Several authors, such as Waardenburg (1932) and Bleyer (1934), suggested that trisomy for one chromosome may be responsible for Down's syndrome. There was, however, no agreement on the correct chromosome number of man at that time: it was reported to be 46 by some authors and 48 by others. Due to this uncertainty and poor techniques of chromosome analysis, no progress in determining the chromosome number of the individuals with Down's syndrome could be made.

Tjio and Levan (1956) developed a greatly improved technique for counting somatic chromosomes. By using cultures of lung fibroblasts derived from human embryos, they established beyond doubt that the normal disomic chromosome number in man is 46. Development of improved technique for human chromosome studies and the establishment of correct chromosome number stimulated studies of the karyotypes in various pathological conditions in man. Several laboratories were engaged in independent studies of human chromosomes when Lejeune *et al.* (1959a, b) showed that the individuals with Down's syndrome (mongolism) had 47 chromosomes and were trisomic for a small chromosome of the complement. This finding gave impetus to the studies of chromosomes of humans with different morphological anomalies. Many laboratories in the United States, Canada, Great Britain, France, Germany, and Japan started investigations of human chromosomes and a large number of papers have since been published on trisomics and suspected trisomics of man.

The improved technique for chromosome studies developed by Tjio and Levan (1956) has been successfully employed in the karyotypic analysis of several other animal species and a few aneuploids have been discovered. However, the number of aneuploids known in the animal kingdom is still very small as compared with the plant kingdom, reflecting the general paucity of polyploids in the animal kingdom as compared with plant kingdom.

Two theories have been put forward to explain the scarcity of polyploidy in the bisexual animals. These theories may explain the paucity of aneuploidy as well. According to the theory offered by Muller (1925), polyploidy upsets the balance of sex determination to such an extent that regular segregation of sex factors is impossible in polyploids. Many aneuploids for sex chromosomes of *Drosophila* and man have been studied to date and most of them have abnormal sexual features. Tolerance of aneuploidy for sex chromosomes in animals, however, is much higher than tolerance for other chromosomes (autosomes) of the complement. Thus, trisomics, tetrasomics, and monosomics for the sex chromosomes of *Drosophila,* man, and mouse are known to be viable. On the other hand, only the trisomics for a few shorter autosomes in these species have been obtained and tetrasomics and monosomics (except for chromosome 4 in *Drosophila*) for any of the autosomes have not been obtained.

Another theory which explains the scarcity of polyploidy in the animal kingdom, was put forward by von Wettstein (1927) and is equally applicable to aneuploids of animals. According to this theory polyploidy upsets the complex processes of animal development by sudden alteration of the genotype. Plants can tolerate this alteration as their developmental processes are much simpler than those of animals. Li (1927) has shown that a great proportion of mono-4 *Drosophila melanogaster* zygotes die in the earlier stages of development. Similarly, there is evidence that many trisomic human embryos abort in the earlier stages of development. Thus Carr (1965) examined the chromosomes of 200 human abortuses and found that 44 of them had chromosomal abnormalities—mainly extra chromosomal materials.

Having reviewed these special features of aneuploidy in animals we can now turn to the specific cases of aneuploidy in animal kingdom.

Aneuploidy in Man

The literature on trisomics of man is so voluminous that it cannot be reviewed adequately in a book of this size and scope. Authoritative reviews on the subject have been made by Smith (1964) and Schöneich (1965). In this section, I have attempted to give only the highlights of the problem.

Human Chromosomes

Besides the XY or sex-determining chromosome pair, there are 22 pairs of autosomes in the human chromosome complement. These 22 pairs have been numbered 1 to 22 according to their size in somatic mitoses, 1 being the longest and 22, the shortest. Even in the best mitotic preparations it is seldom possible to identify more than a few specific pairs individually. Because of the structural similarity between many of the autosomal pairs, subgroups have been classified into more readily identifiable letter categories ranging from A through G (Patau, 1961). Thus individual chromosomes can be assigned to either of the above seven groups on the basis of length and position of centromere but the members of the same group may not always be differentiated from one another. This classification is given in the following tabulation:

Chromosomes	Group
1–3	A
4–5	B
6–12 + X	C
13–15	D
16–18	E
19–20	F
21–22 + Y	G

The chromosomes of each group may also be designated by letter and a numeral as subscript. Thus G_1 is No. 21 and G_2 is No. 22.

When a new trisomic in man is discovered it is assigned to one of the chromosome groups. As more information accumulates the extra chromosome may be identified as to the specific number in the chromosome complement. Thus trisomics with Down's syndrome were called G trisomics, because the extra chromosome belonged to the group G or smallest chromosomes. There is now general agreement that the extra chromosome in the mongols is No. 21 (G_1) although it was believed to be No. 22 by some.

Aneuploids for Autosomes

Three primary trisomics for autosomes are now well known. The most thoroughly investigated is of course triplo-21 or G trisomic. As mentioned earlier this trisomic phenotype was first described by Langdon-Down in 1866. He

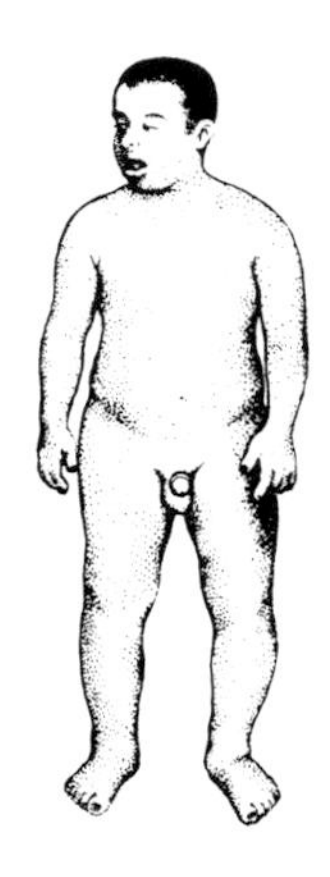

Fig. 12.1. A boy trisomic for chromosome 21 (Down's syndrome). (From V. A. McKusick, *Medical Genetics* 1958–1960, St. Louis, 1961. By permission of C. V. Mosby Company.)

gave the name of mongols to these phenotypes, although the term Down's syndrome has become more popular in modern literature. Numerous reports have been published in the medical literature by the physicians who examined this phenotype. Penrose and Smith (1966) have given detailed description of various clinical features of the mongols. Diagnosis can generally be established within first postnatal week. Mongols are characterized by mental deficiency (I.Q. of 25 to 50), relatively flat occiput, pouting expression when crying, small ears with overlapping upper helices, small nose, protruding tongue, short neck, short hands and fingers, and a small penis in males. In addition, many other clinical defects have been recorded by the physicians (Fig. 12.1).

It has been estimated that one out of 700 babies born is born a mongol. The chances of mongol births increase, the older the mother. Why the frequency of nondisjunction of chromosome 21 increases with increasing maternal age is not fully understood.

Mongols have lived up to 30 years of age or even longer. Male mongols are generally sterile but female mongols have become mothers. Thompson (1961) has reviewed the literature on reproduction in mongols. There are ten documented cases of female mongols who have reproduced. Seven normal and five mongol offspring have been produced by ten mongol mothers in eleven pregnancies including a monozygotic twin pair. This gives a transmission rate of about 40% for the extra chromosome.

In two cases, the fathers of the mongol mothers sired the children (Sawyer, 1949; Schlaug, 1957) while in one case a brother of the mongol was the father of the child (Hanhard, 1960). In some other cases, the fathers were the fellow inmates of the mental institutions or nursing homes where these mongol mothers were confined. Forssman and Thysell (1957) have reported a case of a child born of a mongol mother and an epileptic blind man who grew up to be a

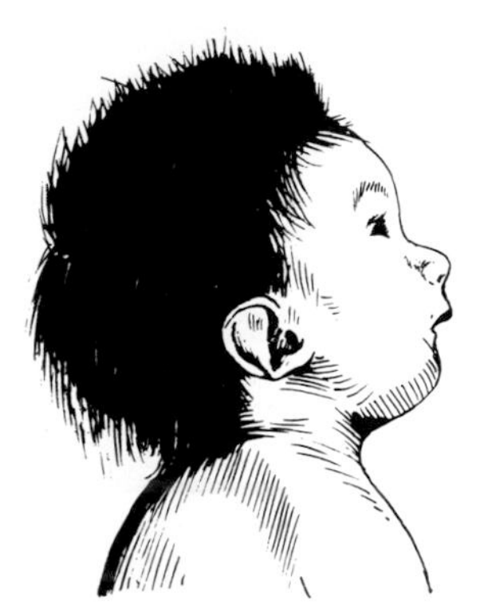

Fig. 12.2. A child trisomic for chromosome 18 (E syndrome). (From *J. Pediat.* **60,** 516, 1962. Reproduced by permission of C. V. Mosby Company.)

normal boy. The mongol woman and the epileptic blind man were being cared for by the wife of a farmer. The mongol woman was generally locked up when she was left alone. However, it turned out that the blind man who lived in the next room knew where the key was kept. He admitted later, that he had gone into her room and had sexual intercourse with her several times. Pregnancy was not suspected by the caretaker. When she noticed the large abdomen in the mongol she attributed it to an abdominal tumor. When, however, the mongol gave birth to a baby boy, police enquiry established the parentage of the child. Both the father and the son had the normal chromosome number, 46, and the son was free of any mental or physical abnormalities (Forssman *et al.,* 1961).

Another well-known trisomic syndrome is the E trisomy or trisomy for chromosome No. 18. This syndrome was first described by Edwards *et al.* (1960). Smith *et al.* (1960) gave another account of this anomaly. Conen and Erkman (1966a) have reviewed the literature on this trisomic. The trisomics for this chromosome do not live long and generally die in infancy. The mean life span of the triplo-18 individuals was reported to be 238 days by Conen and Erkman (1966a), although a few have survived to 15 years of age (Hook *et al.,* 1965).

Triplo-18 individuals are malformed, have poor survival, and are mentally deficient. They have prominent occiput, low set and malformed ears, narrow palatal arch, flexed fingers, dorsiflexed feet, small pelvis, and limited hip abduction (Fig. 12.2.).

The third well-known trisomic of man, that for a D chromosome or triplo-13, was first discovered by Patau *et al.* (1960). Smith (1964) and Conen and Erkman (1966b) have reviewed the literature dealing with this trisomic. Like triplo-21 and triplo-18, the trisomics for this chromosome are mentally defective and grossly malformed and have even lower survival. The mean survival time was calculated to be 131.7 days for 36 cases by Conen and Erkman (1966b) and only two lived up to 199 days after birth. This trisomic is readily identified

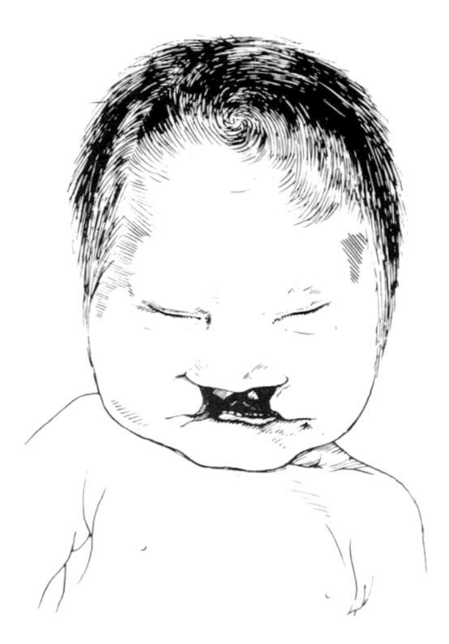

Fig. 12.3. A child trisomic for chromosome 13 (D syndrome). (From *Amer. J. Dis. Child.* **3**, 241, 1966. Reproduced by permission of American Medical Association.)

because of cleft lip and palate, defect of eye development, and sloping forehead with relatively small cranium. The brain is incomplete in its development. None of the patients with this anomaly examined to date have responded to sound. They have short necks and are characterized by polydactyly and narrow hyperconvex fingernails (Fig. 12.3).

Besides these three well known cases of autosomal trisomy, several other anomalies have been described in human beings which have been attributed to chromosomal duplications or deficiencies. Several cases of tertiary trisomics (translocation trisomics) have been reported (see review by Schöneich, 1965). A new case of primary trisomy that for chromosome No. 22 has been reported by Uchida *et al.* (1968). If verified this would be the fourth autosomal trisomic of man to be investigated.

Deficiencies for short arms of human chromosomes have been reported. de Grouchy *et al.* (1966) reported on a 13-month-old girl with severe retardation and with facial and limb deformities. She was found to be deficient for short arm of chromosome 18. Similarly, Lejeune *et al.* (1964) described an anomaly which they named "Cat Cry" syndrome due to high pitched voice of the patients. From karyotic analyses of three patients they concluded that this anomaly was due to deficiency of short arm of chromosome 5.

Aneuploids for Sex Chromosomes

As mentioned earlier, tolerance for aneuploidy of sex chromosomes in animals and man is much greater than tolerance for aneuploidy of autosomes. Thus monosomy is tolerated only for the sex chromosome in man. XO females with 45 chromosomes and lacking a Y chromosome are characterized by a well known phenotype called Turner's syndrome. These individuals have short stature, webbed neck, low-set ears and broad shieldlike chest. Breasts are underdeveloped, nipples are widely spaced, ovaries are represented only by fibrous streaks and the uterus is small. Hence Turner females are sterile. A majority of XO conceptions end in spontaneous abortions. Approximately one out of 3500 female births is a Turner "female" (McKusick, 1964).

The individuals trisomic for an X chromosome, e.g., having XXY chromosomal, constitution occur more frequently (once in every 400–600 male births) and the anomalies associated with this trisomy are referred to as Klinefelter syndrome. Klinefelter individuals have male external genitalia but the testes are consistently very small and body hair is sparse. Most of the cases have femalelike breast development. They have unusually long legs (McKusick, 1964) and are sterile.

It may be noted that the human aneuploids for sex chromosomes are not so malformed and are not as mentally retarded as the aneuploids for autosomes. Even individuals with three extra X chromosomes (XXXXY) have been reported (Fraccaro *et al.*, 1962) and are less malformed than simple trisomics for autosomes.

Aneuploidy in Other Animals

Aneuploids for sex chromosomes and chromosome 4 of *Drosophila melanogaster* have been known for long time (Bridges, 1921a, b, 1922, 1925). Triplo-4 flies are slightly smaller than normal, have narrower wings but are fertile. Even tetra-4 flies have been obtained, although viability is very low (Grell, 1961). Mono-4 flies, on the other hand, have paler body color and have wide wings. These flies are characterized by low productivity, tardy development, heavy mortality, and frequent sterility (Li, 1927). The trisomic and monosomic condition for chromosome 4 is transmitted to the next generation. Aneuploids for sex chromosomes of *Drosophila melanogaster* have been thoroughly studied. Flies of XXX chromosomal constitution are superfemales but the mortality rate of these individuals is very high. XO individuals are sterile males while XXY flies are fertile females.

In mice, XO and XXY individuals have been obtained. The former are fertile females (Welshons and Russell, 1959; Cattanach, 1962) and the latter are viable and phenotypically normal but sterile males (Cattanach, 1961; Russell and Chu, 1961). The first reports of autosomal primary trisomy in mouse were made independently by Cattanach (1964) and Griffen and Bunker (1964). Cattanach (1964) obtained one mouse which was trisomic for one small chromosome of the complement. The animal was phenotypically normal but was completely sterile. The trisomy was suspected to be the result of triethylenemelamine-induced nondisjunction in the meiotic divisions of father. Three primary trisomics reported by Griffen and Bunker (1964) and four additional primary trisomics reported by Griffen and Bunker (1967) were found among the offspring sired by irradiated males. All of these trisomics are phenotypically normal, some are completely sterile, while others are partially sterile. Tertiary trisomics of mouse have also been described by Lyon and Meredith (1966) and Griffen (1967).

Solitary cases of trisomy have been reported in chimpanzee (McClure *et al.*, 1969), wild pigs (McFee *et al.*, 1966), and dog (Shive *et al.*, 1965).

Bibliography

Allan, R. E. (1959). Monosomic analysis studies. *Wheat Newsl.* **6,** 85.

Anderson, L. M., and Driscoll, C. J. (1967). The production and breeding behavior of a monosomic alien substitution line. *Can. J. Genet. Cytol.* **9,** 399-403.

Andrews, G. Y., and McGinnis, R. C. (1964). The artificial induction of aneuploids in *Avena. Can. J. Genet. Cytol.* **6,** 349-356.

Arnold, C. G., and Kressel, M. (1965). Über das gehäufte auf treten trisomer Pflanzen in der Nachkommenschaft diploider Röntgeumutanten bei *Oenothera berteriana. Z. Vererbungsl.* **96,** 83-92.

Arora, O. P., and Khoshoo, T. N. (1969). Primary trisomics in moss *verbena. Euphytica* **18,** 237-248.

Avers, C. J. (1954). Chromosome behavior in fertile triploid Aster hybrids. *Genetics* **39,** 117-126.

Avery, A. G., and Blakeslee, A. F. (1948). Effect of extra chromosomes on the shape of stigmas of *Datura stramonium. Genetics* **33,** 603.

Avery, A. G., Satina, S., and Rietsema, J. (1959). "Blakeslee: The Genus *Datura,*" 289 pp. The Ronald Press, New York.

Avery, P. (1929). Chromosome number and morphology in *Nicotiana.* IV. The nature and effects of chromosomal irregularities in *N. alata* var Grandiflora. *Univ. Calif. Pub. Bot.* **11,** 265-284.

Babcock, E. B., and Navashin, M. (1930). The genus Crepis. Bibliogr. Genet. **6,** 1-90.

Baker, E. P., Sanghi, A. K., McIntosh, R. A., and Luig, N. H. (1970). Cytogenetical studies in wheat. III. Studies of a gene conditioning resistance to stem rust strains with unusual genes for avirulence. *Aust. J. Biol. Sci.* **23,** 369-375.

Baker, R. L., and Morgan, D. T., Jr. (1966). Monosomics in maize induced by X-irradiation of the pollen. *Cytologia* **31,** 172-175.

Bakshi, J. M., and Schlehuber, A. M. (1958). Identification of a substituted chromosome pair in *Triticum-Agropyron* line. *Proc. Okla. Acad. Sci.* **39,** 16-21.

Barton, D. W. (1950). Pachytene morphology of the tomato chromosome complement. *Amer. J. Bot.* **37,** 639-643.

Belling, J. (1921). The behavior of homologous chromosomes in a triploid *Canna. Proc. Nat. Acad. Sci. U. S.* **7,** 197-201.

Belling, J. (1925). Homologous and similar chromosomes in diploid and triploid *Hyacinths. Genetics* **10,** 59-71.

Belling, J. (1927). The attachments of chromosomes at the reduction division in flowering plants. *J. Genet.* **18,** 177-204.

Belling, J., and Blakeslee, A. F. (1922). The assortment of chromosomes in triploid *Daturas*. *Amer. Natur.* **56,** 339-346.

Belling, J., and Blakeslee, A. F. (1924). The configurations and sizes of the chromosomes in the trivalents of 25 chromosome *Daturas*. *Proc. Nat. Acad. Sci. U. S.* **10,** 116-120.

Belling, J., and Blakeslee, A. F. (1926). On the attachment of nonhomologous chromosomes at the reduction division in certain 25 chromosome *Daturas*. *Proc. Nat. Acad. Sci. U. S.* **12,** 7-11.

Bergner, A. D., Avery, A. G., and Blakeslee, A. F. (1940). Chromosomal deficiencies in *Datura stramonium* induced by colchicine. *Amer. J. Bot.* **27,** 676-683.

Bhattacharyya, N. K., and Jenkins, B. C. (1960). Karyotype analysis and chromosome designations for *Secale cereale* L. "Dakold." *Can. J. Genet. Cytol.* **2,** 268-277.

Bhowal, J. G. (1964). An unusual transmission rate of the deficient male gamete in a substitution monosomic of chromosome 3D in wheat. *Can. J. Bot.* **42,** 1321-1328.

Bielig, L. M., and Driscoll, C. J. (1970). Substitution of rye chromosome 5R for its three wheat homoeologues. *Genet. Res.* **16,** 317-324.

Bielig, L. M., and Driscoll, C. J. (1971). Production of alien substitution lines in *Triticum aestivum*. *Can. J. Genet. Cytol.* **13,** 429-436.

Blakeslee, A. F. (1921a). The globe mutant in the Jimson weed (*Datura stramonium*). *Genetics* **6,** 241-264.

Blakeslee, A. F. (1921b). Types of mutations and their possible significance in evolution. *Amer. Natur.* **55,** 254-267.

Blakeslee, A. F. (1921c). The globe a simple trisomic mutant in *Datura*. *Proc. Nat. Acad. Sci. U. S.* **7,** 148-152.

Blakeslee, A. F. (1922). Variations in *Daturas* due to the changes in chromosome number. *Amer. Natur.* **56,** 16-31.

Blakeslee, A. F. (1924). Distinction between primary and secondary chromosomal mutants in *Datura*. *Proc. Nat. Acad. Sci. U. S.* **10,** 109-116.

Blakeslee, A. F. (1927a). The chromosomal constitution of Nubbin, a compound ($2n + 1$) type in *Datura*. *Proc. Nat. Acad. Sci. U. S.* **13,** 79-85.

Blakeslee, A. F. (1927b). Nubbin, a compound chromosomal type in *Datura*. *Ann. N.Y. Acad. Sci.* **30,** 1-29.

Blakeslee, A. F. (1930). Extra chromosomes, a source of variations in Jimson weed. Smithsonian Rep. 3096, pp. 431-450.

Blakeslee, A. F. (1934). New Jimson weeds from old chromosomes. *J. Hered.* **25,** 80-108.

Blakeslee, A. F., and Avery, A. G. (1934). Three genes located in the 21·22 chromosome of the Jimson weed. *J. Hered.* **25,** 393-404.

Blakeslee, A. F., and Avery, A. G. (1938). Fifteen-year breeding records of $2n + 1$ types in *Datura stramonium*. Cooperation in research. *Carnegie Inst. Wash. Publ.* **501,** 315-351.

Blakeslee, A. F., and Avery, B. T. (1919). Mutations in the Jimson weed. *J. Hered.* **10,** 111-120.

Blakeslee, A. F., and Belling, J. (1924a). Chromosomal mutations in the Jimsom weed, *Datura stramonium*. *J. Hered.* **15,** 195-206.

Blakeslee, A. F., and Belling, J. (1924b). Chromosomal chimeras in the Jimson weed. *Science* **60,** 19-20.

Blakeslee, A. F., and Cartledge, J. L. (1926). Pollen abortion in chromosomal types of *Datura*. *Proc. Nat. Acad. Sci. U. S.* **12,** 315-323.

Blakeslee, A. F., and Farnham, M. E. (1923). Trisomic inheritance in the poinsettia mutant of *Datura*. *Amer. Natur.* **57,** 481-495.

Blakeslee, A. F., Avery, A. G., and Bergner, A. D. (1936). A method of isolating tertiary $2n + 1$ forms in *Datura* from prime types by use of double half chromosomes. *Science* **83,** 486.

Blakeslee, A. F., Belling, J., and Farnham, M. E. (1920). Chromosomal duplication and mendelian phenomena in *Datura* mutants. *Science* **52,** [N.S.] 388-390.

Blakeslee, A. F., Bergner, A. D., and Avery, A. G. (1930). Compensating extra chromosomal types in *Datura* and their use as testers. *Science* **71,** 516.

Blakeslee, A. F., Morrison, G., and Avery, A. G. (1927). Mutations in a haploid *Datura* and their bearing on the hybrid-origin theory of mutants. *J. Hered.* **18,** 193-199.

Blakeslee, A. F., Avery, A. G., Bergner, A. D., Satina, S., Warmke, H. E., Buchholz, J. T., Conklin, M. E., Sinnott, E. W., and Van Overbeek, J. (1940). Chromosome investigations. *Carnegie Inst. Wash. Rep. Yearb.* **39,** 203-211.

Bleyer, A. (1934). Indications that mongoloid imbecility is a gametic mutation of degressive type. *Amer. J. Dis. Child.* **47,** 342-348.

Bremer, G. (1923). A cytological investigation of some species and species hybrids within the genus *Saccharum*. *Genetica* **5,** 97-148.

Brewer, G. J., Sing, C. F., and Sears, E. R. (1969). Studies of isozyme patterns in nullisomic-tetrasomic combinations of hexaploid wheat. *Proc. Nat. Acad. Sci. U. S.* **64,** 1224-1229.

Bridges, C. B. (1921a). Genetical and cytological proof of non-disjunction of the fourth chromosome of *Drosophila melanogaster*. *Proc. Nat. Acad. Sci. U. S.* **7,** 186-192.

Bridges, C. B. (1921b). Triploid intersexes in *Drosophila melanogaster*. *Science* **54,** [N.S.] 252-254.

Bridges, C. B. (1922). The origin and variations in sexual and sex-limited characters. *Amer. Natur.* **56,** 51-63.

Bridges, C. B. (1925). Sex in relation to chromosomes and genes. *Amer. Natur.* **59,** 127-137.

Briggle, L. W. (1966). Three loci in wheat involving resistance to *Erysiphe graminis* f. sp. *tritici*. *Crop Sci.* **6,** 461-465.

Briggle, L. W., and Sears, E. R. (1966). Linkage of resistance to *Erysiphe graminis* f. sp. *tritici* (Pm_3) and hairy glume (*Hg*) on chromosome 1A of wheat. *Crop. Sci.* **6,** 559-561.

Brown, M. S. (1966). Attributes of intra- and interspecific aneuploidy in *Gossypium*. *Heredity Suppl.* **20,** 98-112.

Brown, M. S., and Endrizzi, J. E. (1964). The origin, fertility and transmission of monosomics in *Gossypium*. *Amer. J. Bot.* **51,** 108-115.

Brown, S. W. (1949). The structure and meiotic behavior of the differentiated chromosomes of tomato. *Genetics* **34,** 437-461.

Buchholz, J. T., and Blakeslee, A. F. (1922). Studies of the pollen tubes and abortive ovules of the globe mutant of *Datura*. *Science* **55,** 597-599.

Buchholz, J. T., and Blakeslee, A. F. (1927). Abnormalities in pollen-tube growth in *Datura* due to the gene "Tricarpel." *Proc. Nat. Acad. Sci. U. S.* **13,** 242-249.

Buchholz, J. T., and Blakeslee, A. F. (1930a). Pollen tube growth and control of gametophytic selection in Cocklebur, a 25-chromosome *Datura*. *Bot. Gaz.* **90,** 366-383.

Buchholz, J. T., and Blakeslee, A. F. (1930b). Pollen tube growth of the primary mutant of *Datura,* rolled and its two secondaries. *Proc. Nat. Acad. Sci. U. S.* **16,** 190-195.

Buchholz, J. T., and Blakeslee, A. F. (1932). Pollen tube growth in primary and secondary $2n + 1$ *Daturas*. *Amer. J. Bot.* **19,** 604-626.

Buchholz, J. T., Doak, C. C., and Blakeslee, A. F. (1932). Control of gametophytic selection in *Datura* through shortening and splicing of styles. *Bull. Torrey Bot. Cl.* **59,** 109-118.

Buchman, J. (1857). Report of Botanical Garden of Royal Agricultural College at Cirencester. *Rep. 27th Meeting Brit. Ass. Advan. Sci.,* pp. 200-215.

Burnham, C. R. (1930). Genetical and cytological studies of semi-sterility and related phenomena in maize. *Proc. Nat. Acad. Sci. U. S.* **16,** 269-277.

Burnham, C. R. (1934). Cytogenetic studies of an interchange between chromosomes 8 and 9 in maize. *Genetics* **19,** 430-447.

Burnham, C. R. (1946). A gene for "long" chromosomes in barley. *Genetics* **31,** 212-213.

Burnham, C. R. (1948). Cytogenetic studies of a translocation between chromosomes 1 and 7 in maize. *Genetics* **33,** 5-21.

Burnham, C. R. (1956). The use of chromosomal interchanges to test for the independence of linkage groups established genetically. *Proc. Intern. Genet. Symp. Tokyo and Kyoto, Sci. Council Japan, Tokyo,* pp. 453-456.

Burnham, C. R. (1962). "Discussions in Cytogenetics," 375 pp. Burgess, Minneapolis, Minnesota.

Burnham, C. R., and Hagberg, A. (1956). Cytogenetic notes on chromosomal interchanges in barley. *Hereditas* **42,** 467-482.

Burnham, C. R., White, F. H., and Livers, R. (1954). Chromosomal interchanges in barley. *Cytologia* **19,** 191-202.

Butler, L. (1952). The linkage map of the tomato. *J. Hered.* **43,** 25-35.

Caldecott, R. S., and Smith, L. (1952). A study of X-ray-induced chromosomal aberrations in barley. *Cytologia* **17,** 224-242.

Cameron, D. R. (1966). Some experimental applications of aneuploidy in *Nicotiana. Heredity Suppl.* **20,** 1-7.

Cameron, D. R., and Moav, R. (1957). Inheritance in *Nicotiana tabacum.* XXVII. Pollen killer, an alien genetic locus inducing abortion of microspores not carrying it. *Genetics* **42,** 326-335.

Campbell, A. B., and McGinnis, R. C. (1958). A monosomic analysis of stem rust reaction and awn expression in Redman wheat. *Can. J. Plant Sci.* **38,** 184-187.

Carr, D. H. (1965). Chromosome studies in spontaneous abortions. *Obstet. Gynecol.* **26,** 308-326.

Catcheside, D. G. (1954). The genetics of brevistylis in *Oenothera. Heredity* **8,** 125-137.

Cattanach, B. M. (1961). XXY mice. *Genet. Res.* **2,** 156-158.

Cattanach, B. M. (1962). XO mice. *Genet. Res.* **3,** 487-490.

Cattanach, B. M. (1964). Autosomal trisomy in the mouse. *Cytogenetics* **3,** 159-166.

Chapman, V., and Riley, R. (1955). The disomic addition of rye chromosome II to wheat. *Nature (London)* **175,** 1091-1092.

Chapman, V., and Riley, R. (1966). The allocation of the chromosomes of *Triticum aestivum* to the A and B genomes and evidence on genome structure. *Can. J. Genet. Cytol.* **8,** 57-63.

Chang, T. D., and Sadanaga, K. (1964a). Crosses of six monosomics in *Avena sativa* L. with varieties, species and chlorophyll mutants. *Crop. Sci.* **4,** 589-593.

Chang, T. D., and Sadanaga, K. (1964b). Breeding behavior, morphology, karyotype, and interesting results of six monosomics in *Avena sativa* L. *Crop. Sci.* **4,** 609-615.

Chen, C. C., and Grant, W. F. (1968a). Morphological and cytological identification of the primary trisomics of *Lotus pedunculatus (Leguminosae). Can. J. Genet. Cytol.* **10,** 161-179.

Chen, C. C., and Grant, W. F. (1968b). Trisomic transmission in *Lotus pedunculatus. Can. J. Genet. Cytol.* **10,** 648-654.

Chomchalow, N., and Garber, E. D. (1964). The genus *Collinsia.* XXVI. Polyploidy and trisomy in *C. tinctoria. Can. J. Genet. Cytol.* **6,** 488-499.

Clausen, R. E. (1930). Inheritance in *Nicotiana tabacum.* X. Carmine-coral variegation. *Cytologia* **1,** 358-368.

Clausen, R. E. (1931). Inheritance in *Nicotiana tabacum.* XI. The fluted assemblage. *Amer. Natur.* **65,** 316-331.

Clausen, R. E., and Cameron, D. R. (1944). Inheritance in *Nicotiana tabacum.* XVIII. Monosomic analysis. *Genetics* **29,** 447-477.

Clausen, R. E., and Cameron, D. R. (1950). Inheritance in *Nicotiana tabacum.* XXIII. Duplicate factors for chlorophyll production. *Genetics* **35,** 4-10.

Clausen, R. E., and Goodspeed, T. H. (1924). Inheritance in *Nicotiana tabacum.* IV. The trisomic character "enlarged." *Genetics* **9,** 181-197.

Clausen, R. E., and Goodspeed, T. H. (1926a). Inheritance in *Nicotiana tabacum*. VII. The monosomic character "fluted." *Univ. Calif. Pub. Bot.* **11,** 61-82.

Clausen, R. E., and Goodspeed, T. H. (1926b). Interspecific hybridization in *Nicotiana*. III. The monosomic *tabacum* derivative, "corrugated" from the *sylvestris-tabacum* hybrid. *Univ. Calif. Pub. Bot.* **11,** 83-101.

Clayberg, C. D. (1959). Cytogenetic studies of precocious meiotic centromere division in *Lycopersicon esculentum* Mill. *Genetics* **44,** 1335-1346.

Clayberg, C. D., Butler, L., Kerr, E. A., Rick, C. M., and Robinson, R. W. (1966). Third list of known genes in the tomato. *J. Hered.* **57,** 188-196.

Cole, K. (1956). The effect of various trisomic conditions in *Datura stramonium* on crossability with other species. *Amer. J. Bot.* **43,** 794-801.

Conen, P. E., and Erkman, B. (1966a). Frequency and occurrence of chromosomal syndromes. II. E-trisomy. *Amer. J. Hum. Genet.* **18,** 387-398.

Conen, P. E., and Erkman, B. (1966b). Frequency and occurrence of chromosomal syndromes. I. D-trisomy. *Amer. J. Hum. Genet.* **18,** 374-386.

Costa-Rodriguez, L. (1954). Chromosomal aberrations in oats, *Avena sativa* L. *Agron. Lusitana* **16,** 49-65.

Darlington, C. D. (1930). Studies in Prunus. III. *J. Genet.* **22,** 65-93.

Darlington, C. D. (1934). The origin and behavior of chiasmata. VII. Zea mays. *Z. Indukt. Abstamm. Vererbungsl.* **67,** 96-114.

Darlington, C. D. (1939). Misdivision and the genetics of the centromere. *J. Genet.* **37,** 341-364.

Darlington, C. D. (1940). The origin of iso-chromosomes. *J. Genet.* **39,** 351-361.

Darlington, C. D., and Mather, K. (1932). The origin and behavior of chiasmata. III. Triploid *Tulipa*. *Cytologia* **4,** 1-15.

Darlington, C. D., and Mather, K. (1944). Chromosome balance and interaction in *Hyacinthus*. *J. Genet.* **46,** 52-61.

Das, K., and Goswami, B. C. (1967). A note on translocation-trisomic plant from gamma-irradiated barley. *Cytologia* **32,** 55-58.

Das, K., and Srivastava, H. M. (1969). Interchange-trisomics in barley. *Genetica* **40,** 555-565.

Davenport, C. B. (1924). Rate of pollen-tube growth in *Datura* mutant. *Annu. Rep. Director Dept. Genet. Yearb. Carnegie Inst. Wash.* **22,** 92.

Dermen, H. (1931). Polyploidy in *Petunia*. *Amer. J. Bot.* **18,** 250-261.

de Grouchy, J., Bonnette, J., and Salmon, C. (1966). Deletion du bras court du chromosome 18. *Ann. de Genet.* **9,** 19-26.

de Vries, H. (1901). "Die Mutation theorie Bd 1 Die Entstehung der Arten durch Mutation," 648 pp. Leipzig.

de Vries, H., and Boedijn, K. (1923). On the distribution of mutant characters among the chromosomes of *Oenothera lamarckiana*. *Genetics* **8,** 233-238.

Dhillon, T. S., and Garber, E. D. (1960). The genus *Collinsia*. X. Aneuploidy in *C. heterophylla*. *Bot. Gaz.* **121,** 125-133.

Douglas, C. R. (1968). Abortive pollen: A phenotypic marker of monosomics in upland cotton, *Gossypium hirsutum* L. *Can. J. Genet. Cytol.* **10,** 913-915.

Driscoll, C. J., and Baker, E. P. (1965). Location of genes for resistance to stem rust race 126-Anz-1 in four varieties of wheat. *Wheat Inf. Serv.* **19, 20,** 47-49.

Driscoll, C. J., and Jensen, N. F. (1964a). Chromosomes associated with waxlessness, awnedness and time of maturity of common wheat. *Can. J. Genet. Cytol.* **6,** 324-333.

Driscoll, C. J., and Jensen, N. F. (1964b). Characteristics of leaf rust resistance transferred from rye to wheat. *Crop Sci.* **4,** 372-374.

Driscoll, C. J., and Sears, E. R. (1965). Mapping of a wheat-rye translocation. *Genetics* **51,** 439-443.

Dubuc, J. P., and McGinnis, R. C. (1970). The identification and meiotic behavior of a ditelosomic line in *Avena sativa* L. *Can. J. Genet. Cytol.* **12,** 876-881.

Dyck, P. L. (1964). Desynapsis in diploid oats, *Avena strigosa,* and its use in the production of trisomics. *Can. J. Genet. Cytol.* **6,** 238.

Dyck, P. L., and Rajhathy, T. (1963). Cytogenetics of a hexaploid oat with an extra pair of chromosomes. *Can. J. Genet. Cytol.* **5,** 408-413.

Dyck, P. L., and Rajhathy, T. (1965). A desynaptic mutant in *Avena strigosa. Can. J. Genet. Cytol.* **7,** 418-421.

East, E. M. (1933). The behavior of a triploid in *Nicotiana tabacum* L. *Amer. J. Bot.* **20,** 269-289.

Edwards, J. H., Harnden, G. D., Cameron, A. H., Crosse, V. M., and Wolfe, O. H. (1960). A new trisomic syndrome. *Lancet* **i,** 787-789.

Einset, J. (1943). Chromosome length in relation to transmission frequency of maize trisomes. *Genetics* **28,** 349-364.

Ekstrand, H. (1932). Ein Fall von erblicher Asyndese bei *Hordeum. Sv. Bot. Tidskr.* **26,** 292-302.

Ellis, J. R., and Janick, J. (1960). The chromosomes of *Spinacea oleracea. Amer. J. Bot.* **47,** 210-214.

Emerson, S. H. (1935). The genetic nature of de Vries's mutations in *Oenothera lamarckiana. Amer. Natur.* **69,** 545-559.

Emerson, S. H. (1936). The trisomic derivatives of *Oenothera lamarckiana. Genetics* **21,** 200-224.

Endrizzi, J. E. (1963). Genetic analysis of six primary monosomes and one tertiary monosome in *Gossypium Hirsutum. Genetics* **48,** 1625-1633.

Endrizzi, J. E. (1966). Use of haploids in *Gossypium barbadense* L. as a source of aneuploids. *Curr. Sci.* **2,** 34-35.

Endrizzi, J. E., and Brown, M. S. (1964). Identification of monosomes for six chromosomes in *Gossypium hirsutum. Amer. J. Bot.* **51,** 117-120.

Endrizzi, J. E., and Kohel, R. J. (1966). Use of telosomes in mapping three chromosomes in cotton. *Genetics* **54,** 535-550.

Endrizzi, J. E., and Morgan, T. D. (1955). Chromosomal interchanges and evidence for duplication in haploid *Sorghum vulgare. J. Hered.* **46,** 201-208.

Endrizzi, J. E., and Taylor, T. (1968). Cytogenetic studies of $N\ Lc_1\ Yg_2\ R_2$ marker genes and chromosome deficiencies in cotton. *Genet. Res.* **12,** 295-304.

Endrizzi, J. E., McMichael, S. C., and Brown, M. S. (1963). Chromosomal constitution of "Stag" plants of *Gossypium hirsutum,* Acala 4-42. *Crop Sci.* **3,** 1-3.

Erlanson, E. W. (1929). Cytological conditions and evidences for hybridity in North American wild roses. *Bot. Gaz.* **87,** 443-506.

Evans, L. E., and Jenkins, B. C. (1960). Individual *Secale cereale* chromosome additions to *Triticum aestivum.* I. The addition of individual "Dakold" fall rye chromosomes to "Kharkov" winter wheat and their subsequent identification. *Can. J. Genet. Cytol.* **2,** 205-215.

Fedak, G., Tsuchiya, T., and Helgason, S. B. (1971). Cytogenetics of some monotelotrisomics in barley. *Can. J. Genet. Cytol.* **13,** 760-770.

Federley, H. (1931). Chromosome analyse der reziproken Bastarde zwischen *Pygaera pigra* and *P. curtula* sowie ihrer Ruckkreuzungbastarde. *Z. Zellforsch.* **12,** 772-816.

Feldman, M. (1966). The effect of chromosomes 5B, 5D and 5A on chromosomal pairing in *Triticum aestivum. Proc. Nat. Acad. Sci. U. S.* **55,** 1447-1453.

Feldman, M., Mello-Sampayo, T., and Sears, E. R. (1966). Somatic association in *Triticum aestivum. Proc. Nat. Acad. Sci. U. S.* **56,** 1192-1199.

Florell, V. H. (1931). A cytologic study of wheat × rye hybrids and back crosses. *J. Agr. Res.* **42,** 341-362.

Forssman, H., and Thysell, T. (1957). A woman with mongolism and her child. *Amer. J. Ment. Def.* **62,** 500-503.

Forssman, H., Lehmann, O., and Thysell, T. (1961). Reproduction in mongolism: Chromosome studies and re-examination of a child. *Amer. J. Ment. Def.* **65,** 495-498.

Fraccaro, M., Klinger, H. P., and Schutt, W. (1962). A male with XXXXY sex chromosomes. *Cytogenetics* **1,** 52-64,

Frandsen, N. O. (1967). Haploid production in potato breeding material with intensive back crossing to wild species. *Zuechter* **37,** 120-134.

Frost, H. B. (1927). Chromosome-mutant types in stocks *(Matthiola incana* R. Br.). I. Characters due to extra chromosomes. *J. Hered.* **18,** 474-486.

Frost, H. B. (1931). Uncomplicated trisomic inheritance of purple versus red in *Matthiola incana. Proc. Nat. Acad. Sci. U. S.* **17,** 434-436.

Frost, H. B., and Mann, M. C. (1924). Mutant forms of *Matthiola* resulting from non-disjunction. *Amer. Natur.* **58,** 569-572.

Frost, S., and Ising, G. (1964). Cytogenetics of fragment chromosomes in barley. *Hereditas* **52,** 176-180.

Gager, C. S., and Blakeslee, A. F. (1927). Chromosomes and gene mutations in *Datura* following exposure to radium rays. *Proc. Nat. Acad. Sci. U. S.* **13,** 75-79.

Gaines, E. F., and Aase, H. C. (1926). A haploid wheat plant. *Amer. J. Bot.* **13,** 373-385.

Galen, D. F., and Endrizzi, J. R. (1968). Induction of monosomes and mutations in cotton by gamma irradiation of pollen. *J. Hered.* **59,** 343-346.

Garber, E. D. (1964). The genus *Collinsia.* XXII. Trisomy in *C. heterophylla. Bot. Gaz.* **125,** 46-50.

Gates, R. R. (1909). The behavior of the chromosomes in *Oenothera lata* × *Oenothera gigas. Bot. Gaz.* **48,** 179-199.

Gates, R. R. (1923). The trisomic mutations in *Oenothera. Ann. Bot.* **37,** 543-563.

Gauthier, F. M., and McGinnis, R. C. (1965). The association of "Kinky" neck with chromosome 20 of *Avena sativa* L. *Can. J. Genet. Cytol.* **7,** 120-125.

Gerassimova, H. (1940). A translocation between the B- and D-chromosomes and the trisomic effect of the B-chromosome in *Crepis tectorum* L. *Akad. Nauk SSSR Izv. Ser. Biol.* **1,** 31-44.

Gerstel, D. U. (1943). Inheritance in *Nicotiana tabacum.* XVII. Cytogenetical analysis of glutinosa-type resistance to mosaic disease. *Genetics* **28,** 533-536.

Gerstel, D. U. (1945). Inheritance in *Nicotiana tabacum.* XX. The addition of *Nicotiana glutinosa* chromosomes to tobacco. *J. Hered.* **36,** 197-206.

Gerstel, D. U., and Burk, L. G. (1960). Controlled introgression in *Nicotiana*: A cytological study. *Tobacco Sci.* **4,** 147-150.

Gill, B. S., Virmani, S. S., and Minocha, J. L. (1970a). Primary simple trisomics in pearl millet. *Can. J. Genet. Cytol.* **12,** 474-483.

Gill, B. S., Virmani, S. S., and Minocha, J. L. (1970b). Aneuploids in pearl millet. *Experientia* **26,** 1021.

Goodspeed, T. H., and Avery, P. (1939). Trisomic and other types in *Nicotiana sylvestris. J. Genet.* **38,** 381-458.

Goodspeed, T. H., and Avery, P. (1941). The twelfth primary trisomic type in *Nicotiana sylvestris. Proc. Nat. Acad. Sci. U. S.* **27,** 13-14.

Goodspeed, T. H., and Clausen, R. E. (1928). Interspecific hybridization in *Nicotiana.* VIII. The *sylvestris-tomentosa-tabacum* hybrid triangle and its bearing on the origin of *tabacum. Univ. Calif. Publ. Bot.* **11,** 245-256.

Gottschalk, W. (1951). Untersuchungen am Pachytan normaler und rontgenbestrahlter Pollen mutterzellen von *Solanum lycopersicum. Chromosoma* **4,** 298-341.

Gottschalk, W. (1954). Die Grundzahl der Gattung *Solanum* und einiger *Nicotiana* arten. *Ber. Deut. Bot. Ges.* **67,** 369-376.

Goulden, C. H. (1926). A genetic and cytological study of dwarfing in wheat and oats. *Tech. Bull. Minn. Agr. Exp. Sta.* **33,** 1-37.

Greenleaf, W. H. (1941). The probable explanation of low transmission ratios of certain monosomic types of *Nicotiana tabacum*. *Proc. Nat. Acad. Sci. U. S.* **27,** 427-430.

Grell, E. H. (1961). The tetrasomic for chromosome 4 in *Drosophila melanogaster*. *Genetics* **46,** 1177-1183.

Griesinger, R. (1937). Über hypo- und hyperdiploide Formen von *Petunia, Hyoscyamus, Lamium* und einige andere chromosomenzahlungen. *Ber. Deut. Bot. Ges.* **55,** 556-571.

Griffen, A. B. (1967). A case of tertiary trisomy in the mouse, and its implications for the cytological classification of trisomics in other mammals. *Can. J. Genet. Cytol.* **9,** 503-510.

Griffen, A. B., and Bunker, M. C. (1964). Three cases of trisomy in the mouse. *Proc. Nat. Acad. Sci. U. S.* **52,** 1194-1198.

Griffen, A. B., and Bunker, M. C. (1967). Four further cases of autosomal primary trisomy in the mouse. *Proc. Nat. Acad. Sci. U. S.* **58,** 1446-1452.

Griffiths, D. J., and Thomas, P. T. (1953). Genotypic control of chromosome loss in *Avena*. *Proc. 9th Intern. Congr. Genet. Bellagio (Como)* **II,** 1172-1175.

Gupta, S. B. (1968). Chlorophyll variegation caused by unstable behaviour of an alien chromosome in hybrid derivatives of *Nicotiana* spp. *Genetica* **39,** 193-208.

Gupta, P. K. (1968). Homeology of a rye (*Secale cereale* var. Dakold) chromosome. *Wheat Inf. Serv.* **27,** 13-15.

Gupta, P. K. (1969). Studies on transmission of rye substitution gametes in common wheat. *Indian J. Genet. Plant Breed.* **29,** 163-172.

Gupta, P. K. (1971). Homoeologous relationship between wheat and rye chromosomes. Present status. *Genetica* **42,** 199-213.

Hacker, J. B. (1965). The inheritance of chromosome deficiency in *Avena sativa* monosomics. *Canad. J. Genet. Cytol.* **7,** 316-327.

Hacker, J. B. (1966). The inheritance of virescent albinism in *Avena sativa* var. Sun II. *Canad. J. Genet. Cytol.* **8,** 387-392.

Hacker, J. B., and Riley, R. (1963). Aneuploids in oat varietal populations. *Nature (London)* **197,** 924-925.

Hacker, J. B., and Riley, R. (1965). Morphological and cytological effects of chromosome deficiency in *Avena sativa*. *Can. J. Genet. Cytol.* **7,** 304-315.

Hagberg, A. (1954). Cytogenetic analysis of erectoides mutations in barley. *Acta Agr. Scand.* **4,** 472-490.

Hagemann, R. (1969). Somatic conversion (Para mutation) at the sulfurea locus of *Lycopersicon esculentum* Mill. III. Studies with trisomics. *Can. J. Genet. Cytol.* **11,** 346-358.

Håkansson, A. (1936). Die Zytologie eines trisomischen *Pisum*-Typus. *Hereditas* **21,** 223-226.

Håkansson, A. (1940). Eine tertiare trisome von *Godetia whitneyi*. *Bot. Notis.* pp. 395-398.

Håkansson, A. (1945). Zytologische Studien an Monosomischen Typen von *Godetia whitneyi*. *Hereditas* **31,** 129-162.

Haldane, J. B. S. (1930). Theoretical genetics of autopolyploids. *J. Genet.* **22,** 359-372.

Halloran, G. M., and Boydell, C. W. (1967). Wheat chromosomes with genes for vernalization response. *Can. J. Genet. Cytol.* **9,** 632-639.

Hanhart, E. (1960). Mongoloid idiocy in mother and two children from an incestuous relationship. *Acta Genet. Med. Gemellol.* **9,** 112-130.

Hanna, W. W., and Schertz, K. F. (1970). Inheritance and trisomic linkage of seedling characters in *Sorghum bicolor* (L.) Moench. *Crop Sci.* **10,** 441-443.

Hanna, W. W., and Schertz, K. F. (1971). Trisome identification in *Sorghum bicolor* (L.) Moench by observing progeny of triploid × translocation stocks. *Can. J. Genet. Cytol.* **13,** 105-109.

Hanson, W. D. (1952). An interpretation of the observed amount of recombination in interchange heterozygotes of barley. *Genetics* **37,** 90-100.

Hanson, W. D., and Kramer, H. H. (1949). The genetic analysis of two chromosome interchanges in barley. from F_2 data. *Genetics* **34,** 687-700.

Hanson, W. D., and Kramer, H. H. (1950). The determination of linkage intensities from F_2 and F_3 genetic data involving chromosomal interchanges in barley. *Genetics* **35,** 559-569.

Haunold, A. (1968). A trisomic hop, *Humulus lupulus* L. *Crop. Sci.* **8,** 503-505.

Haunold, A. (1970). Fertility studies and cytological analysis of the progeny of a triploid × diploid cross in hop, *Humulus lupulus* L. *Can. J. Genet. Cytol.* **12,** 582-588.

Hermsen, J. G. Th. (1963). The localization of two genes for dwarfing in the wheat variety Timstein by means of substitution lines. *Euphytica* **12,** 126-129.

Hermsen, J. G. Th. (1969). Induction of haploids and aneuploids in colchicine-induced tetraploid *Solanum chacoense* Bitt. *Euphytica* **18,** 183-189.

Hermsen, J. G. Th., Wagenvoort, M., and Ramana, M. S. (1970). Aneuploids from natural and colchicine-induced autotetraploids of *Solanum*. *Can. J. Genet. Cytol.* **12,** 601-613.

Heyne, E. G., and Livers, R. W. (1953). Monosomics analysis of leaf rust reaction, awnedness, winter injury and seed color in Pawnee wheat. *Agron. J.* **45,** 54-58.

Hiorth, G. (1948a). Über das Wesen der Monosomen und der disomen Anordnung 3-Katte + Univalent bei *Godetia whitneyi*. Vorlaüfige *Mitt. Z. Vererbungsl.* **82,** 1-11.

Hiorth, G. (1948b). Zur Genetik der Monosomen von *Godetia whitneyi*. *Z. Vererbungsl.* **82,** 230-275.

Hook, E. B., Lehrke, R., Roesner, A., and Yunis, J. J. (1965). Trisomy-18 in a 15-year-old female. Lancet **ii,** 910-911.

Howard, H. W. (1948). Meiotic irregularities in hexaploid oats. II. A cytological examination of the variety Picton. *J. Agr. Sci.* **38,** 332-338.

Hu, C. H. (1968). Studies on the development of twelve types of trisomics in rice with reference to genetic study and breeding program. *J. Agr. Ass. China* **63,** [N.S.], 53-71.

Hurd, E. A., and McGinnis, R. C. (1958). Note on the location of genes for dwarfing in Redman wheat. *Can. J. Plant Sci.* **38,** 506.

Huskins, C. L. (1927). On the genetics and cytology of fatuoid or false wild oats. *J. Genet.* **18,** 315-364.

Huskins, C. L., and Hearne, E. M. (1933). Meiosis in asynaptic dwarf oats and wheat. *J. Roy. Microsc. Soc.* **53** [Ser. III], 109-117.

Hyde, B. B. (1953). Addition of individual *Haynaldia villosa* chromosomes to hexaploid wheat. *Amer. J. Bot.* **40,** 174-182.

Iwata, N., Omura, T., and Nakagahra, M. (1970). Studies on the trisomics in rice plants (*Oryza sativa* L.). I. Morphological classification of trisomics. *Jap. J. Breed.* **20,** 230-236.

Iyer, R. D. (1968). Towards evolving a trisomic series in jute. *Curr. Sci.* **37,** 181-183.

Jachuck, P. J. (1963). Isolation of chromosome variants in cultivated rice. *Oryza* **1**(2), 131-132.

Jagathesan, D., and Swaminathan, M. S. (1961). Viability and fertility of monosomics in *Gossypium hirsutum*. *Curr. Sci.* **30,** 155.

Jakob, K. M. (1963). A trisomic male castor bean plant. *J. Hered.* **54,** 292-296.

Janick, J., Mahony, D. L., and Pfahler, P. L. (1959). The trisomics of *Spinacea oleracea*. *J. Hered.* **50,** 47-50.

Jenkins, B. C. (1966). *Secale* additions and substitutions to common wheat. *Proc. 2nd Intern. Wheat Genet. Symp. Hereditas Suppl.* **2,** 301-312.

Jha, K. K. (1964). The association of a gene for purple coleoptile with chromosome 7D of common wheat. *Can. J. Genet. Cytol.* **6,** 370-372.

Johnson, R. (1966). The substitution of a chromosome from *Agropyron elongatum* for chromosomes of hexaploid wheat. *Can. J. Genet. Cytol.* **8,** 279-292.

Johnson, R., Wolfe, M. S., and Scott, P. R. (1969). *Annu. Rep. Plant Breed. Inst.*, Cambridge, p. 134.

Johnsson, H. (1942). Cytological studies of triploid progenies of *Populus tremula*. *Hereditas* **28,** 306-312.

Johnsson, H. (1946). Progeny of triploid *Betula verrucosa* Erh. *Bot. Notis.*, 285-290.

Joshi, A. B., and Howard, H. W. (1954). Meiotic irregularities in hexaploid oats. III. Further observations on the frequency of univalents and other meiotic irregularities in spring × winter variety hybrids of *Avena sativa*. *J. Agr. Sci.* **45**, 380-387.

Kamanoi, M., and Jenkins, B. C. (1962). Trisomics in common rye. *Secale cereale* L. *Seiken Ziho* **13**, 118-123.

Karibasappa, B. K. (1961). An auto-triploid of rice and its progeny. *Curr. Sci.* **30**, 432-433.

Karpechenko, G. D. (1927). The production of polyploid gametes in hybrids. *Hereditas* **9**, 349-368.

Kasha, K. J., and McLennan, H. A. (1967). Trisomics in diploid alfalfa. I. Production, fertility and transmission. *Chromosoma* **21**, 232-242.

Katayama, T. (1963). Study on the progenies of autotriploid and asynaptic rice plants. *Jap. J. Breed.* **13**, 83-87.

Kattermann, G. (1938). Uber konstante, halmbehaarte Stamime aus Weizenroggenbastardierung mit $2n = 42$ chromosomen. *Z. Indukt. Abstamm. Vererbungsl.* **74**, 354-375.

Kattermann, G. (1939). Sterilitats studien bei *Hordeum distichum*. *Z. Indukt. Abstamm. Vererbungsl.* **77**, 63-103.

Kerber, E. R. (1954). Trisomics in barley. *Science* **120**, 808-809.

Khan, S. I. (1962). Monosomes and their influence on meiosis in "Holdfast." *Cytologia* **27**, 386-397.

Khush, G. S. (1962). Cytogenetic and evolutionary studies in *Secale*. II. Interrelationships of wild species. *Evolution* **16**, 484-496.

Khush, G. S., and Rick, C. M. (1966a). The use of tertiary trisomics in linkage mapping. *Genetics* **54**, 343.

Khush, G. S., and Rick, C. M. (1966b). The origin, identification and cytogenetic behavior of tomato monosomics. *Chromosoma* **18**, 407-420.

Khush, G. S., and Rick, C. M. (1967a). Novel compensating trisomics of the tomato: Cytogenetics, monosomic analysis, and other applications. *Genetics* **55**, 297-307.

Khush, G. S., and Rick, C. M. (1967b). Tomato tertiary trisomics: Origin, identification, morphology and use in determining position of centromeres and arm location of markers. *Can. J. Genet. Cytol.* **9**, 610-631.

Khush, G. S., and Rick, C. M. (1967c). Studies on the linkage map of chromosome 4 of the tomato and on the transmission of induced deficiencies. *Genetica* **38**, 74-94.

Khush, G. S., and Rick, C. M. (1967d). Haplo-triplo-disomics of the tomato: Origin, cytogenetics, and utilization as a source of secondary trisomics. *Biol. Zentrabl.* **86**, 257-265.

Khush, G. S., and Rick C. M. (1968a). Cytogenetic analysis of the tomato genome by means of induced deficiencies. *Chromosoma* **23**, 452-484.

Khush, G. S., and Rick, C. M. (1968b). Tomato telotrisomics: Origin, identification, and use in linkage mapping. *Cytologia* **33**, 137-148.

Khush, G. S., and Rick, C. M. (1968c). The use of secondary trisomics in the cytogenetic analysis of the tomato genome. *Proc. 12th Intern. Congr. Genet. Tokyo,* **1**, 117.

Khush, G. S., and Rick, C. M. (1969). Tomato secondary trisomics: Origin, identification, morphology and use in cytogenetic analysis of tomato genome. *Heredity* **24**, 129-146.

Khush, G. S., and Stebbins, G. L. (1961). Cytogenetic and evolutionary studies in *Secale*. I. Some new data on the ancestry of *S. cereale*. *Amer. J. Bot.* **48**, 723-730.

Khush, G. S., Rick, C. M., and Robinson, R. W. (1964). Genetic activity in a heterochromatic chromosome segment of the tomato. *Science* **145**, 1432-1434.

Kihara, H. (1924). Cytologische und genetische Studien bei weichtigen Getreidearten mit besonderer Rucksicht auf das Verhaltan der Chromosomen und die Sterilitat in den Bastarden. *Mem. Coll. Sci. Kyoto Imp. Univ. Ser.* **B1**, 1-200.

Kihara, H. (1932). Chromosomale Aberranten bei *Pharbitis nil*. *Kwagaku* **2,** 196-198.

Kihara, H. (1939). Polyploidy in genus *Triticum*. *Bot. Zool.* **7**.

Kihara, H., and Nishiyama, I. (1937). Possibility of crossing over between semi homologous chromosomes from two different genomes. *Cytologia* (Fuji Jubilee Vol.), Part II, 654-666.

Kihara, H., and Tsunewaki, K. (1962). Polyploids and aneuploids of *Triticum dicoccum*, var. Khapli produced by N_2O-treatment. *Wheat Inf. Serv.* **14,** 1-3.

Kihara, H., and Wakakuwa, S. (1930). Dwarf plants with 40 chromosomes in the progenies of a pentaploid *Triticum*-hybrid. *Jap. J. Genet.* **5,** 134–137.

Kihara, H., and Wakakuwa, S. (1935). Veraenderung von Wuchs, fertilitaet und chromosomenzahl in den Folgegenerationen der 40-chromosomigen Zwerge bei Weizen. *Jap. J. Genet.* **11,** 102-108.

Kimber, G. (1967). The addition of the chromosomes of *Aegilops umbellulata* to *Triticum aestivum* (var. Chinese Spring). *Genet. Res.* **9,** 111-114.

Kimber, G., and Sears, E. R. (1968). Nomenclature for the description of aneuploids in the *Triticinae*. *Proc. 3rd Intern. Wheat Genet. Symp., Aust. Acad. Sci. Canberra,* pp. 468-473.

Knott, D. R. (1958). The effect on wheat of an *Agropyron* chromosome carrying rust resistance. *Proc. 10th Intern. Congr. Genet. Montreal* **2,** 148.

Knott, D. R. (1959). The inheritance of rust resistance. IV. Monosomic analysis of rust resistance and some other characters in six varieties of wheat including Gabo and Kenya Farmer. *Can. J. Plant Sci.* **39,** 215-228.

Knott, D. R. (1961). The inheritance of rust resistance. VI. The transfer of stem rust resistance from *Agropyron elongatum* to common wheat. *Can. J. Plant Sci.* **41,** 109-123.

Knott, D. R. (1964). The effect on wheat of an *Agropyron* chromosome carrying rust resistance. *Can. J. Genet. Cytol.* **6,** 500-507.

Kohel, R. J. (1966). Identification of super cup mutant in cotton, *Gossypium hirsutum* L. *Crop Sci.* **6,** 86-87.

Koller, P. C. (1938). Asynapsis in *Pisum sativum*. *J. Genet.* **36,** 275-306.

Konzak, C. F., and Heiner, R. E. (1959). Progress in the transfer of resistance to bunt (*Tilletia caries* and *T. foetida*) from *Agropyron* to wheat. *Wheat Inf. Serv.* **9-10,** 31.

Kramer, H. H., Veyl, R., and Hanson, W. D. (1954). The association of two genetic linkage groups in barley with one chromosome. *Genetics* **39,** 159-168.

Kuspira, J. (1966). Intervarietal chromosome substitution in hexaploid wheat. *Proc. 2nd Intern. Wheat Genet. Symp. Hereditas Suppl.* **2,** 355-369.

Kuspira, J., and Millis, L. A. (1967). Cytogenetic analysis of tetraploid wheats using hexaploid wheat aneuploids. *Can. J. Genet. Cytol.* **9,** 79-86.

Kuspira, J., and Unrau, J. (1957). Genetic analysis of certain characters in common wheat using whole chromosome substitution lines. *Can. J. Plant Sci.* **37,** 300-326.

Kuspira, J., and Unrau, J. (1958). Determination of the number and dominance relationships of genes on substituted chromosomes in common wheat—*Triticum aestivum* L. *Can. J. Plant Sci.* **38,** 199-205.

Kuspira, J., and Unrau, J. (1960). Determination of gene chromosome associations and establishment of chromosome markers by aneuploid analysis in common wheat. I. F_2 analysis of glume pubescence, spike density and culm color. *Can. J. Genet. Cytol.* **2,** 301-310.

Lafever, H. N., and Patterson, F. L. (1964a). Cytological and breeding behavior of an aneuploid line of *Avena sativa* L. *Crop Sci.* **4,** 535-539.

Lafever, H. N., and Patterson, F. L. (1964b). Aneuploids for commercial hybrid oats? *Crop Sci.* **4,** 635-638.

Lam, S. L., and Erickson, H. T. (1971). The nucleolar trisomic and trisomic transmission in a diploid potato. *J. Hered.* **62,** 375-376.

Lammerts, W. E. (1932). Inheritance of monosomics in *Nicotiana rustica*. *Genetics* **17,** 689-696.

Langdon-Down, J. (1866). Observations on an ethnic classification of idiots. *Clin. Lect. Rep. London Hosp.* **3,** 259-262.

Larson, R. I. (1952). Aneuploid analysis of inheritance of solid stem in common wheat. *Genetics* **37,** 597-598.

Larson R. I. (1959). Cytogenetics of solid stem in common wheat. I. Monosomic F_2 analysis of the variety S-615. *Can. J. Bot.* **37,** 135-156.

Larson, R. I. (1966). Aneuploid analysis of quantitative characters in wheat. *Proc. 2nd Intern. Wheat Genet. Symp: Hereditas Suppl.* **2,** 345-354.

Larson, R. I., and Atkinson, T. G. (1970). Identity of the wheat chromosomes replaced by *Agropyron* chromosomes in a triple alien chromosome substitution line immune to wheat streak mosaic. *Can. J. Genet. Cytol.* **12,** 145-150.

Larson, R. I., and MacDonald, M. D. (1959). Cytogenetics of solid stem in common wheat. II. Stem solidness of monosomic lines of the variety S-615. *Can. J. Bot.* **37,** 365-378.

Larson, R. I., and MacDonald, M. D. (1962). Cytogenetics of solid stem in common wheat. IV. Aneuploid lines of the variety Rescue. *Can. J. Genet. Cytol.* **4,** 97-104.

Larson, R. I., and MacDonald, M. D. (1966). Cytogenetics of solid stem in common wheat. V. Lines of S-615 with whole chromosome substitutions from Apex. *Can. J. Genet. Cytol.* **8,** 64-70

Law, C. N. (1966a). Biometrical analysis using chromosome substitutions within a species. *Heredity Suppl.* **20,** 59-85.

Law, C. N. (1966b). The location of genetic factors affecting a quantitative character in wheat. *Genetics* **53,** 487-498.

Law, C. N. (1967). The location of genetic factors controlling a number of quantitative characters in wheat. *Genetics* **56,** 445-461.

Law C. N., and Wolfe, M. S. (1966). Location of genetic factors for mildew resistance and ear emergence time on chromosome 7B of wheat. *Can. J. Genet. Cytol.* **8,** 462-470.

Lee, Y. H., Larter, E. N., and Evans, L. E. (1969). Homoeologous relationship of rye chromosome VI with two homoeologous groups from wheat. *Can. J. Genet. Cytol.* **11,** 803-809.

Lee-Chen, S, and Steinitz-Sears, L. M. (1967). The location of linkage groups in *Arabidopsis thaliana*. *Can. J. Genet. Cytol.* **9,** 381-384.

Leighty, C. E., and Taylor, J. W. (1924). "Hairy Neck" wheat segregates from wheat-rye hybrids. *J. Agr. Res.* **28,** 567-576.

Lejeune, J., Gautier, M., and Turpin, R. (1959a). Les chromosomes humaines en culture de tissus. *C. R. Acad. Sci. (Paris)* **248,** 602-603.

Lejeune, J., Gautier, M., and Turpin, R. (1959b). Etudes des chromosomes somatiques de neuf enfants mongoliens. *C. R. Acad. Sci. (Paris)* **248,** 1721-1722.

Lejeune, J., La Fourcade, J., Berger, R., Viallate, J., Boeswillwald, M., Seringe, P., and Turpin, R. (1964). Trois cas de deletion partielle du bras court d'un chromosome 5. *C. R. Acad. Sci. (Paris)* **257,** 3098-3102.

Lesley, J. W. (1926). The genetics of *Lycopersicon esculentum* Mill. I. The trisomic inheritance of "Dwarf". *Genetics* **11,** 352-354.

Lesley, J. W. (1928). A cytological and genetical study of progenies of triploid tomatoes. *Genetics* **13,** 1-43.

Lesley, J. W. (1932). Trisomic types of the tomato and their relations to the genes. *Genetics* **17,** 545-559.

Lesley, J. W. (1937). Crossing over in tomatoes trisomic for the "A" or first chromosome. *Genetics* **22,** 297-306.

Lesley, M. M., and Frost, H. B. (1927). Mendelian inheritance of chromosome shape in *Matthiola*. *Genetics* **12,** 449-460.

Lesley, M. M., and Lesley, J. W. (1941). Parthenocarpy in a tomato deficient for a part of a chromosome and its aneuploid progeny. *Genetics* **26,** 374-386.

Levan, A. (1937). Chromosome numbers in *Petunia. Hereditas* **23,** 99-112.

Levan, A. (1942). The effect of chromosomal variation in sugar beets. *Hereditas* **28,** 345-399.

Lewitsky, G. A. (1940). A cytological study of the progeny of X-rayed *Crepis capillaris* wallr. *Cytologia* **11,** 1-29.

Li, H. W., Hsia, C. A., and Lee, C. L. (1948). An inquiry into the nature of speltoid and compactoid types of wheat as a function of genic dosage. *Bot. Bull. Acad. Sinica* **2,** 243-264.

Li, J. C. (1927). The effect of chromosome aberrations on development in *Drosophila melanogaster. Genetics* **12,** 1-58.

Lin, P. S., and Ross, J. G. (1969a). Morphology and cytological behavior of aneuploids of *Sorghum bicolor. Can. J. Genet. Cytol.* **11,** 908-918.

Lin, P. S., and Ross, J. G. (1969b). Ovular tumors in a trisomic sorghum plant. *J. Hered.* **60,** 183-185.

Little, T. M. (1945). Gene segregation in autotetraploids. *Bot. Rev.* **11,** 60-85.

Little, T. M. (1958). Gene segregation in autotetraploids. II. *Bot. Rev.* **24,** 318-339.

Livers, R. W. (1949). Genetic analysis of wheat using monosomics. *Abstr. Annu. Meet. Amer. Soc. Agron.* **41,** 8.

Loegering, W. Q., and Sears, E. R. (1966). Relationships among stemrust genes on wheat chromosomes 2B, 4B and 6B. *Crop Sci.* **6,** 157-160.

Longwell, J. H., and Sears, E. R. (1963). Nullisomics in tetraploid wheat. *Amer. Natur.* **97,** 401-403.

Love, R. M. (1938). A cytogenetic study of white chaff off-types occurring spontaneously in Dawson's Golden Chaff winter wheat. *Genetics* **23,** 157.

Love, R. M. (1940). Chromosome number and behaviour in a plant breeder's sample of pentaploid wheat hybrid derivatives. *Can. J. Res.* **C18,** 415-434.

Love, R. M. (1941). Chromosome behavior in F_1 wheat hybrids. I. Pentaploids. *Can. J. Res.* **C19,** 351-369.

Love, R. M. (1943). A cytogenetic study of off-types in a winter wheat, Dawson's Golden Chaff, including a white chaff mutant. *Can. J. Res.***C21,** 257-264.

Lucov, Z., Cohen, S., and Moav, R. (1970). Effects of low temperature on the somatic instability of an alien chromosome in *Nicotiana tabacum. Heredity* **25,** 431-440.

Luig, N. H., and McIntosh, R. A. (1968). Location and linkage of genes on wheat chromosome 2D. *Can. J. Genet. Cytol.* **10,** 99-105.

Lutkov, A. N. (1937). Reciprocal translocations and gene mutations in *Pisum sativum* induced by X-radiation of pollen. *Bull. Appl. Bot. Ser. II* **7,** 377-416.

Lutz, A. M. (1909). Notes on the first generation hybrid of *Oenothera lata* × *O. gigas. Science* **29,** [N.S.], 263-267.

Lyon, M. F., and Meredith, R. (1966). Autosomal translocations causing male sterility and viable aneuploidy in the mouse. *Cytogenetics* **5,** 335-354.

Macer, R. C. F. (1966). The formal and monosomic genetic analysis of stripe rust (*Puccinia striiformis*) resistance in wheat. *Proc. 2nd Intern. Wheat Genet. Symp. Hereditas Suppl.* **2,** 127-143.

McClintock, B. (1929a). A cytological study of triploid maize. *Genetics* **14,** 180-222.

McClintock, B. (1929b). A $2n - 1$ chromosomal chimera in maize. *J. Hered.* **20,** 218.

McClintock, B. (1941). The association of mutants with homozygous deficiencies in *Zea mays. Genetics* **26,** 542-571.

McClintock, B. (1944). The relation of homozygous deficiencies to mutations and allelic series in maize. *Genetics* **29,** 478-502.

McClintock, B., and Hill, H. E. (1931). The cytological identification of the chromosome associated with the *R-G* linkage group in *Zea mays*. *Genetics* **16,** 175-190.

McClure, H. M., Belden, K. H., Pieper, W. A., and Jacobson, C. B. (1969). Autosomal trisomy in a Chimpanzee: Resemblance to Down's syndrome. *Science* **165,** 1010-1011.

McDaniel, R. G., and Ramage, R. T. (1970). Genetics of a primary trisomic series in barley: Identification by protein electrophoresis. *Can. J. Genet. Cytol.* **12,** 490-495.

MacFadden, E. S., and Sears, E. R. (1944). The artificial synthesis of *Triticum spelta*. *Genetics* **30,** 14.

McFee, A. F., Bonner, M. W., and Rory, J. M. (1966). Variation in chromosome number among European wild pigs. *Cytogenetics* **5,** 75.

McGinnis, R. C. (1962a). The occurrence of spontaneous aneuploids in common oats, *Avena sativa*. *Genetics* **47,** 969.

McGinnis, R. C. (1962b). Aneuploids in common oats *Avena sativa*. *Can. J. Genet. Cytol.* **4,** 296-301.

McGinnis, R. C. (1966). Establishing a monosomic series in *Avena sativa* L. *Heredity Suppl.* **20,** 86-97

McGinnis, R. C., and Andrews, G. Y. (1962). The identification of a second chromosome involved in chlorophyll production in *Avena sativa*. *Can. J. Genet. Cytol.* **4,** 1-5.

McGinnis, R. C., and Campbell, A. B. (1958). Fertile triple monosomics from a *Triticum vulgare* cross. *Proc. 10th Intern. Congr. Genet. Montreal* **2,** 184.

McGinnis, R. C., and Campbell, A. B. (1960). A case of maintainable hypoploidal variability in *Triticum aestivum* L. cross. *Can. J. Genet. Cytol.* **2,** 47-56.

McGinnis, R. C., and Campbell, A. B. (1961). Differential inheritance in crosses involving three phenotypically similar wheat varieties with redman monosomics. *Can. J. Genet. Cytol.* **3,** 10-12.

McGinnis, R. C., and Lin, C. C. (1966). A monosomic study of panicle shape in *Avena sativa*. *Can. J. Genet. Cytol.* **8,** 96-101.

McGinnis, R. C., and Taylor, D. K. (1961). The association of a gene for chlorophyll production with a specific chromosome in *Avena sativa*. *Can. J. Genet. Cytol.* **3,** 436-443.

McGinnis, R. C., Andrews, G. Y., and McKenzie, R. I. H. (1963). Determination of Chromosome arm carrying a gene for chlorophyll production in *Avena sativa*. *Can. J. Genet. Cytol.* **5,** 57-59.

McGinnis, R. C., Dyck, P. L., Hildebrandt, S. G., and Lin, C. C. (1968). The association of a third chromosome with chlorophyll production in *Avena sativa*. *Can. J. Genet. Cytol.* **10,** 228-231.

McIntosh, R. A., and Baker, E. P. (1966). Chromosome location of mature plant leaf rust resistance in Chinese Spring wheat. *Aust. J. Biol. Sci.* **19,** 943-944.

McIntosh, R. A., and Baker, E. P. (1968). A linkage map for chromosome 2D. *Proc. 3rd Intern. Wheat Genet. Symp. Aust. Acad. Sci. Canberra,* pp. 305-309.

McIntosh, R. A., and Baker, E. P. (1969a). Chromosome location and linkage studies involving the Pm_3 locus for powdery mildew resistance in wheat. *Proc. Linnean Soc. N. S. Wales, Aust.* **93,** (Part 2), 232-238.

McIntosh, R. A., and Baker, E. P. (1969b). Telocentric mapping of a second gene for grass-clump dwarfism. *Wheat Inf. Serv.* **29,** 6-7.

McIntosh, R. A., and Baker, E. P. (1970). Cytogenetic studies in wheat. V. Monosomic analysis of Vernstein stem rust resistance. *Can. J. Genet. Cytol.* **12,** 60-65.

McIntosh, R. A., Baker, E. P., and Driscoll, C. J. (1965). Cytogenetical studies in wheat. I. Monosomic analysis of leaf rust resistance in the cultivars Uruguay and Transfer. *Aust. J. Biol. Sci.* **18,** 971-977.

McIntosh, R. A., Luig, N. H., and Baker, E. P. (1967). Genetic and cytogenetic studies of stem rust, leaf rust and powdery mildew resistances in Hope and related wheat cultivars. *Aust.*

J. Biol. Sci. **20,** 1181-1192.

McKusick, V. A. (1964). "Human Genetics," 221 pp. Prentice-Hall, Englewood Cliffs, New Jersey.

McLennan, H. A. (1947). Cytogenetic studies of a strain of barley with long chromosomes. M. Sc. Thesis, Univ. Minnesota, Minneapolis, (Cited by R. Tsuchiya, 1963).

McLennan, H. A., and Kasha, K. J. (1964). The production of trisomics in diploid alfalfa. *Can. J. Genet. Cytol.* **6,** 240-241.

Maguire, M. P. (1962). Pachytene and diakinesis behavior of the isochromosome 6 of maize. *Science* **138,** 445-446.

Maguire, M. P. (1963). High transmission frequency of a *Tripsacum* chromosome in corn. *Genetics* **48,** 1185-1194.

Maneephong, C., and Sadanaga, K. (1967). Induction of aneuploids in hexaploid oats, *Avena sativa* L. through X-irradiation. *Crop Sci.* **7,** 522-523.

Mann, T. J., Gerstel, D. U., and Apple, J. L. (1963). The role of interspecific hybridization in tobacco disease control. *Proc. 3rd World Tobacco Sci. Congr. Salisbury,* pp. 201-207.

Mather, K. (1935). Reductional and equational separation of the chromosome in bivalents and multivalents. *J. Genet.* **30,** 53-78.

Mather, K. (1936). Segregation and linkage in autotetraploids. *J. Genet.* **32,** 287-314.

Mather, K. (1939). Chiasma frequencies in trisomic maize. *Genetics* **24,** 104.

Matsumura, S. (1940). Weitere Untersuchungen uber die pentaploiden *Triticum*-Bastarde X. *Jap. J. Bot.* **10,** 477-487.

Matsumura, S. (1947). Chromosomen analyse de Dinkelgenoms auf Grund zytogenetischer Untersüchungen an pentaploiden Weizenbastarden. *La Kromosomo* **3-4,** 113-132.

Matsumura, S. (1952a). Chromosome analysis of the Dinkel genome in the offspring of a pentaploid wheat hybrid. I. Nullisomics deficient for a pair of D-chromosomes. *Cytologia* **16,** 265-287.

Matsumura, S. (1952b). Chromosome analysis of the Dinkel genome in the offspring of a pentaploid wheat hybrid. III. 29-chromosome D-haplosomics and their relation to nullisomics. *Cytologia* **17,** 35-49.

Matsumura, S. (1954). Chromosome analysis of the Dinkel genome in the offspring of a pentaploid wheat hybrid. V. Gigas-plants in the offspring of nullisomic dwarfs and analysis of their additional chromosome pair. *Cytologia* **19,** 273-285.

Matsuura, H. (1937). Chromosome studies in *Trillium kamtschaticum* Pall. V. Abnormal meiotic divisions due to high temperatures. *Cytologia* (Fuji Jubilee Vol.), Part I, 20-34.

Mavor, J. W. (1924). The production of non-disjunction by X-rays. *J. Exp. Zool.* **39,** 381-432.

Menzel, M. Y., and Brown, M. S. (1952). Viable deficiency duplications from a translocation in *Gossypium hirsutum*. *Genetics* **37,** 678-692.

Metzger, R. J., Rhode, C. R., and Trione, E. J. (1963). Inheritance of genetic factors which condition resistance of the wheat variety Columbia to selected races of smut, *Tilletia caries* and their association with red glumes. *Agron. Abstr.,* p. 85.

Meurman, O. (1928). Cytological studies in the genus *Ribes* L. *Hereditas* **11,** 289-356.

Moav, R. (1961). Genetic instability in *Nicotiana* hybrids. II. Studies of the *Ws (pbg)* locus of *N. plumbaginifolia* in *N. tabacum* nuclei. *Genetics* **46,** 1069-1087.

Mochizuki, A. (1962). *Agropyron* addition lines of *durum* wheat. *Seiken Ziho* **13,** 133-138.

Mochizuki, A. (1968a). Production of monosomics in *durum* wheat. *Wheat Inf. Serv.* **26,** 8-10.

Mochizuki, A. (1968b). The monosomics of *durum* wheat. *Proc. 3rd Intern. Wheat Genet. Symp. Aust. Acad. Sci. Canberra,* pp. 310-315.

Mochizuki, A. (1971). Production of three monosomic series in Emmer and common wheat. *Seiken Ziho* **22**, 39-49.

Mochizuki, A., and Shigenaga, S. (1964). Some notes from cytogenetical observations in the nullisomics of Chinese Spring wheat at Sasayama, Japan. *Sci. Rep., Hyogo Univ. Agr.* **6,** 53-58.

Moens, P. (1965). The transmission of a heterochromatic isochromosome in *Lycopersicon esculentum*. *Can. J. Genet. Cytol.* **7,** 296-303.

Morgan, D. T., Jr. (1956). X-ray induced deficiencies and multiple embryo formation. *Maize Genet. Coop. Newsl.* **30,** 83.

Morinaga, T., and Fukushima, E. (1935). Cytogenetical studies on *Oryza sativa* L. II. Spontaneous autotriploid mutants in *Oryza sativa* L. *Jap. J. Bot.* **7,** 207-225.

Morris, R. (1955). Induced reciprocal translocations involving homologous chromosomes in maize. *Amer. J. Bot.* **42,** 546-550.

Morris, R., and Sears, E. R. (1967). The cytogenetics of wheat and its relatives. *Amer. Soc. Agron. No.* **13,** 19-87.

Morris, R., Schmidt, J. W., Mattern, P. J., and Johnson, V. A. (1966). Chromosomal location of genes for flour quality in the wheat variety Cheyenne using substitution lines. *Crop. Sci.* **6,** 119-122.

Morrison, J. W. (1953). Chromosome behavior in wheat monosomics. *Heredity* **7,** 203-217.

Morrison, J. W., and Unrau, J. (1952). Frequency of micronuclei in pollen quartets of common wheat monosomics. *Can. J. Bot.* **30,** 371-378.

Moseman, H. G., and Smith, L. (1954). Gene location by three point test and telocentric half chromosome fragment in *Triticum monococcum*. *Agron. J.* **46,** 120-124.

Müntzing, A. (1930). Einige Beobachten uber die Zytologie der Speltoidmutanten. *Bot. Natur.*, 35-47.

Muller, H. J. (1914). A new mode of segregation in Gregory's tetraploid *Primulas*. *Amer. Natur.* **48,** 508-512.

Muller, H. J. (1925). Why polyploidy is rarer in animals than in plants. *Amer. Natur.* **59,** 346-353.

Muramatsu, M. (1968). Studies on translocations between hairy neck chromosome of rye and chromosomes which belong to homoeologous group 5 of common wheat. *Proc. 12th Intern. Congr. Genet. Tokyo* **1,** 180.

Nakamori, E. (1932). On the appearance of the triploid plant of rice, *Oryza sativa* L.. *Proc. Imp. Acad. Japan* **8,** 528-529.

Nandi, H. K. (1937). Trisomic mutations in jute. *Nature (London)* **140,** 973-974.

Nebel, B. R. (1933). Chromosome numbers in aneuploid apple seedlings. *N. Y. State Agr. Exp. Sta. Bull.* **209,** 1-12.

Newton, W. C. F., and Darlington, C. D. (1929) Meiosis in polyploids. I. *J. Genet.* **21,** 1-16.

Nilsson-Ehle, H. (1907). Om hafresorters constans. *Sv. Utsoderforenings Tidskr.*, pp. 227-239.

Nilsson-Ehle, H. (1911). Uber Falle spontanen Wegfallens eines Hemmungsfaktor beim Hafer. *Z. Indukt. Abstamm. Vererbungsl.* **5,** 1-37.

Nishiyama, I. (1928). On hybrids between *Triticum spelta* and two dwarf wheat plants with 40 somatic chromosomes. *Bot. Mag.* **42,** 154-177.

Nishiyama, I. (1931). The genetics and cytology of certain cereals. II. Karyo-genetic studies of fatuoid oats with special reference to their origin. *Jap. J. Genet.* **7,** 49-102.

Nishiyama, I. (1933). The genetics and cytology of certain cereals. IV. Further studies on fatuoid oats. *Jap. J. Genet.* **8,** 107-124.

Nishiyama, I. (1951). A genic analysis of the grain characters of oats by means of the monosomic inheritance. *Bull. Res. Inst. Food Sci. Kyoto Univ.* **4,** 67-85.

Nishiyama, I., Forsberg, R. A., Shands, H. L., and Tabata, M. (1968). Monosomics of Kanota oats. *Can. J. Genet. Cytol.* **10,** 601-612.

Noronha-Wagner, M., and Mello-Sampayo, T. (1966). Aneuploids in *durum* wheat. *Proc. 2nd Intern. Wheat Genet. Symp. Hereditas Suppl.* **2,** 382-392.

Nyquist, W. E. (1957). Monosomic analysis of stem rust resistance of a common wheat strain derived from *Triticum timopheevi*. *Agron. J.* **49,** 222-223.

Ohta, T., and Matsumura, S. (1961). Mechanism of appearance of gigas-plants from nullisomic dwarf wheat. *Cytologia* **26,** 226-235.

Okamoto, M. (1957). Further information on identification of the chromosomes in the A and B genones. *Wheat Inf. Serv.* **6,** 3-4.

Okamoto, M. (1960). An awn suppressor located on chromosome 5B. *Wheat Inf. Serv.* **11,** 2-3.

Okamoto, M. (1962). Identification of the chromosomes of common wheat belonging to the A and B genomes. *Can. J. Genet. Cytol.* **4,** 31-37.

Okamoto, M., and Sears, E. R. (1962). Chromosomes involved in translocations obtained from haploids of common wheat. *Can. J. Genet. Cytol.* **4,** 24-30.

Olmo, H. P. (1935). Genetical studies of monosomic types of *Nicotiana tabacum*. *Genetics* **20,** 286-300.

Olmo, H. P. (1936). Cytological studies of monosomic and derivative types of *Nicotiana tabacum*. *Cytologia* **7,** 143-159.

O'Mara, J. G. (1940). Cytogenetic studies on *Triticinae*. I. A method for determining the effects of individual *Secale* chromosomes on *Triticum*. *Genetics* **25,** 401-408.

O'Mara, J. G. (1947). The substitution of a specific *Secale cereale* chromosome for a specific *Triticum vulgare* chromosome. *Genetics* **32,** 99-100.

O'Mara, J. G. (1951). Cytogenetic studies in *Triticale*. II. The kinds of intergeneric chromosome additions. *Cytologia* **16,** 225-232.

O'Mara, J. G. (1953). The cytogenetics of *Triticale*. *Bot. Rev.* **19,** 587-605.

O'Mara, J. G. (1961). Cytogenetics. *Amer. Soc. Agron. No.* **8,** 112-124.

Pal, B. P., and Ramanujam, S. (1940). Asynapsis in chilli *(Capsicum annum* L.*)*. *Curr. Sci.* **9,** 126-128.

Parry, D. C., and Gerstel, D. U. (1967). On the occurrence of monosomics among diploid species of *Solanaceous* plants. *Tobacco Sci.* **11,** 87-88.

Parthasarathy, N. (1938). Cytogenetical studies in *Oryzeae* and *Phalarideae*. I. Cytogenetics of some X-ray derivatives in rice. *J. Genet.* **37,** 1-40.

Patau, K. (1961). Chromosome identification and the Denver report. *Lancet* **i,** 933-934.

Patau, K., Smith, D. W., Therman, E., Inhorn, S. L., and Wagner, H. P. (1960). Multiple congenital anomaly caused by an extra autosome. *Lancet* **i,** 790-793.

Patil, S. H. (1968). Cytogenetics of X-ray induced aneuploids in *Arachis hypogea* L. *Can. J. Genet. Cytol.* **10,** 545-550.

Patterson, E. B. (1952). The use of functional duplicate-deficient gametes in locating genes in maize. *Genetics* **37,** 612.

Penrose, L. S. and Smith, G. F. (1966). "Down's Anomaly," 218 pp. Little Brown, Boston, Massaschusetts.

Person, C. (1956). Some aspects of monosomic wheat breeding. *Can. J. Bot.* **34**, 60-70.

Philp, J. (1935). Aberrant albinism in polyploid oats. *J. Genet.* **30,** 267-302.

Philp, J. (1938). Aberrant leaf width in polyploid oats. *J. Genet.* **36,** 405-429.

Philp, J., and Huskins, C. L. (1931). The cytology of *Matthiola incana* R. Br., especially in relation to the inheritance of double flowers. *J. Genet.* **24,** 359-404.

Piech, J. (1968). Monosomic and conventional genetic analyses of semi-dwarfism and grass clump dwarfism in common wheat. *Euphytica Suppl. No.* **1,** 153-170.

Pochard, E. (1968). Isolation of ten primary trisomics in the progeny of a haploid pepper plant *(Capsicum annum)*. *Proc. 12th Intern. Congr. Genet. Tokyo.* **1,** 266.

Poon, N. H., and Wu, H. K. (1967). Identification of involved chromosomes in trisomics of *Sorghum vulgare* Pers. *J. Agr. Ass. China.* **58** [N.S.], 18-32.

Powers, L. (1932). Cytologic and genetic studies of variability of strains of wheat derived from interspecific crosses. *J. Agr. Res.* **44,** 797-831.

Prakken, R. (1942). A new trisomic *Matthiola* type. *Hereditas* **28,** 297-305.

Price, M. E., and Ross, W. M. (1955). The occurrence of trisomics and other aneuploids in a cross of triploid × diploid *Sorghum vulgare*. *Agron. J.* **47,** 591-592.

Price, M. E., and Ross, W. M. (1957). A cytological study of a triploid × diploid cross of *Sorghum vulgare*. *Agron. J.* **49,** 237-240.

Propach, H. (1935). Studien über heteroploide Formen von *Antirrhinum majus* L.II. Die Reifetelongen in der Pollenmutterzellen der Trisomen Anaemica, Fusca, Purpurea, Rotunda, Candida. *Planta* **23,** 349-357.

Punyasingh, K. (1947). Chromosome numbers in crosses of diploid, triploid, and tetraploid maize. *Genetics* **32,** 541-554.

Quinn, C. J., and Driscoll, C. J. (1967). Relationship of the chromosomes of common wheat and related genera. *Crop Sci.* **7,** 74-75.

Rai, K. S., and Garber, E. D. (1961). The genus *Collinsia*. XI. Trisomic inheritance in *C. heterophylla*. *Bot. Gaz.* **122,** 109-117.

Rajhathy, T., and Dyck, P. L. (1964). Methods for aneuploid production in common oats, *Avena sativa*. *Can. J. Genet. Cytol.* **6,** 215-220.

Rajhathy, T., and Fedak, G. (1970). A secondary trisomic in *Avena strigosa*. *Can. J. Genet. Cytol.* **12,** 358-360.

Ramage, R. T. (1960). Trisomics from interchange heterozygotes in barley. *Agron. J.* **42,** 156-159.

Ramage, R. T. (1963). Genetic recombination and chromosome disjunction in a balanced tertiary trisomic of barley. *Genet. Today Proc. 11th Intern. Congr.* **1,** 122.

Ramage, R. T. (1965). Balanced tertiary trisomics for use in hybrid seed production. *Crop Sci.* **5,** 177-178.

Ramage, R. T., and Day, A. D. (1960). Separation of trisomic and diploid barley seeds produced by interchange heterozygotes. *Agron. J.* **52,** 590-591.

Ramage, R. T., and Humphrey, D. F. (1964). Frequency of orientation of normal and interchanged heterozygotes. *Crop Sci.* **4,** 539-540.

Ramage, R. T., and Suneson, C. A. (1958a). A gene marker for the *g* chromosome of barley. *Agron. J.* **50,** 114.

Ramage, R. T., and Suneson, C. A. (1958b). A nullisomic oat. *Agron. J.* **50,** 52-53.

Ramage, R. T., and Tuleen, N. A. (1964). Balanced tertiary trisomics in barley serve as a pollen source homogeneous for a recessive lethal gene. *Crop Sci.* **4,** 81-82.

Ramanujam S. (1937). Cytogenetical studies in the *Oryzeae*. II. Cytological behaviour of an autotriploid in rice *(Oryza sativa* L.*)*. *J. Genet.* **35,** 183-221.

Rana, R. S. (1965a). A cytological evaluation of temperature sensitivity in diploid wheat. *Caryologia* **18,** 117-125.

Rana, R. S. (1965b). Monosomic interchange heterozygote of diploid *Chrysanthemum*. *Nature (London)* **206,** 532-533.

Rana, R. S. (1967). A nullisomic plant in diploid *Chrysanthemum*. *Experientia* **23,** 119.

Randolph, L. F., and Fischer, H. E. (1939). The occurrence of parthenogenetic diploid in tetraploid maize. *Proc. Nat. Acad. Sci. U.S.* **25,** 161-164.

Rao, P. N., and Stokes, G. W. (1963). Cytogenetic analysis of the F_1 of haploid × diploid tobacco. *Genetics* **48,** 1423-1433.

Ray, M., and Swaminathan, M. S. (1959). Monosomic analysis in bread wheat. IV. Morphology and pairing of chromosomes in some monosomics and nullisomics of Chinese Spring and Redman. *Indian J. Genet. Plant Breed.* **19,** 176-185.

Reddy, R. P. (1969). Monosomic analysis of leaf rust resistance to races 9 and 15 in Parker wheat. *J. Hered.* **60,** 341-343.

Reeves, A. F. (1968). A cytogenetic analysis of an atypical tertiary trisomic of tomato. Ph.D. Thesis, University of California, Davis, California.

Reeves, A. F., Khush, G. S., and Rick, C. M. (1968). Segregation and recombination in trisomics: A reconsideration. *Can. J. Genet. Cytol.* **10,** 937-940.

Rhoades, M. M. (1933a). An experimental and theoretical study of chromatid crossing over. *Genetics* **18,** 535-555.

Rhoades, M. M. (1933b). A secondary trisome in maize. *Proc. Nat. Acad. Sci. U.S.* **19,** 1031-1038.

Rhoades, M.M. (1936). A cytogenetic study of a chromosome fragment in maize. *Genetics* **21,** 491-502.

Rhoades, M. M. (1938). On the origin of a secondary trisome through the doubling of a half-chromosome fragment. *Genetics* **23,** 163-164.

Rhoades, M. M. (1940). Studies of a telocentric chromosome in maize, with reference to the stability of its centromere. *Genetics* **25,** 483-520.

Rhoades, M. M. (1942). Preferential segregation in maize. *Genetics* **27,** 395-407.

Rhoades, M. M. (1955). The cytogenetics of maize. *In* "Corn and Corn Improvement" (G. F. Sprague,ed.). pp. 123-219. Academic Press, New York.

Rhoades, M. M., and McClintock, B. (1935). The cytogenetics of maize. *Bot. Rev.* **1,** 292-325.

Rick, C. M. (1943). Cytogenetic consequences of X-ray treatment of pollen in *Petunia*. *Bot. Gaz.* **104,** 528-540.

Rick, C. M. (1945). A survey of cytogenetic causes of unfruitfulness in the tomato. *Genetics* **30,** 347-362.

Rick, C. M., and Barton, D. W. (1954). Cytological and genetical identification of the primary trisomics of the tomato. *Genetics* **39,** 640-666.

Rick, C. M., and Khush, G. S. (1961). X-ray induced deficiencies of chromosome 11 in the tomato. *Genetics* **46,** 1389-1393.

Rick, C. M., and Khush, G. S. (1966). Chromosome engineering in *Lycopersicon*. *Heredity Suppl.* **20,** 8-20.

Rick, C. M., and Khush, G. S. (1969). Cytogenetic explorations in the tomato genome. *In* "Genetics Lectures"(R. Bogart, ed.), Vol. 1, 45-68. Oregon State University Press, Oregon.

Rick, C. M., and Notani, N. K. (1961). The tolerance of extra chromosomes by primitive tomatoes. *Genetics* **46,** 1231-1235.

Rick, C. M., Dempsey, W. H., and Khush, G. S. (1964). Further studies on the primary trisomics of the tomato. *Can. J. Genet. Cytol.* **6,** 93-108.

Riley, R. (1955). The cytogenetics of the differences between some *Secale* species. *J. Agr. Sci.* **46,** 377-383.

Riley, R. (1958). Chromosome pairing and haploids in wheat. *Proc. 10th Intern. Congr. Genet. Montreal* **2,** 234-235.

Riley, R. (1960a). The meiotic behavior, fertility, and stability of wheat-rye chromosome addition lines. *Heredity* **14,** 89-100.

Riley, R. (1960b). The diploidisation of polyploid wheat. *Heredity* **15,** 407-429.

Riley, R. (1965). Cytogenetics and Plant Breeding. *Genet. Today Proc. 11th Intern. Congr. The Hague* **3**, 681-688.

Riley, R. (1966a). Genotype-environment interaction affecting chiasma frequency in *Triticum aestivum*. *In* "Chromosomes Today" Vol. 1, pp. 57-65. (C. D. Darlington and K. R. Lewis, eds.). Oliver and Boyd, Edinburgh.

Riley, R. (1966b). The genetic regulation of meiotic behavior in wheat and its relatives. *Proc. 2nd Intern. Wheat Genet. Symp. Hereditas Suppl.* **2,** 395-408.

Riley, R. (1967). Theoretical and practical aspects of chromosome pairing. *Genet. Agr.* **21,** 111-128.

Riley R., and Chapman, V. (1958a). Genetic control of the cytologically diploid behaviour of hexaploid wheat. *Nature, (London)* **182,** 713-715.

Riley R., and Chapman, V. (1958b). The production and phenotypes of wheat-rye chromosome addition lines. *Heredity* **12,** 301-315.

Riley, R., and Chapman, V. (1963). The effects of the deficiency of chromosome V (5B) of *Triticum aestivum* on the meiosis of synthetic amphiploids. *Heredity* **18,** 473-484.

Riley R., and Chapman, V. (1964). The effect of the deficiency of the long arm of chromosome 5B on meiotic pairing in *Triticum aestivum. Wheat Inf. Serv.* **17,** 12-15.

Riley, R., and Chapman, V. (1967). Effect of 5BS in suppressing the expression of altered dosage of 5BL on meiotic chromosome pairing in *Triticum aestivum. Nature (London)* **216,** 60-62.

Riley, R., and Ewart, J. A. D. (1970). The effect of individual rye chromosomes on the amino acid content of wheat grains. *Genet. Res.* **15,** 209-219.

Riley, R., and Kempanna, C. (1963). The homoeologous nature of the non-homologous meiotic pairing in *Triticum aestivum* deficient for chromosome V (5B). *Heredity* **18,** 287-306.

Riley, R., and Kimber, G. (1961). Aneuploids and the cytogenetic structure of wheat varietal populations. *Heredity* **16,** 275-290.

Riley, R., and Kimber, G. (1966). The transfer of alien genetic variation to wheat. *Rep. Plant Breed. Inst. Cambridge,* pp. 6-36.

Riley R., Unrau, J., and Chapman, V. (1958). Evidence on the origin of the B genome of wheat. *J. Hered.* **49,** 91-98.

Riley, R., Kimber, G., and Chapman, V. (1961). Origin of genetic control of diploid-like behavior of polyploid wheat. *J. Hered.* **52,** 22-25.

Riley R., Chapman, V., and Macer, R. C. F. (1966a). The homoeology of an *Aegilops* chromosome causing stripe rust resistance. *Can. J. Genet. Cytol.* **8,** 616-630.

Riley, R., Chapman, V., and Belfield, A. M. (1966b). Induced mutation affecting the control of meiotic chromosome pairing in *Triticum aestivum. Nature (London)* **211,** 368-369.

Riley, R., Chapman, V., Young, R. M., and Belfield, A. M. (1966c). Control of meiotic chromosome pairing by the chromosomes of homoeologous group 5 of *Triticum aestivum. Nature (London)* **212,** 1475-1477.

Riley, R., Chapman, V., and Johnson, R. (1968). The incorporation of alien disease resistance in wheat by genetic interference with the regulation of meiotic chromosome synapsis. *Genet. Res.* **12,** 199-219.

Robbelen, G., and Kribben, F. J. (1966). Erfahrungen bei der Auslese von Trisomen. *Arabid. Inf. Serv.* **3,** 16-17.

Robertson, L. D., and Curtis, B. C. (1967). Monosomic analysis of fertility-restoration in common wheat *(Triticum aestivum* L.*). Crop Sci.* **7,** 493-495.

Rudorf-Lauritzen, M. (1958). The trisomics of *Antirrhinum majus* L. *Proc. 10th Intern. Congr. Genet. Montreal* **2,** 243-244.

Russell, L. B., and Chu, E.H.Y. (1961). An XXY male in the mouse. *Proc. Nat. Acad. Sci. U.S.* **47,** 571-575.

Sachar, K., Upadhaya, L. P., and Swaminathan, M. S. (1967). Trisomics of jute isolated in progenies of *Olitorius capsularis* hybrids. *Indian J. Genet. Plant Breed.* **27,** 334-348.

Sadanaga, K. (1957). Cytological studies of hybrids involving *Triticum durum* and *Secale cereale.* I. Alien addition races in tetraploid wheat. *Cytologia* **22,** 312-321.

Sadanaga, K. (1970). Genetics and test of association of the flourescence genes in hexaploid oats. *Crop Sci.* **10,** 103-104.

Sakamura, T. (1918). Kurze Mitteilung über die chromosomenzahlen und die Verwandtschaftsverhaltnisse de *Triticum*-Arten. *Bot. Mag.* **32,** 150-153.

Sakanaga, K. (1956). Cytological studies of hybrids involving *Triticum durum* and *Secale cereale. Wheat Inf. Serv.* **3,** 23-24.

Sampson, D. R., Hunter, A. S. W., and Bradley, E. C. (1961). Triploid × diploid progenies and primary trisomics of *Antirrhinum majus. Can. J. Genet. Cytol.* **3,** 184-194.

Sanchez-Monge, E. (1950). Two types of misdivision of the centromere. *Nature (London)* **165,** 80-81.

Sanchez-Monge, E. (1951). The stability of isochromosomes. *Anal. Estac. Exp. Aula Dei* **2,** 168-173.

Sanchez-Monge, E., and MacKey, J. (1948). On the origin of sub-compactoides in *Triticum vulgare*. *Hereditas* **34,** 321-337.

Sansome, E. R. (1933). Segmental interchange in *Pisum* II. *Cytologia* **5,** 15-30.

Sansome, F. W., and Philp, J. (1939). "Recent Advances in Plant Genetics," 411 pp. Blakiston, Philadelphia, Pennsylvania.

Sarkar, P., and Stebbins, G. L. (1956). Morphological evidence concerning the origin of the B genome of wheat. *Amer. J. Bot.* **43,** 297-304.

Sasaki, M., Morris, R., Schm'dt, J. W., and Gill, B. S. (1963). Metaphase I studies of F_1 monosomics of crosses between the Chinese Spring and Cheyenne common wheat varieties. *Can. J. Genet. Cytol.* **5,** 318-325.

Sasaki, M., Moriyasu, M., Morris, R., and Schmidt, J. W. (1968). Chromosomal location of genes for some quantitative characters of wheat using chromosome substitution lines. *Proc. 3rd Intern. Wheat Genet. Symp. Aust. Acad. Sci. Canberra,* pp. 343-349.

Satina, S., and Blakeslee, A. F. (1937a). Chromosome behavior in triploids of *Datura stramonium*. I. The male gametophyte. *Amer. J. Bot.* **24,** 518-527.

Satina, S., and Blakeslee, A. F. (1937b). Chromosome behavior in triploid *Datura*. II. The female gametophyte. *Amer. J. Bot.* **24,** 621-627.

Satina, S., Blakeslee, A. F., and Avery, A. G. (1938). Chromosomal behavior in triploid *Datura*. III. The seed. *Amer. J. Bot.* **25,** 595-602.

Sawyer, G. M. (1949). Case report: Reproduction in a mongoloid. *Amer. J. Ment. Def.* **54,** 204-206.

Sax, K. (1922). Sterility in wheat hybrids. II. Chromosome behavior in partially sterile hybrids. *Genetics* **7,** 513-552.

Sax, K. (1937a). Chromosome behavior and nuclear development in *Tradescantia*. *Genetics* **22,** 523-533.

Sax, K. (1937b). Effect of variations in temperature on nuclear and cell division in *Tradescantia*. *Amer. J. Bot.* **24,** 218-225.

Schertz, K. F. (1963). Chromosomal, morphological, and fertility characteristics of haploids and their derivative in *Sorghum vulgare*. Pers. *Crop Sci.* **3,** 445-447.

Schertz, K. F. (1966). Morphological and cytological characteristics of five trisomics of *Sorghum vulgare*. Pers. *Crop Sci.* **6,** 519-523.

Schlaug, R. (1957). A mongolian mother and her child: A case report. *Acta Genet. Stat. Med.* **7,** 533-540.

Schmidt, J. W., Morris, R., and Johnson, V. A. (1969). Monosomic analysis for bunt resistance in derivatives of Turkey and Oro wheats. *Crop Sci.* **9,** 286-288.

Schöneich, V. J. (1965). Genon-und Chromosomen-aberrationen de Menschen. *Biol. Zentralbl.* **84,** 409-445.

Sears, E. R. (1939). Cytogenetic studies with polyploid species of wheat. I. Chromosomal aberrations in the progeny of a haploid of *Triticum vulgare*. *Genetics* **24,** 509-523.

Sears, E. R. (1941). Nullisomics in *Triticum vulgare*. *Genetics* **26,** 167-168.

Sears, E. R. (1944). Cytogenetic studies with polyploid species of wheat. II. Additional chromosomal aberrations in *Triticum vulgare*. *Genetics* **29,** 232-246.

Sears, E. R. (1947). The *sphaerococcum* gene in wheat. *Genetics* **32,** 102-103.

Sears, E. R. (1952a). Misdivision of univalents in common wheat. *Chromosoma* **4,** 535-550.

Sears, E. R. (1952b).Homoeologous chromosomes in *Triticum aestivum*. *Genetics* **37,** 624.

Sears, E. R. (1953). Nullisomic analysis in common wheat. *Amer. Natur.* **87,** 245-252.

Sears, E. R. (1954). The aneuploids of common wheat, *M. Agr. Exp. Sta. Res. Bull.* **572,** 58 pp.

Sears, E. R. (1956a). Neatby's virescent. *Wheat Inf. Serv.* **3,** 5.

Sears, E. R. (1956b). The transfer of leaf-rust resistance from *Aegilops umbellulata* to wheat. *Brookhaven Symp. Biol.* **92,** 1-22.

Sears, E. R. (1957). Effect of chromosomes XII and XVI on the action of Neatby's virescent. *Wheat Inf. Serv.* **6,** 1.

Sears, E. R. (1958). The aneuploids of common wheat. *Proc. 1st Intern. Wheat Genet. Symp.,* pp. 221-229.

Sears, E. R. (1962). The use of telocentric chromosomes in linkage mapping. *Genetics* **47,** 983.

Sears, E. R. (1966a). Chromosome mapping with the aid of telocentrics. *Proc. 2nd Intern. Wheat Genet. Symp. Hereditas Suppl.* **2,** 370-381.

Sears, E. R. (1966b). Nullisomic-tetrasomic combinations in hexaploid wheat. *Heredity Suppl.* **20,** 29-45.

Sears, E. R. (1967a). Genetic suppression of homoeologous pairing in wheat breeding. *Cienc. Cult. Sao Paulo* **19,** 175-178.

Sears, E. R. (1967b). Induced transfer of hairy neck from rye to wheat. *Z. Pflanzenzuecht.* **57,** 4-25.

Sears, E. R. (1968). Relationships of chromosomes 2A, 2B, 2D with their rye homoeologue. *Proc. 3rd Intern. Wheat Genet. Symp. Aust. Acad. Sci. Canberra,* pp. 53-61.

Sears, E. R. (1969). Wheat cytogenetics. *Annu. Rev. Genet.* **3,** 451-468.

Sears, E. R., and Briggle, L. W. (1969). Mapping the gene *Pml* for resistance to *Erysiphe graminis* f. sp. *tritici* on chromosome 7A of wheat. *Crop Sci.* **9,** 96-97.

Sears, E. R., and Loegering, W. Q. (1968). Mapping of stem rust genes *Sr* 9 and *Sr* 16 of wheat. *Crop Sci.* **8,** 371-373.

Sears, E. R., and Okamoto, M. (1956). Genetic and structural relationships of non-homologous chromosomes in wheat. *Proc. Intern. Genet. Symp. Tokyo* and *Kyoto, Sci. Council, Jap. Tokyo,* pp. 332-335.

Sears, E. R., and Okamoto, M. (1958). Intergenomic chromosome relationships in hexaploid wheat. *Proc. 10th Intern. Congr., Genet. Montreal* **2,** 258-259.

Sears, E. R., and Rodenhiser, H. A. (1948). Nullisomic analysis of stem-rust resistance in *Triticum vulgare* var. Timstein. *Genetics* **33,** 123-124.

Sears, E. R., Loegering, W. Q., and Rodenhiser, H. A. (1957). Identification of chromosomes carrying genes for stem rust resistance in four varieties of wheat. *Agron. J.* **49,** 208-212.

Sears, E. R., Schaller, C. W., and Briggs, F. N. (1960). Identification of the chromosome carrying the Martin gene for resistance of wheat to bunt. *Can. J. Genet. Cytol.* **2,** 262-267.

Sears, L. M. S., and Lee-Chen, S. (1970). Cytogenetic studies in *Arabidopsis thaliana. Can. J. Genet. Cytol.* **12,** 217-223.

Sears, L. M. S., and Sears, E. R. (1968). The mutants chlorina-1 and Hermsen's virescent. *Proc. 3rd Intern. Wheat Genet. Symp. Aust. Acad. Sci. Canberra,* pp. 299-304.

Sen, N. K. (1952). Isochromosomes in tomato. *Genetics* **37,** 227-241.

Sen, S. K. (1965). Cytogenetics of trisomics in rice. *Cytologia* **30,** 229-238.

Semeniuk, Wm. (1947). Chromosomal stability in certain rust resistant derivatives from a *T. vulgare* × *T. timopheevi* cross. *Sci. Agr.* **27,** 7-20.

Shah, S. S. (1964). Studies on a triploid, a tetrasomic triploid and a trisomic plant of *Dactylis glomerata. Chromosoma* **15,** 469-477.

Sharma, D., and Knott, D. R. (1966). The transfer of leaf-rust resistance from *Agropyron* to *Triticum* by irradiation. *Can. J. Genet. Cytol.* **8,** 137-143.

Shaver, D. L. (1965). A new maize monosomic. *Maize Genet. Coop. Newsl.* **39,** 24.

Sheen, S. J., and Snyder, L. A. (1964). Studies on the inheritance of resistance to six stem rust cultures using chromosome substitution lines of a Marquis wheat selection. *Can. J. Genet. Cytol.* **6,** 74-82.

Sheen, S. J., and Snyder, L. A. (1965). Studies on the inheritance of resistance to six stem rust cultures using chromosome substitution lines of a Kenya wheat. *Can. J. Genet. Cytol.* **7**, 374-387.

Shive, R. J., Hare, W. C. D., and Patterson, D. F. (1965). Chromosome studies in dogs with congenital cardiac defects. *Cytogenetics* **4,** 340.

Singh, M. P. (1967a). Monosomic analysis in wheat. *Heredity* **22,** 591-596.

Singh, M. P. (1967b). Identification of chromosomes carrying genes for resistance to race 15, 122 and 34 of stem rust *(Puccinia graminis tritici)* in wheat variety Yaqui-53. *Caryologia* **20,** 51-60.

Singh, M. P., and Swaminathan, M. S. (1959). Monosomic analysis in bread wheat. III. Identification of chromosomes carrying genes for resistance to two races of yellow rust in Cometa Klein. *Indian J. Genet. Plant Breed.* **19,** 171-175.

Singh, M. P., and Swaminathan, M. S. (1960). V. Identification of chromosomes carrying genes for resistance to two races of stem rust in the variety N.P. 790. *Indian J. Genet. Plant Breed.* **20,** 160-165.

Singh, R. M., and Wallace, A. T. (1967a). Monosomics of *Avena byzantina* C. Koch. I. Karyotype and chromosome pairing studies. *Can. J. Genet. Cytol.* **9,** 87-96.

Singh, R. M., and Wallace, A. T. (1967b). Monosomics of *Avena byzantina* C. Koch. II. Breeding behavior and morphology. *Can. J. Genet. Cytol.* **9,** 97-106.

Singleton, W. R., and Mangelsdorf, P. C. (1940). Gametic lethals on the fourth chromosome of maize. *Genetics* **25,** 366-390.

Sinnott, E. W., and Blakeslee, A. F. (1922). Structural changes associated with factor mutations and with chromosome mutations in *Datura*. *Proc. Nat. Acad. Sci. U.S.* **8,** 17-19.

Sinnott, E. W., Houghtaling, H., and Blakeslee, A. F. (1934). The comparative anatomy of extra-chromosomal types in *Datura stramonium*. *Carnegie Inst. Wash. Publ.* **451**.

Skovsted, A. (1933). Cytological studies in cotton. I. The mitosis and the meiosis in diploid and triploid Asiatic cotton. *Ann. Bot.* **47,** 227-251.

Smith, D. W. (1964). Autosomal abnormalities. *Amer. J. Obstet. Gynecol.* **90,** 1055-1077.

Smith D. W., Patau, K., Therman, E., and Inhorn, S. L. (1960). A new autosomal trisomy syndrome: Multiple congenital anomalies caused by an extra chromosome. *J. Pediat.* **57,** 338-345.

Smith, H. H. (1943). Studies on induced heteroploids of *Nicotiana*. *Amer. J. Bot.* **30,** 121-130.

Smith, J. D. (1963). The effect of chromosome number on competitive ability of hexaploid wheat gametophytes. *Can. J. Genet. Cytol.* **5,** 220-226.

Smith, L. (1941). An inversion, a reciprocal translocation, trisomics, and tetraploids in barley. *J. Agr. Res.* **63,** 741-750.

Smith, L. (1947). A fragmented chromosome in *Triticum monococcum* and its use in studies of inheritance. *Genetics* **32,** 341-349.

Snow, R. (1964). Cytogenetic studies in *Clarkia*, section *Primigenia*. III. Cytogenetics of monosomics in *Clarkia amoena*. *Genetica* **35,** 205-235.

Soriano, J. D. (1957). The genus *Collinsia*. IV. The cytogenetics of colchicine-induced reciprocal translocations in *C. heterophylla*. *Bot. Gaz.* **118,** 139-145.

Stadler, L. J. (1933). On the genetic nature of induced mutations in plants. II. A haplo-viable deficiency in maize. *Mo. Agr. Exp. Sta. Res. Bull.* **204,** 1-29.

Stadler, L. J. (1935). Genetic behaviour of a haplo-viable internal deficiency in maize. *Amer. Natur.* **69,** 80-81.

Stebbins, G. L. (1950). "Variation and Evolution in Plants," 643 pp. Columbia Univ. Press, New York.

Steinitz-Sears, L. M. (1963). Chromosome studies in *Arabidopsis thaliana*. *Genetics* **48,** 483-490.

Stephens, S. G. (1947). Cytogenetics of *Gossypium* and the problem of the New World cottons. *Advan. Genet.* **1,** 431-442.

Stettler, R. F. (1964). Dosage effects of the Lanceolate gene in tomato. *Amer. J. Bot.* **51,** 253-264.

Stino, K. R. (1940). Inheritance in *Nicotiana tabacum*. XV. Carmine-white variegation. *J. Hered.* **31,** 19-24.

Stubbe, H. (1934). Studien über heteroploide Formen von *Antirrhinum majus* L. I. Mitteilung: zur Morphologie und Genetik der Trisomen Anaemica, Fusca, Purpurea, Rotunda. *Planta* **22,** 153-170.

Suto, T., and Sugiyama, S. (1961a). Sex expression and determination in spinach. II. Inheritance and breeding of intersexuality. *Jap. J. Breed.* **10,** 215-222.

Suto, T., and Sugiyama, S. (1961b). Sex expression and determination in spinach. III. Inheritance and breeding of intersexuality (continued). *Jap. J. Breed.* **11,** 29-36.

Sutton, E. (1939). Trisomics in *Pisum sativum* derived from an interchange heterozygote. *J. Genet.* **38,** 459-476.

Swaminathan, M. S., and Natarajan, A. T. (1959). Cytological and genetic changes induced by vegetable oils in *Triticum. J. Hered.* **50,** 177-187.

Sybenga, J. (1966). The quantitative analysis of chromosome pairing and chiasma formation based on the relative frequencies of M 1 configuration. V. Interchange trisome. *Genetica* **37,** 481-510.

Tabushi, J. (1958). Trisomics of spinach. *Seiken Ziho* **9,** 49-57.

Tagenkamp, T. R., and Finkner, V. C. (1954). Inheritance of albinism in oats and the effects of carbohydrates on rust sori development of albino oat leaves. *Annu. Meet. Amer. Soc. Agron.*, pp. 74-75.

Takagi, F. (1935). Karyogenetical studies on rye. I. A trisomic plant. *Cytologia* **6,** 496-501.

Taylor, J. W. (1934). Irregularities in the inheritance of the hairy neck character transposed from *Secale* to *Triticum. J. Agr. Res.* **48,** 603-617.

Thomas, H. (1966). Hypoploid variability in hexaploid oats, *Avena sativa. Can. J. Genet. Cytol.* **8**, 568-574.

Thomas, H. (1968). The addition of single chromosomes of *Avena hirtula* to the cultivated hexaploid oat *A. sativa. Can. J. Genet. Cytol.* **10,** 551-563.

Thomas, H., and Mytton, J. (1970). Monosomic analysis of fatuoids in cultivated oat *Avena sativa. Can. J. Genet. Cytol.* **12,** 32-35.

Thomas, H., and Rajhathy, T. (1966). A gene for desynapsis and aneuploidy in tetraploid *Avena. Can. J. Genet. Cytol.* **8,** 506-515.

Thompson, M. W. (1961). Reproduction in two female mongols. *Can. J. Genet. Cytol.* **3,** 351-354.

Tikhomirova, M. M. (1965). On the mechanism of X-ray induction of aneuploid gametes. *Genetics (USSR)* **4,** 63-68

Tjio, J. H., and Levan, A. (1956). The chromosome number of man. *Hereditas* **42,** 1-6.

Tsuchiya, T. (1952). Cytogenetics of a hypo-triploid barley and its progeny. *Mem. Beppu. Univ.* **2,** 19-42.

Tsuchiya, T. (1954). Trisomics in barley. *Jap. J. Genet.* **29**, 179.

Tsuchiya, T. (1956). Studies on the relationships between chromosomes and genetic linkage groups in trisomic barley. *Jap. J. Genet.* **31,** 313-314.

Tsuchiya, T. (1958). Studies on the trisomics in barley. I. Origin and the characteristics of primary simple trisomics in *Hordeum spontaneum* C. Koch. *Seiken Ziho* **9,** 69-86, (English Resume, p. 84.)

Tsuchiya, T. (1959). Genetic studies in trisomic barley. I. Relationships between trisomics and genetic linkage groups of barley. *Jap. J. Bot.* **17,** 14-28.

Tsuchiya, T. (1960). Cytogenetic studies of trisomics in barley. *Jap. J. Bot.* **17,** 177-213.

Tsuchiya, T. (1961). Studies on the trisomics in barley. II. Cytological identification of the extra chromosomes in crosses with Burnham's translocation testers. *Jap. J. Genet.* **36,** 444-451.

Tsuchiya, T. (1963). Chromosome aberrations and their use in genetics and breeding in barley —trisomics and aneuploids. *Barley Genet. Proc. 1st Intern. Symp.*, pp. 116-150.Center for Agricultural Publications and Documentation, Wageningen.

Tsuchiya, T. (1967). Establishment of a trisomic series in a two-rowed cultivated variety of barley. *Can. J. Genet. Cytol.* **9,** 667-682.

Tsuchiya, T. (1969). Status of studies of primary trisomics and other aneuploids in barley. *Genetica* **40,** 216-232.
Tsuchiya, T., Hayashi, J., and Takahashi, R. (1960). Genetic studies in trisomic barley. II. Further studies on the relationships between trisomics and the genetic linkage groups. *Jap. J. Genet.* **35,** 153-160.
Tsunewaki, K. (1960). Monosomic and conventional analyses in common wheat. III. Lethality. *Jap. J. Genet.* **35,** 71-75.
Tsunewaki, K. (1964a). The transmission of the monosomic condition in a wheat variety, Chinese Spring. II. A critical analysis of nine-year records. *Jap. J. Genet.* **38,** 270-281.
Tsunewaki, K. (1964b). Transmission of monosomes and trisomes in an Emmer wheat. *Wheat Inf. Serv.* **17, 18,** 34-35.
Tsunewaki, K. (1966a). Comparative gene analysis of common wheat and its ancestral species. III. Glume hairiness. *Genetics* **53,** 303-311.
Tsunewaki, K. (1966b). Comparative gene analysis of common wheat and its ancestral species. II. Waxiness, growth habit and awnedness. *Jap. J. Bot.* **19,** 175-229.
Tsunewaki, K., and Jenkins, B. C. (1961). Monosomic and conventional gene analysis in common wheat. II. Growth habit and awnedness. *Jap. J. Genet.* **36,** 428-443.
Tsunewaki, K., and Kihara, H. (1961). F_1 monosomic analysis of *Triticum macha*. *Wheat Inf. Serv.* **12,** 1-3.
Uchida, I. A., Ray, M., McRae, K. N., and Besant, D. F. (1968). Familial occurrence of trisomy 22. *Amer. J. Hum. Genet.* **20,** 107-118.
Uchikawa, I. (1941). Genetic and cytological studies of speltoid wheat. II. Origin of speltoid wheat. *Mem. Coll. Agr. Kyoto Imp. Univ.* **50,** 1-64.
Unrau, J. (1950). The use of monosomes and nullisomes in cytogenetic studies of common wheat. *Sci. Agr.* **30,** 66-89.
Unrau, J. (1958). Genetic analysis of wheat chromosomes. I. Description of proposed methods. *Can. J. Plant Sci.* **38,** 415-418.
Unrau, J., Person, C., and Kuspira, J. (1956). Chromosome substitution in hexaploid wheat. *Can. J. Bot.* **34,** 629-640.
Van Overeem, C. (1921). Uber Formen mit abweichender chromosomenzahal bei *Oenothera*. *Beih. Bot. Zentralbl.* **38,** 73-113.
Vasek, F. C. (1956). Induced aneuploidy in *Clarkia unguiculata (Onagraceae)*. *Amer. J. Bot.* **43,** 366-371.
Vasek, F. C. (1961). Trisomic transmission in *Clarkia unguiculata*. *Amer. J. Bot.* **48,** 829-833.
Vasek, F. C. (1963). Phenotypic variation in trisomics of *Clarkia unguiculata*. *Amer. J. Bot.* **50,** 308-314.
Venkateswarlu, J., and Reddi, V. R. (1968). Cytological studies of *Sorghum* trisomics. *J. Hered.* **59,** 179-182.
Virmani, S. S. (1969). Trisomics of *Pennisetum typhoides*. Ph.D. Thesis Punjab Agricultural University, Ludhiana, Punjab, India.
Vogt, G. E., and Rowe, P. R. (1968). Aneuploids from triploid-diploid crosses in the series *Tuberosa* of the genus *Solanum*. *Can. J. Genet. Cytol.* **10,** 479-486.
Von Berg, K. (1935). Cytologische Untersüchungen an den Bastarden des *Triticum turgidovillosum* und an einer F_1 *Triticum turgidum* × *villosum*. III. Weitere Studien am fertilen Knostanten Artbastard *Triticum turgidovillosum* und seinen Verwandten. *Z. Indukt. Abstamm. Vererbungsl.* **68,** 94-126.
Von der Schulenburg, H. (1965). Beitrage zur Aufstellung von Monosomen-Sortimenten bein Hafer. (Contributions towards building a monosomic collection in oats.) *Z. Pflanzenzuecht.* **53,** 247-265.
Von Wettstein, F. (1927). Die Erscheinung der Heteroploide, besonders im Pflanzenreich. *Ergeb. Biol.* **2,** 311-356.

Waardenburg, P. J. (1932). Mongolismus. "Das Menschliche Auge und seine Erbaulagen," pp. 44-48. Nijhoff, The Hague.

Wang, S., Yeh, P., Lee, S. S. Y., and Li, H. W. (1965). Effect of low temperature on desynapsis in rice. *Bot. Bull. Acad. Sinica* **6,** 197-207.

Washington, W. J., and Sears, E. R. (1970). Ethyl methanesulfonate-induced chlorophyll mutations in *Triticum aestivum*. *Can. J. Genet. Cytol.* **12,** 851-859.

Watanabe, Y., Ono, S., Mukai, Y., and Koga, Y. (1969). Genetic and cytogenetic studies on the trisomic plants of rice, *Oryza sativa* L. *Jap. J. Breed.* **19,** 12-18.

Welshons, W. J., and Russell, L. B. (1959). The Y chromosome as the bearer of male determining factors in the mouse. *Proc. Nat. Acad. Sci. U.S.* **45,** 560-566.

White, T. G. (1966). Monosomic analyses of some marker loci in cotton. *Abstr. Annu. Meet. Amer. Soc. Agron. 1966,* p. 16.

White, T. G., and Endrizzi, J. E. (1965). Tests for the association of marker loci with chromosomes in *Gossypium hirsutum* by the use of aneuploids. *Genetics* **51,** 605-612.

Wienhues, A. (1965). Cytogenetische Untersüchungen uber die chromosomale Grundlage der Rostresistenz der Weizensorte Weique. *Zuchter* **35,** 352-354.

Wienhues, A. (1966). Transfer of rust resistance of *Agropyron* to wheat by addition, substitution and translocation. *Proc. 2nd Intern. Wheat Genet. Symp. Hereditas Suppl.* **2,** 328-341.

Winge, O. (1924). Zytologische Untersüchungen uber Speltoide und andere mutantenahnliche Aberranten beim Weizen. *Hereditas* **5,** 241-286.

Winkler, H. (1916). Uber die experimentelle Erzeugung von pflanzen mit abweichenden chromosomenzahlen. *Z. Bot.* **8,** 417-531.

Yarnell, S. H. (1931). A study of certain polyploid and aneuploid forms in *Fragaria*. *Genetics* **16,** 455-489.

Yen, Fu-sun, Evans, L. E., and Larter, E. N. (1969). Monosomic analysis of fertility restoration in three restorer lines of wheat. *Can. J. Genet. Cytol.* **11,** 531-546.

Young, P. A., and MacArthur, J. W. (1947). Horticultural characters of tomatoes. *Texas Agr. Exp. Sta. Bull.* **698,** 1-61.

SUBJECT INDEX

I

K

L

M

T